Controlling the Morphology of Polymers

Geoffrey R. Mitchell • Ana Tojeira
Editors

Controlling the Morphology of Polymers

Multiple Scales of Structure and Processing

Editors
Geoffrey R. Mitchell
Product
Center for Rapid and Sustainable
  Product Development
Institute Polytechnic of Leiria
Marinha Grande, Portugal

Ana Tojeira
Product
Center for Rapid and Sustainable
  Product Development
Institute Polytechnic of Leiria
Marinha Grande, Portugal

ISBN 978-3-319-39320-9   ISBN 978-3-319-39322-3   (eBook)
DOI 10.1007/978-3-319-39322-3

Library of Congress Control Number: 2016947449

© Springer International Publishing Switzerland 2016

This work is subject to copyright. All rights are reserved by the Publisher, whether the whole or part of the material is concerned, specifically the rights of translation, reprinting, reuse of illustrations, recitation, broadcasting, reproduction on microfilms or in any other physical way, and transmission or information storage and retrieval, electronic adaptation, computer software, or by similar or dissimilar methodology now known or hereafter developed.

The use of general descriptive names, registered names, trademarks, service marks, etc. in this publication does not imply, even in the absence of a specific statement, that such names are exempt from the relevant protective laws and regulations and therefore free for general use.

The publisher, the authors and the editors are safe to assume that the advice and information in this book are believed to be true and accurate at the date of publication. Neither the publisher nor the authors or the editors give a warranty, express or implied, with respect to the material contained herein or for any errors or omissions that may have been made.

Printed on acid-free paper

The registered company is Springer International Publishing AG Switzerland
The Springer imprint is published by Springer Nature

To our families, for all the joy and inspiration.

Dedicated to the memory of Jill Mitchell (1948–2015).

Preface

The possibility of long-chain molecules was established at the start of the twentieth century through the pioneering work of Hermann Staudinger (1920), who was awarded the Nobel Prize in Chemistry in 1953 "for his discoveries in the field of macromolecular chemistry." Much of the twentieth century was dominated by the ingenious work of synthetic organic chemists in producing polymers with particular chemical configurations and molecular lengths. The first to dominate the field were Giulio Natta and Karl Zeigler whose work led to the commercial production of stereoregular alpha olefins such as polypropylene (Natta 1967). The contribution of Karl Ziegler was the discovery of the first titanium-based catalysts. Giulio Natta used such catalysis to prepare stereoregular polymers from propylene. Both were awarded the Nobel Prize in Chemistry in 1963.

Other developments have built on these early steps. In the 1950s, scientists at Phillips Petroleum discovered that chromium catalysts are very effective for the low temperature polymerization of ethylene, which launched major industrial technologies. A little later, Ziegler discovered that the combination of titanium chloride and ethyl aluminum sesquichloride gave comparable activities for the production of polyethylene. Natta used crystalline α-titanium chloride in combination with triethyl aluminum to produce the first isotactic polypropylene. In the 1960s, BASF developed a gas phase, a mechanically stirred polymerization process for making polypropylene which led to the UNIPOL process, which was commercialized by Union Carbide to produce polyethylene. Later in the 1970s, magnesium chloride was discovered to greatly enhance the activity of the Ti-based catalysts. These catalysts are so active that the residual Ti is no longer removed from the product. They enabled the commercialization of linear low-density polyethylene (LLDPE) resins and allowed the development of noncrystalline copolymers. These developments and similar research have transformed plastics from an oddity only suited to Hula-Hoops into a major technological materials industry with a total global demand of over 200 million tonnes.

Functional polymers appeared in the second half of the twentieth century. Although polyaniline was first described in the mid-nineteenth century by Henry Letheby and polypyrrole derivatives were reported to be electrically conducting in 1963 by B.A. Bolto et al. (1963), substantial progress was not made with intrinsically conducting polymers until the pioneering work of Hideki Shirakawa, Alan J. Heeger, and Alan MacDiarmid who reported similar high conductivity in oxidized iodine-doped polyacetylene in 1977 (Shirakawa 1977). For this research, they were awarded the 2000 Nobel Prize in Chemistry "for the discovery and development of conductive polymers."

Liquid crystal polymers and elastomers (Donald et al. 2006), dendrimers (Hawker and Fréchet 1990), and block copolymers (Szwarc 1956) also appeared in the second half of the twentieth century. Today we can select polymers with many different properties and functions. Of particular current note is the development of photovoltaic devices which are beginning to show considerable promise. However, in the field of photovoltaics, researchers have found that it is not just the ingenuity of the molecule makers that delivers the efficiency but also controlling the morphology. Controlling the morphology is a major tool in polymer science and engineering. It enables chemically the same material such as polyethylene to be exploited to produce low-cost plastic bags as well as bulletproof vests.

This book sets out to embrace this control of morphology recognizing from the outset that the different scales of structure are connected. As Natta discovered the precise positioning of the methyl group in polypropylene leads to a high crystallinity and without precise positioning to a poor quality amorphous polymer used in roofing membranes, others have identified that the control of that crystallization process leads to clear packaging material which constitutes about half of the 60 million tonnes of polypropylene-based materials used in 2013.

Like the scales of structure, the contributors to this book are all interconnected in a myriad of ways. At the core, connecting many is the highly successful Reading Polymer Physics Group, later morphing into the Polymer Science Centre at the University of Reading. It was there that Alison Hodge, Robert Olley, and David Bassett optimized the permanganic etching process for revealing the morphology of semicrystalline polymers (Olley 1979). The advent of time-resolving X-ray scattering at synchrotron-based beamlines (Keates et al. 1994) has greatly added to the ability to follow in real time the development of structure and morphology during processing. Not all good things go on forever, and the University brought the curtain down on polymer physics at Reading in 2010 with the closure of the Department of Physics. With every closing door, another opens and so please check out controlling morphology during 3D printing in Chap. 7. The future is direct digital manufacturing as Mr. McGuire[1] might say.

---

[1] Mr. McGuire (played by Walter Brooke) is a character in the motion picture "The Graduate" who said in the movie "Plastics," followed by "There's a great future in plastics," to the graduate Ben Braddock played by Dustin Hoffman, which is one of all-time top 100 quotes judged by the American Film Institute.

We would like to thank all the authors for their contributions and their positive approach to this crosscutting project. It has been a pleasure to work with them all.

This work was supported by the Fundação para a Ciência e a Tecnologia (Portugal) through project PTDC/CTM-POL/7133/2014.

Marinha Grande, Portugal   Geoffrey R. Mitchell
                                                          Ana Tojeira

References

Bolto BA, McNeil R, Weiss DE (1963) Electronic conductions in polymers III electronic properties of polypyrrole. Aust J Chem 16:1090–1103

Donald AM, Windle AH, Hanna S (2006) Liquid crystal polymers. Cambridge University Press

Hawker, CJ, Fréchet JMJ (1990) Preparation of polymers with controlled molecular architecture. A new convergent approach to dendritic macromolecules. J Am Chem Soc 112(21): 7638–7647

Keates P, Mitchell GR, Peuvrel-Disdier E, Riti JB, Navard P (1994) A novel X-ray rheometer for in situ studies of polymer melts and solutions during shear flow. J Nonnewton Fluid Mech 52 (2):197–215

Keller A (1957) A note on single crystals in polymers: evidence for a folded chain configuration. Philos Mag Ser 8(2): 1171–1175

Letheby H (1862) J Chem Soc 15: 161

Natta G, Danusso F (eds) (1967) Stereoregular polymers and stereospecific polymerizations. Pergamon Press

Olley RH, Hodge AM, Bassett DC (1979) A permanganic etchant for polyolefines. J Polym Sci 17: 627–643

Shirakawa H, Louis EJ, MacDiarmid AG, Chiang CK, Heeger AJ (1977). Synthesis of electrically conducting organic polymers: halogen derivatives of polyacetylene, $(CH)_x$. J Chem Soc Chem Commun 16: 578

Staudinger H (1920) Über Polymerisation. Ber Deut Chem Ges 53(6): 107.

Szwarc M (1956) Living polymers. Nature 178: 1168–1169

Contents

1. **Scales of Structure in Polymers** 1
   Geoffrey R. Mitchell, Fred J. Davis, and Robert H. Olley

2. **Evaluating Scales of Structures** 29
   Saeed Mohan, Robert H. Olley, Alun S. Vaughan,
   and Geoffrey R. Mitchell

3. **Crystallization in Nanocomposites** 69
   Geoffrey R. Mitchell, Donatella Duraccio, Imran Khan,
   Aurora Nogales, and Robert Olley

4. **Theoretical Aspects of Polymer Crystallization** 101
   Wenbing Hu and Liyun Zha

5. **Controlling Morphology Using Low Molar
   Mass Nucleators** 145
   Geoffrey R. Mitchell, Supatra Wangsoub, Aurora Nogales,
   Fred J. Davis, and Robert H. Olley

6. **Crystallization in Nanoparticles** 163
   Aurora Nogales and Daniel E. Martínez-Tong

7. **Controlling Morphology in 3D Printing** 181
   Ana Tojeira, Sara S. Biscaia, Tânia Q. Viana,
   Inês S. Sousa, and Geoffrey R. Mitchell

8. **Electrically Conductive Polymer Nanocomposites** 209
   Thomas Gkourmpis

9. **Nanodielectrics: The Role of Structure in Determining
   Electrical Properties** 237
   Alun S. Vaughan

10	**Block Copolymers and Photonic Band Gap Materials** 263
	Dario C. Castiglione and Fred J. Davis
11	**Relationship Between Molecular Configuration and Stress-Induced Phase Transitions** 287
	Finizia Auriemma, Claudio De Rosa, Rocco Di Girolamo, Anna Malafronte, Miriam Scoti, Geoffrey R. Mitchell, and Simona Esposito
12	**Summary** 329
	Geoffrey R. Mitchell

Index 331

Chapter 1
Scales of Structure in Polymers

Geoffrey R. Mitchell, Fred J. Davis, and Robert H. Olley

1.1 Introduction

The multi-valency of the carbon atom leads immediately to a rich variety of molecular structures including long-chain polymers. At the smallest scale, the chemical bonds are anisotropic but without appropriate molecular ordering, that anisotropy is not conveyed to the macroscopic scale which has a substantial impact on the properties. This chapter focuses on detailing these scales of structures and the types of ordering processes which are observed in polymer-based materials. We consider the mechanisms for the transformation of polymer melts by these ordering processes. Finally, we consider 'top-down' manufacturing processes which lead to 'ordering' on a scale larger than that intrinsic to the polymer material.

1.2 Types of Bonds

Although many inorganic systems may be considered to be polymeric, the term polymer is generally taken to refer to long chains of organic polymers. The importance of the high molecular weight is that it is this that controls their mechanical properties. The useful mechanical properties of a polymer arise from the increased magnitude of the bonding between molecules. This intermolecular

G.R. Mitchell (✉)
Center for Rapid and Sustainable Product Development, Institute Polytechnic of Leiria, Marinha Grande, Portugal
e-mail: geoffrey.mitchell@ipleiria.pt

F.J. Davis
Department of Chemistry, University of Reading Whiteknights, Reading RG6 6AD, UK

R.H. Olley
EMLab, JJ Thomson Building, University of Reading Whiteknights, Reading RG6 6AF, UK

bonding is collectively classified as Van der Waals bonding although this terminology is often used just to describe the bonding from induced dipole-induced dipole forces (London Forces). In addition to the London dispersion forces, intermolecular bonding is provided by dipole-induced dipole forces, dipole–dipole forces and hydrogen bonding. In polymers with ionic groups, there may be ionic interactions. While London forces are generally considered to be the weakest of the interaction, because of their ubiquitous nature they tend to contribute more towards the overall intermolecular bonding (except in small molecules such as water). However, the presence of interactions additional to London forces results in increased strength; thus up to a molecular weight of 10,000 Da, polyethylene which has no permanent dipole is a waxy solid, while polyamides (with hydrogen bonding) are hard solids at weights as low as 1000 Da (Stevens 1990).

1.3 Types of Polymers

There are a number of different ways to classify polymers but perhaps the simplest division is between natural polymers (biopolymers) which include proteins (polypeptides), polysaccharides, and poly(nucleotides) and synthetic polymers which include polyethylene, poly(vinyl chloride) and nylon; some natural polymers are synthetically modified as in the formation of viscose rayon from cellulose or vulcanized rubber from natural rubber [largely poly(isoprene)].

Chemists generally classify synthetic polymers by their mode of synthesis: chain growth polymers are formed by the addition of single monomers to a growing chain, while step growth polymers are formed by the reaction of (for example) bifunctional monomers to form dimers, trimers, etc. which may react together to form larger molecules. An example of a chain growth polymerization would be the formation of polystyrene [Reaction (1.1)], while an example of the second is provided by poly(ethylene terephthalate) as shown in Reaction (1.2).

Reaction 1.1

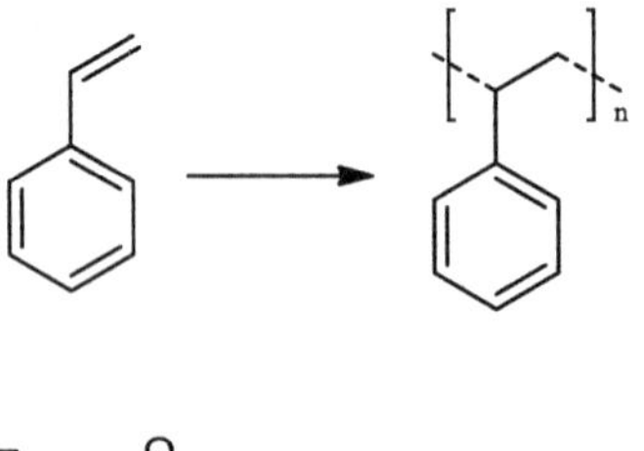

Reaction 1.2

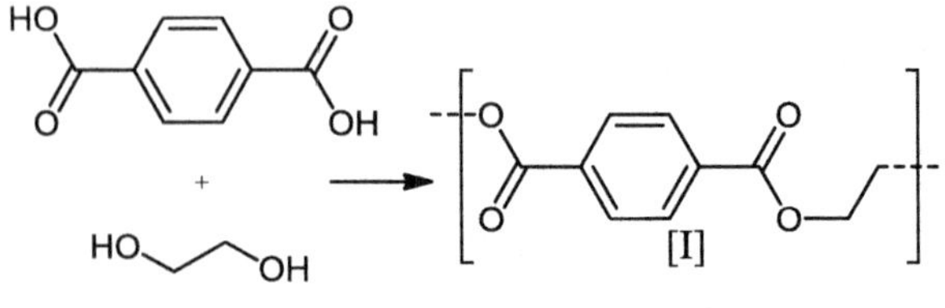

Chain-growth polymers are often formed from vinyl systems ($CH_2=CHX$) and represent some of the most common commercial polymers; examples include

1 Scales of Structure in Polymers

Fig. 1.1 Polymer tacticities for vinyl polymers (**a**) isotactic (**b**) syndiotactic (**c**) atactic (heterotactic)

Fig. 1.2 Schematic of copolymer systems: (**a**) homopolymer; (**b**) alternating copolymer; (**c**) random copolymer; (**d**) block copolymer; and (**e**) graft copolymer

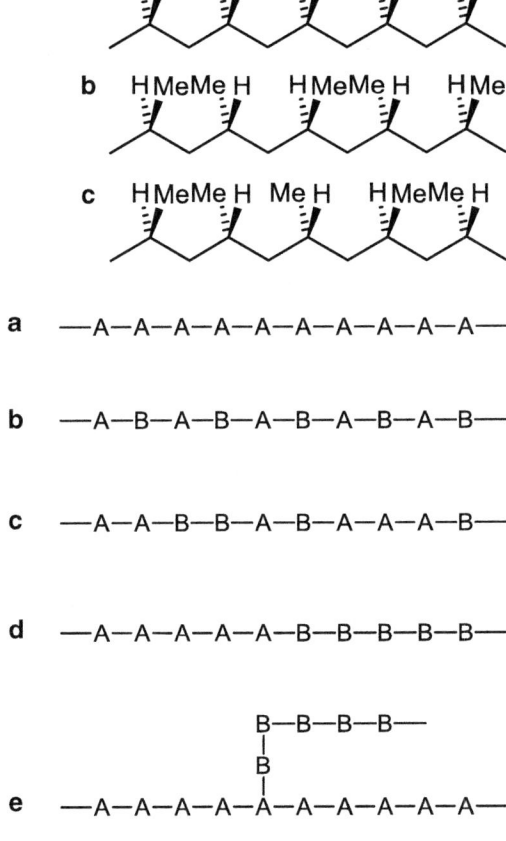

polyethene, polypropene, poly(methyl methacrylate) and poly (vinyl chloride). One particular feature of these polymers is that for all systems except ethene (X=H), the resulting polymers have a stereogenic centre, resulting in a range of arrangements of the side groups, such arrangements are described in Fig. 1.1. The normal atactic (or random) arrangement might be expected from simple processes such as free-radical polymerization, more regular structures such as isotactic and syndiotactic poly(propene), are usually produced by organometallic catalysts such as the Ziegler–Natta catalyst (Stevens 1990). In general, the more stereoregular polymers have a greater tendency towards crystallization, and as such are often valued for their enhanced mechanical properties.

The repeat unit for poly(ethylene terephthalate) ([I] Reaction (1.2) above) is built up from two chemical units, but the polymer is a continuous repeat of this structure. In many cases, the structure of a polymer is modified by the addition of another monomer unit during the polymerization process, a process known as copolymerization. This for a vinyl polymer system, for example, styrene could be copolymerized with methyl methacrylate. Such copolymers can be arranged in different as shown in Fig. 1.2 and each type of material may show interesting or unique properties. For example, block copolymers may show microphase separation of the incompatible blocks of copolymer, and this can produce for example materials useful as photonic band-gap materials (Urbas et al. 1999) (See Chap. 10.)

1.4 Types of Materials

In addition to copolymerization described above, there is considerable interest in the control of polymer architecture. As this may have a considerable influence on the behaviour of polymers as materials. Thus, polymers may be linear (Fig. 1.3a), or branched (Fig. 1.3b) or may for more complex structures such as a star arrangement (Fig. 1.3c) or more sophisticated dendrimer arrangements. Some of these more complex architectures pose considerable challenges to the organic chemist in terms of reagents and equipment (Hadjichristidis et al. 2000), while others, for example, the introduction of cross-links, can be achieved using technically quite simple methodologies. Cross-linked polymers are an important class of materials in themselves and provide another classification, namely thermoplastics and thermosets; The former are those which melt and flow, the latter are materials which cannot melt or dissolve and are built up of cross-linked polymer chains.

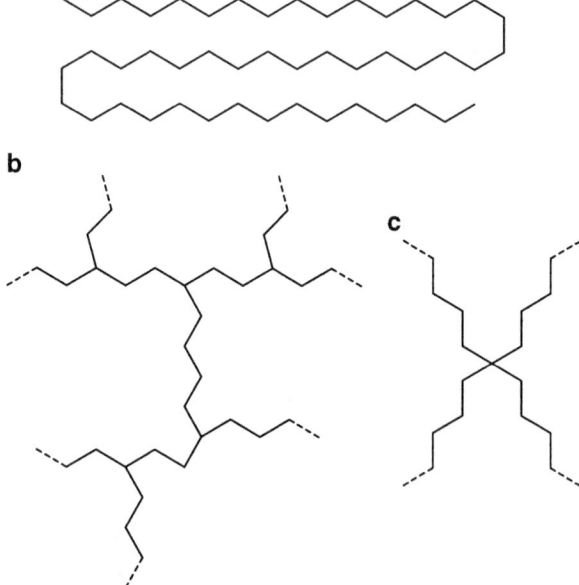

Fig. 1.3 Examples of polymer architecture: (**a**) linear; (**b**) branched; and (**c**) star structure

1.4.1 Thermoplastics

In terms of production of polymeric materials poly(ethene) dominates with typically 30 % of polymer production worldwide, other significant materials are poly (propene), and poly(styrene) and poly (vinyl chloride), though the latter has prompted considerable environmental debate, in terms of the toxicity of the monomer, the use of plasticisers in the products (particularly those which come into contact with children) and the production of dioxins in the ultimate disposal by pyrolysis. The major advantage of the general class of thermoplastics is that they soften and flow on heating and thus can be processed using a range of simple methodologies, which rely on the solidification of the melt into a particular form. Examples of processing methodologies include blow moulding, injection moulding, extrusion and spinning (Stevens 1990). The ability to process the polymer through melting or (less usually) through dissolution also allows for the polymer to be recycled although in this regard additives either added for appearance as stabilizers or for improved properties may influence the range of options possible.

As thermoplastics do not contain covalent bonds between chains the mechanical properties of the resultant polymers are highly dependent on the intermolecular forces holding neighbouring chains together (vide supra). That being said it is the presence of crystalline regions which often produce the high mechanical strength of many polymeric materials. Thus, careful control of polyethene morphology can result in extremely high modulus materials (Ward and Hine 2004). It should also be noted that while crystallinity can impart high mechanical strength through the close regular alignment of neighbouring molecules, the presence of amorphous regions also provides some elasticity and impact resistance.

1.4.2 Thermosets

Thermosetting polymers are those systems where a network of cross-linking between chains means that the polymer cannot be dissolved or liquefied on heating. In lightly cross-linked system, there may be considerable deformation allowed at the right conditions as in the stretching of an everyday rubber band, but ultimately the bonding of neighbouring chains restricts their translational motion. In highly cross-linked systems, the polymer may have considerable dimensional stability. In both cases, the ultimate form (leaving aside issues of applied forces) is determined at the time of cross-linking and to form useful products the cross-linking process must form part of the manufacturing process. One of the first commercially important synthetic polymers was Bakelite® (Baekeland 1909) which is a phenol formaldehyde resin, since these are the two components involved in its manufacture (Fig. 1.4). This material was extremely popular in the first half of the twentieth century for the production of a range of domestic and electrical goods.

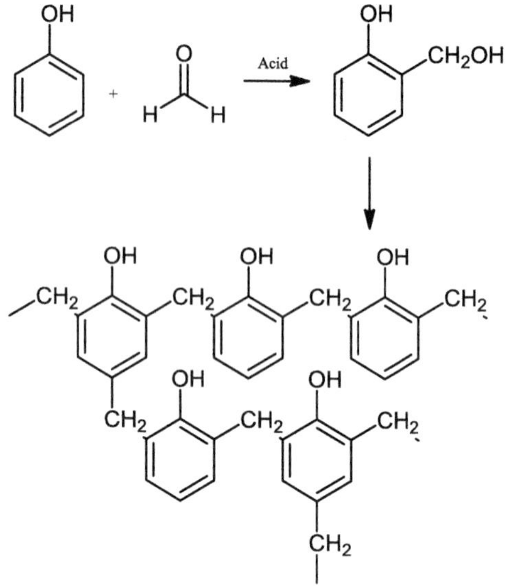

Fig. 1.4 Synthesis of the thermosetting material Bakelite®

1.4.3 Composites, Micro-Fillers, Nano-Fillers

Increasingly a range of applications require polymers with enhanced performance. This has led to a further classification, namely between engineering and commodity polymers. The latter are those that are produced in high quantities as discussed in Sect. 1.4.1; the former are generally more expensive materials, which are manufactured on a smaller scale but are valued for certain superior properties, such as increased mechanical strength or thermal stability. Perhaps the best known example of an engineering plastic is Kevlar [II], which finds use in a range of applications including body protection.

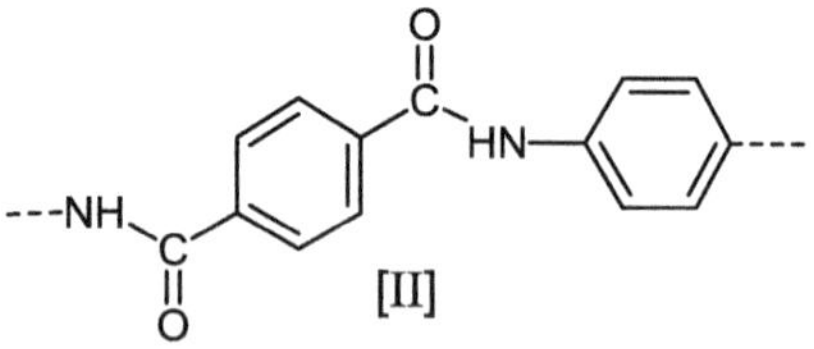

Notwithstanding the remarkable properties of materials such as Kevlar, modern engineers are constantly looking for improved performance. One of the most successful areas in this regard is the development of composites; such materials are manufactured in such a way that the final material has superior properties to the sum of its constituents. For example, fibres of Kevlar, glass or carbon are used to reinforce epoxy resins. Increasingly developments are focusing on more sophisticated materials; for example, poly (ether ether ketone) can replace epoxy as the matrix, and more complex reinforcing materials have been used for example carbon nanotubes (Coleman et al. 2006). Indeed, the new Boeing 'Dreamliner' airplane introduced in 2014 is structural based around composite materials.

1.5 Types of Order

1.5.1 Crystalline

The crystalline state is a state of matter ideally characterized by three-dimensional, long-range order on an atomic scale. A crystallizable polymer is polymer that is able to partially crystallize (IUPAC 1997). Large single crystals of polymers are not observed except in the case of polydiacetylenes as they can be prepared via a topochemical polymerization of the single crystals of the monomer. Single crystals of polymers such as polyethylene can be prepared from solution (Geil 1963), whereas polymers crystallized from the melt phase contain both crystalline regions and uncrystallized amorphous material. The crystalline regions are in the form of chain folded lamellae (see Sect. 1.6).

1.5.2 Liquid Crystalline

A liquid crystal polymer is a material that, under suitable conditions of temperature, pressure, and concentration, exists as a liquid crystal mesophase (IUPAC). Liquid crystal phases are those characterized by a high level of orientational order in the absence of positional order as observed in the crystalline state. The nematic state is observed with rod-like molecules with a common axis of orientation as shown in Fig. 1.5.

Fig. 1.5 Schematic representation of the molecular order present in the nematic liquid crystal phase. The *line* represents the rod-line molecules

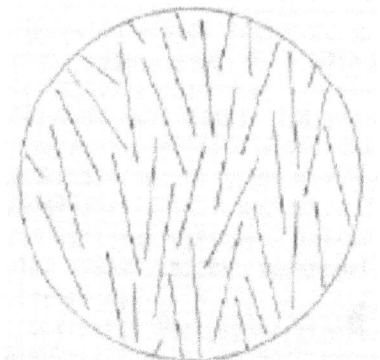

In a polymer, the rod-like structures can be attached as side groups—side-chain liquid crystal polymers or with the skeletal backbone—main chain liquid crystal polymers (Donald et al. 2006). The latter usually exhibit liquid crystal characteristics at elevated temperatures, while some side-chain liquid crystal polymers exhibit liquid crystalline order at room temperature. A number of more ordered smectic phases can be observed as well as chiral, nematic, and smectic phases (Donald et al. 2006).

1.5.3 Amorphous

The amorphous state is a state of matter characterized by the absence of long-range molecular order. Polymers exhibit this disordered but solid state either as a consequence of a thermal history which includes cooling from the melt state at a rate which inhibits crystallization or through the presence of disorder along the polymer chain such as present in random copolymers or atactic systems. Typically glassy polymers are optically clear and a key characteristic is the glass transition temperature. The transformation in molecular dynamics from that of a polymer melt to those of the glassy state does not occur over a very short temperature range as in the case of a first-order phase transition such as melting. Rather, it extends over several Kelvins and a plot of heat capacity of thermal expansion coefficient exhibits a smooth step. The glass transition temperature is defined by convention in which the heating rate is a critical factor.

1.5.4 Blends and Mixtures

As with other material systems, different types of polymers can be mixed to provide control of properties. A polymer blend is a macroscopically homogeneous mixture of two or more different species of polymer [IUPAC]. The low entropy of mixing of polymers due to the high molecular weight means that different polymers do not readily form a homogeneous blend; rather they form mixtures with a phase separated structure whose morphology will depend on the preparation route. The presence of a solvent further complicates matters.

1.6 Structuring Processes

1.6.1 Crystallization

It may come as a surprise to many that a large number of the common plastic items in everyday use are semicrystalline. As the IUPAC definition states, 'A crystallizable polymer ... is able to partially crystallize.'

1 Scales of Structure in Polymers

Crystallinity in polymers was first observed by X-Ray Diffraction even before these materials were understood to be polymers. Prominent examples were polysaccharides, especially cellulose (Herzog et al. 1920) and stretched natural rubber (Katz 1925). This was before Staudinger's macromolecular concept was widely accepted (Morawetz 1995). By the time synthetic crystalline polymers such as polyethylene (in 1933) and nylons (in 1935) were developed, it was generally acknowledged that the unit cell of a polymer crystal was based on repeating monomer units, rather than on the whole molecules.

Polymer crystallization is dominated by the process of untangling molecules and then straightening them onto the crystal growth face. Because of this, it is a comparatively slow process compared with crystallization of simple molecular species, and it may require supercooling by tens of Kelvins to occur at a significant rate.

Not all polymers can crystallize. Polyethylene does, but atactic polystyrene (the common form) does not. Though the specialized syndiotactic and the more common isotactic forms do, albeit into completely different crystal forms (de Rosa and Auriemma 2013). Isotactic polypropylene and many nylons commonly crystallize into more than one crystal type, and polyethylene terephthalate (PET), a polyester, is commonly found in a glassy form in soft drinks bottles but crystallizes readily on appropriate thermal treatment.

1.6.1.1 Lamellae to Spherulites

Up to about 1955, the only widely held model of semicrystalline polymer morphology was the **fringed micelle model**. In this, polymer crystallites were composed of small stacks of parallel chains, outside of which chains wandered between crystallites rather than folding back into the same bundles. This model, however, did not fit well with the observation (Bunn and Alcock 1945) that in polyethylene spherulites, the molecular chains lay perpendicular to the radial growth direction of the spherulites, nor with their suggestion that the crystallites might grow radially like the thin plate-like crystals of high molecular weight n-paraffins.

A great advance in the understanding of polymer crystals came with the discovery that polyethylene would crystallize from solution in hot xylene to give thin crystals, known as **lamellae**, which could be observed under the electron microscope. Electron diffraction showed that the chains ran more-or-less perpendicular to the plane of the crystals, and subsequent investigation has shown that this is the normal mode of crystallization for almost all polymers. These were reported by several workers, among whom Keller (1957) deduced that the chains must fold at the top and bottom surfaces of the crystals, so that the chain re-entered, going in and coming out of both surfaces. This had been suggested (Storks 1938) from observations of thin films of gutta-percha, which however did not contain isolated individual crystals.

Polymers on the trigonal and tetragonal systems tend to form lamellae based on their crystal system, hexagonal in the case of iPS and POM, square in the case of

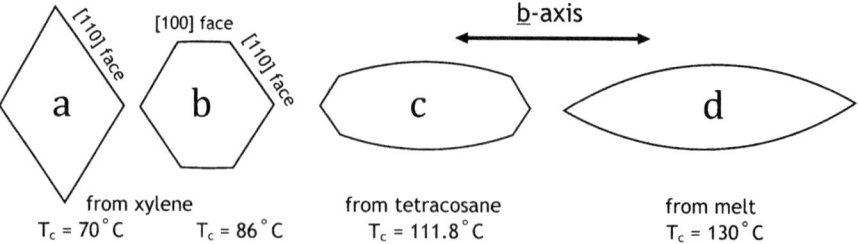

Fig. 1.6 Shapes of polyethylene crystals grown from solution or melt under the conditions indicated

poly4-methyl pentene-1. Polyethylene on the orthorhombic system, when grown from solution, does produce lozenges with [110] faces as at **a** in Fig. 1.6. When grown at the higher temperatures truncated lozenges **b** are formed with [100] faces at each end. For progressively poorer solvents and at higher temperatures **c**, the [110] faces become increasingly dominant and curved, while from the melt **d** only these [100] faces are observed. This is consistent with the observation that in polyethylene, the radial growth direction of spherulites is along the b-axis (Bassett et al 1979).

At the same time, as solution crystals were observed by electron microscopy, Fischer observed screw dislocations in electron micrographs of melt crystallized polyethylene-banded spherulites (Fischer 1957), which have since been shown to be the primary mode of branching in polymer lamellae. This, together with splaying out of the plane of the original lamellae, forms the mechanisms of three-dimensional growth into spherulites. From small initial lamellae, there develop objects which are generally called axialites because they were thought to be axially symmetrical like a wheatsheaf with splaying stalks. Where the initial crystal is of high symmetry, as in the hexagonal lamellae of isotactic polystyrene, intermediate stages of growth are often called hedrites because of their resemblance to mathematical polyhedra. These early objects, however, appear very different according to which direction they are viewed: this is particularly clear in the case of isotactic polystyrene (Bassett and Vaughan 1985) and polypropylene (Olley and Bassett 1989).

The wide variety of the morphologies observed, both in crystals formed from solution and from the melt, including precursors to spherulites and more developed versions especially banded spherulites, are presented in 'Structure of Polycrystalline Aggregates', together with discussion of current theories of their formation, by Crist (2013).

1.6.1.2 Crystal Growth Rates

Much effort has gone into the determination of crystal growth rates in polymers. In practice, this involves determining how fast spherulites grow, so what is determined is the fast growth direction for the polymer in question, for example, the b-axis in polyethylene and the a*-axis for the common α-form of polypropylene. This has led to a large number of theories of crystal growth in polymers.

1.6.1.3 Primary and Secondary Nucleation

One early theory which is still widely applied is that of Hoffman and Lauritzen, in which a nucleus forms on a flat crystal surface. First proposed in 1961, in 1973 it was extended to include regimes of crystallization, initially two (Lauritzen and Hoffman 1973), later extended to three. In Regime 1 (Fig. 1.7), a single molecular chain overcomes a nucleation barrier to settle on the growth surface, producing a step from which further chains can quickly be added. In Regime II at lower temperature, several chains and sideways development are simultaneously present on the same surface, and in Regime III growth is extending outward before completion of the first new layer. In the initial presentation of regime theory, there was observed an apparent correlation of 'axialitic' growth with Regime I and more classical spherulitic growth in Regime II; this was later shown to be a consequence of the different times available for chain fold reorganization in the two regimes (Abo El Maaty and Bassett 2006). The theory was modelled on flat crystal growth surfaces, and received a strong challenge from the discovery of curved growth (Organ 1986). A strong attempt was made to bring both nucleation and curvature into line (Point and Villers 1992) and modifications continued (Armistead and Hoffman 2002). A state-of-the-art review (Cheng and Lotz 2005) provides an excellent starting point for further reading. Some recent theories dealing with both growth rate and crystal thickness involve the intervention of a mesophase as the first stage of crystallization: (Keller et al. 1994; Strobl 2005).

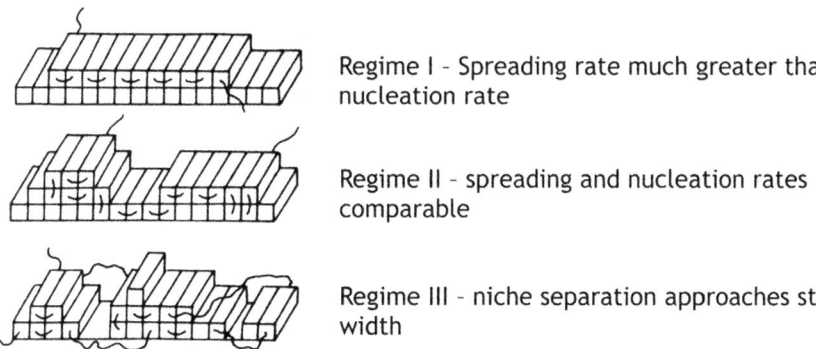

Fig. 1.7 Diagram of crystal growth in Regimes I to III according to the Hoffman–Lauritzen theory

1.6.1.4 Folding Theories

The question of folding resolves itself into three parts: (1) the topology of folding, whether chain folds back adjacent to itself, (2) the packing in the fold surface region, particularly how liquid or solid it may be, and (3) how frequently does the chain fold, a reciprocal space way of looking at how is lamellar thickness determined?

Regarding the first question, adjacent re-entry of stems into the same lamella is only approximated in solution crystals and material crystallized very slowly from the melt. On faster crystallization, stems are more likely to re-enter the same basal surface at some distance from where they emerged, known as the 'switchboard' model. Moreover, unless crystallization is slow enough for a molecule to reconfigure themselves in the melt by the process known as reptation, only local reorganization is possible as in the solidification model pictured below (Fig. 1.8) (Dettenmaier et al. 1980). The shape of the molecule as a whole will not be grossly changed, and so its radius of gyration will not be very different from what it was in the melt. Neutron scattering experiments do indeed show that radius of gyration is very little changed on fast crystallization from the melt, but in solution crystallized lamellae it is of the same order as the lamellar thickness, even for long molecules (Sadler 1984).

From the melt, faster crystallization is known to produce a greater proportion of 'amorphous' in relation to crystalline material, to which an change to switchboard-like rather than adjacent re-entry is only one of several contributory factors. Where there is a large amount of inter-lamellar material, there is the possibility of much of this being similar in properties to the melt. However, near the crystal surface its motion is constrained, and its reduced mobility compared to the rubbery behaviour

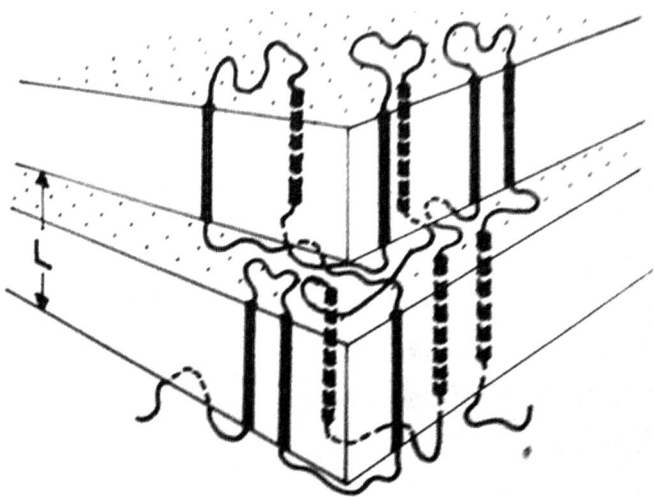

Fig. 1.8 Crystallization according to the solidification theory (Dettenmaier et al. 1980)

of material further away from the lamellae is to be expected. It is observed as an 'interphase' measurable by ^{13}C NMR (Kitamaru et al. 1986), and its existence as a rigid-amorphous' phase in poly(oxy-l,4-phenyleneoxy-l,4-phenylenecarbonyl-1,4-phenylene) (PEEK) though thermal analysis (Cheng et al. 1986). Fold surfaces do not necessarily remain as they are laid down, but in polyethylene are able to reorder subsequent to initial crystallization, this process determining the characteristic S-shaped cross section of lamellae formed at all but the highest crystallization temperatures (Abo El Maaty and Bassett 2001).

1.6.1.5 Computer Modelling Theories

Lamellar thickness, folding, and growth rate are all interconnected phenomena, and in order to understand the crystallization mechanism as a whole, a wide variety of methods, mostly based on molecular dynamics and related methods have been developed to simulate polymer crystallization. For an excellent discussion, on these the reader is directed to Chap. 4 and (Rutledge 2013).

1.6.1.6 The Bell Curve

In terms of bulk polymers, we have talked so far about crystallization 'from the melt': however, if a polymer is cooled rapidly enough from the melt, it may not crystallize but instead form a glass. For polymers in general, the crystallization rate as a function of temperature passes through a maximum (Fig. 1.9). The first report

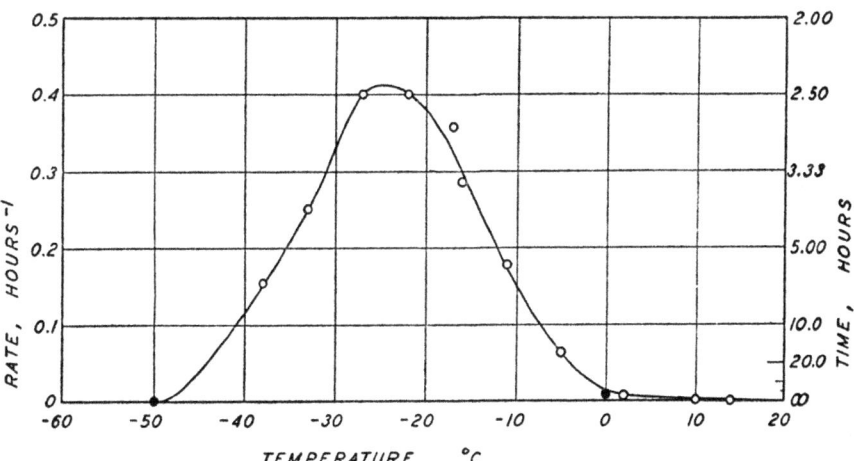

Fig. 1.9 Rate of crystallization of natural rubber over a range of temperatures, according to Wood and Bekkedahl (1946)

of this is attributed to Wood and Bekkedahl (1946) of dilatometric measurements of the crystallization of unvulcanized rubber in an unstretched state.

The best known case of this behaviour is polyethylene terephthalate (PET), which is most commonly encountered in the amorphous state as bottles. Whether heated from the glass or cooled from the melt, it will crystallize more rapidly the further below the melting temperature or the higher above the glass transition temperature, reaching a maximum from either side. Crystallization in polymers is a nucleation-controlled phenomenon: the graph above refers to overall degree of crystallinity, but if growth rate of spherulites is plotted, the maximum will be moved to somewhat lower temperature, since nucleation is denser at lower temperatures. The increase in growth rate while cooling from the melt is a thermodynamically based phenomenon, whereas the increase from the glass transition temperature is a result of decreasing viscosity.

In polymers which crystallize very rapidly, particularly polyethylene and isotactic polypropylene homopolymers, it is not feasible to cool quickly enough to obtain a glass (though with very thin films for some copolymers may be) so the full crystallization curve cannot be obtained for these materials.

1.6.1.7 Nucleation

When crystallizing a polymer such as polyethylene or polypropylene isothermally, it is generally found that the number of spherulites nucleated in a given volume at close to the melting temperature at low supercooling is relatively few, but if the temperature of crystallization is reduced, more spherulites are nucleated. A similar effect is observed where the polymer is crystallized non-isothermally during cooling, with faster cooling rates leading to overall lower crystallization temperatures. One advantage of higher crystallization temperature that may be desired is the greater lamellar thickness and higher degree of crystallinity leading to a stiffer material. Nucleating additives may be used to bring this about, along with economy due faster processing times. The lamellae which constitute spherulites are generally birefringent, so the radial orientation of lamellae in spherulites generally makes polymer objects translucent or even opaque. (Poly 4-methyl pentene-1 is an exception since the refractive indices of its optic axes are almost the same; moreover, the density of crystal and amorphous phases is very similar.) Since spherulites are optically anisotropic reduction of spherulite sizes to sub-micron levels will reduce optical scattering: in such uses the nucleating agents are known as clarifiers: clarified PP articles may be effectively as clear as glassy polystyrene (see Chap. 5). For a comparison of different nucleating agents, see Fillon et al. 1993.

1.6.1.8 Crystallization in Practice

Polymer melts are not molecularly homogeneous. Even in homopolymers, there is polydispersity of molecular weights, while with copolymers there is almost always a variation of comonomer content between different molecules. In homopolymers, fractionation by molecular weight during crystallization can occur, with longer molecules from the original dominant structure and short ones the subsequent infilling subsidiary structure. Much more pronounced effects are found in copolymers, where on continuous cooling molecules with less comonomer content crystallize first at higher temperatures, a range of subsequent melting points can be generated (Chen et al. 2004).

1.6.1.9 Crystallization and Orientation

One factor not taken into account so far is orientation of the polymers. The effect on crystallization can occur in many ways. Transient crystallization on stretching of rubbers, observed for natural rubber by Katz (1925), is also found in certain synthetic rubbers (Toki et al. 2004).

A study by Zachmann and Gehrke (1986) of crystallization of oriented amorphous films shows that in PET orientation leads to faster crystallization and in polyamide-6 leads to formation of the γ-rather than the common α-phase.

The subject of crystallization from stressed melts has been reviewed (Kumaraswamy 2005) and one frequent occurrence in the time sequence of this crystallization is the formation of row nuclei which give rise to structures known as 'shish-kebabs'. Polypropylene is the polymer best known for its remarkable propensity for forming row structures, which often contain considerable amounts of β-phase material (Olley et al. 2014), and are a major feature of the skin-core structure of injection-moulded polypropylene (Shinohara et al. 2012), but they are also found in polyethylene (An et al. 2006), isotactic polystyrene (Azzurri and Alfonso 2008), poly(phenylene sulfide) (Zhang et al. 2008) and polylactide (Xu et al. 2013).

Row structures are an instance of amplification of orientation in a specimen by crystallization onto oriented nuclei, but the other extreme is represented by highly oriented fibres, especially those of ultra-high modulus polyethylene (Abo El Maaty et al. 1999).

1.6.2 Microphase Separation and Block Copolymers (BCP)

In 1956, it was found (Szwarc 1956) that anionic 'living' polymerization could generate block copolymers. (These were not the first block copolymers to be prepared: in 1937 Otto Bayer had synthesized polyurethanes, and Pluronic® surfactants consisting of blocks of polyethylene oxide and polypropylene oxide were patented in 1954.) This method of synthesis went commercial when in 1961 Shell scientists, working to increase the green strength of polyisoprene rubber for tyre applications, discovered that by adding styrene monomer sequentially to the anionic polymerization of polyisoprene, strong thermoplastic elastomers were produced which required no vulcanization yet could be moulded into different shapes with heat. The (typically) 20 % of polystyrene (PS) would segregate into small spherical regions, effectively vulcanizing the material, in place of chemical cross-links. (These materials were called Kraton® after Kratos, the Greek god of strength and brother of Nike.) In a sense, these were 'effectively composite' materials, similar in principle to semicrystalline polyethylene but with rubber elasticity in the 'soft' phase and the glass transition of the PS allowing the material to be stable up to the boiling point of water, whereas a PE of 20 % crystallinity would melt well below this.

Although we do not know the details of commercial synthesis, in the laboratory such a polymer would be synthesized in solution in a solvent such as tetrahydrofuran (Douy and Gallot 1972) which would be a one-phase system. On removal of solvent, a simple mixture of polystyrene and polyisoprene would segregate into a two-phase blend with a coarse structure and poor mechanical cohesion, due to low entropy of mixing. When the two materials are joined in one molecule, however, the two phases are very limited in migration, and form a microscopic structure with the two segregated phases in close connection. An increase in domain size is promoted by the drive to minimize the surface area–volume ratio of the structure, which would reduce the interfacial free energy. However, growth is limited by the localization of the junction between chain segments at the domain boundary, and increasing domain size would also stretch the chains from their random coil (Gaussian) configurations leading to a reduction in entropy.

From 1970 reports started appearing that the segregated regions in block copolymers were organized in lattices of spheres, cylinders or lamellae, depending on the relative volumes of the components. Electron micrographs showing patterned spheres (Lewis and Price 1969), lamellae and cylinders (Mayer 1974) and (Douy and Gallot 1972). Such techniques are usable only with thin films or thin cryosections, and generally require selective staining with reagents such as osmium tetroxide or ruthenium tetroxide.

The electron micrographs in Fig. 1.10 show the (left) (100) and (right) (111) projections of the BCC macrolattice for a styrene-butadiene diblock

Fig. 1.10 Electron micrographs of two projection of the BCC microlattice of a styrene-butadiene block copolymer (Thomas et al. 1987)

1 Scales of Structure in Polymers

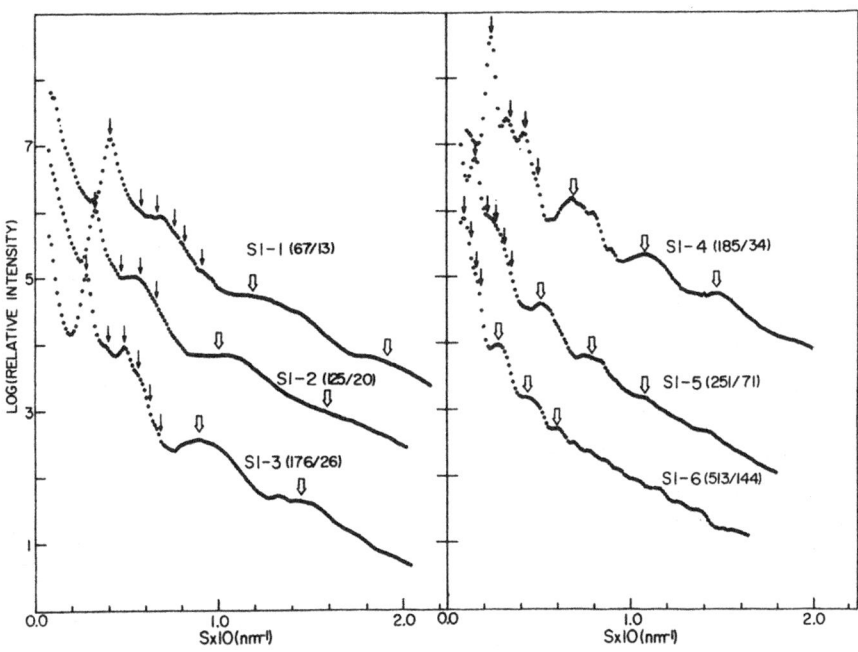

Fig. 1.11 Small angle X-ray scattering of a series of polystyrene-isoprene diblocks (Hashimoto et al. 1980)

(Thomas et al. 1987). The dark circular regions labelled 1, 2, and 3 are polystyrene spheres deposited on the section to confirm the axis and magnitude of tilt between the two projections. Optical diffraction patterns are given in the bottom right inset of each figure.

The importance of projection can be seen here: a hexagonal pattern or an apparent lamellae pattern by itself could also arise from packed cylinders, but tilting would reveal the specimen for what it is.

Graft copolymers can also produce segregated domains, similar in appearance to block copolymers, but less regular (Price et al. 1974).

Periodic structures can also be imaged by phase contrast (defocus) techniques, but these require care as artefacts are easily produced, especially from non-periodic structures, and it has been suggested that some reports concerning polyurethanes have been misinterpreted (Roche and Thomas 1981).

Figure 1.11 shows small angle X-ray scattering from a series of polystyrene-isoprene diblocks (labelling shows PS/PI segment molecular weights $\times 10^{-4}$). Shows peaks both from the lattice arrangement of the PI spheres (simple arrows at lower S) and scattering function of the original spheres (open arrows at higher S). The lower S-values at higher molecular weights derive from the ability of longer Gaussian chains to fit into larger spheres, and from the smaller proportion of chain junctions in a given volume of material.

Originally, only the 'classical' set of phases was observed, spheres, cylinders and lamellae. With time and increasing refinement of both theory and experiment other configurations, stable and metastable, were predicted and observed. The ordered bicontinuous double diamond was observed in a star block copolymer (Thomas et al. 1986) and the gyroid in a diblock copolymer (Hajduk et al. 1994). A recent theoretical calculation (Matsen 2012) is shown at left in the figure below, and compared with experimental observations complied from different sources (Matsen 2002) (Figs. 1.12 and 1.13).

Models predict that block copolymers are not totally segregated, but that there is an interphase in which both components interpenetrate, which increases the conformational entropy of the two parts. In highly dissimilar polymers, this will be very limited in extent, tending towards the strong segregation limit, whereas if they are quite similar, in a 50:50 system the composition profile may be almost sinusoidal, towards the weak segregation limit.

Crystallization in BCP where one or more of the segments are crystallizable can and does occur. A particularly interesting early study was of polycaprolactone/ polyoxyethylene triblock copolymers PCL-PEO-PCL. Where one of the components constitutes a distinct majority, that one crystallizes (Perret and Skoulios 1972a),

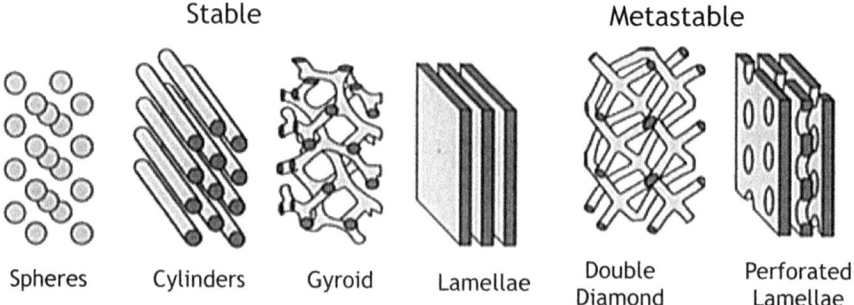

Fig. 1.12 Classical and non-classical block copolymer phases (Matsen and Bates 1996)

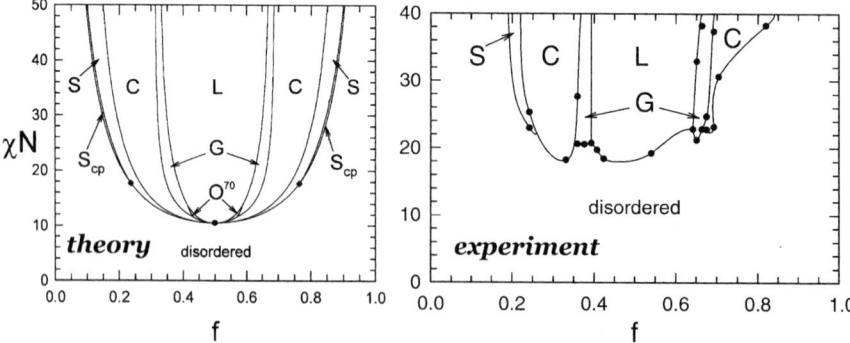

Fig. 1.13 Theoretical and experimental phase diagrams for an AB-type block copolymer melt (*left-hand figure* reproduced from Matsen (2012) and the *right-hand figure* from Matsen (2002))

but where the two are in close to equal proportion, below 51°, PEO crystallizes first: above 52 °C, PCL crystallizes first: between 51 and 52 °C, they crystallize simultaneously (Perret and Skoulios 1972b). In some systems, there is reported a conflict between microphase separation and crystallization, and spherulites of the crystallizable component may or may not form, depending on thermal treatment regime (Kim et al. 2001; Balsamo et al. 1998, 2003).

The ability to control the morphology of block copolymers renders them suitable for applications in fuel cells: especially ion-containing block copolymers, where one block is non-ionic and one or two blocks are based on aromatic sulfonic acids, hold promise as next-generation proton exchange membranes in hydrogen and methanol fuel cells (Elabd and Hickner 2011). Connected ion-rich nanodomains are particularly helpful in increasing conductivity. Reviews of other applications include: optoelectronics (Segalman et al. 2009) nanolithography (Jeong et al. 2013) and solar cells (Topham et al. 2011), which often employ donor–acceptor BCPs (Nakabayashi and Mori 2014).

1.6.2.1 Photonic Crystals

BCPs of low enough molecular weight to self-assemble easily generally only attain scales able to reflect green wavelengths. However, a block copolymer made of two similar brush copolymer segments, the only difference being a phenyl substituent at the end of one of the brushes, is able to self-assemble easily. The brush configuration allows the high molecular weight required for the appropriate scale to be achieved without significant entanglement, so that reflection onto the near infrared is achievable, the exact wavelength depending on (Miyake et al. 2012). See Chap. 10. Block copolymers can form not only photonic crystals, but phononic crystals, and even the possibility of the so-called phoxonic crystals combining both properties may be possible: a most instructive and informative review of recent developments gives some idea of the versatility of these materials (Lee et al. 2014).

1.6.2.2 Micelles

Block (and graft) copolymers can form colloidal aggregates, often in the form of micelles, in organic solvents if these are selectively poor for one of the polymeric blocks. Amphiphilic systems show surfactant behaviour in water and can exhibit micelle formation. The variety of block copolymer micelles is much too large to be explored here, but today they especially find widespread application in drug delivery systems (Gaucher et al. 2005) Amphiphilic block copolymers are particularly important in this regard, and a wide variety of nanostructures are formed (Letchford and Burt 2007). An interesting historical note is that Pluronics, among the first block copolymers to be invented, are still the topic of active research (Alakhova and Kabanov 2014), in the example cited their use to sensitize cancer cells to chemotherapy.

1.6.3 Phase Separation Mixtures

Whereas metals easily form alloys, polymers are generally encountered either unmixed, except for additives in small proportions or with completely different additions such as pigments or fillers or as composites. Single phase mixtures of two or more polymers are quite rare, though materials with two or more separate polymer phases are frequently used. Polymers generally are reluctant to mix intimately at the molecular level, for thermodynamic reasons.

Simple molecular organic compounds ('solvents' and 'monomers' in polymer parlance) generally have unfavourable (positive) enthalpy of mixing, but entropy of mixing is always positive and with a large number of molecules is comparatively large. So, chemically dissimilar species (as measured by the Hildebrand solubility parameter) are nevertheless able to mix. However, with oligomers and even more so with high polymers the number of molecules is considerably smaller and so is the entropy of mixing, tending to zero with 'infinite' molecular weight. Thus, even chemically quite similar species such as polyethylene and polypropylene will form a two-phase system in the melt. Even so, there are a considerable number of polymer pairs that are miscible. One common type is a polyester which has lone pairs of electrons with an electronegative halogenated polymer. Copolymers such as poly(styrene-acrylonitrile) also tend to be compatible with other polymers because of lower cohesive energy within the copolymer itself. Blends such as these display either upper or lower critical solution temperatures, depending on the specific kind of interaction between the two polymers (Tambasco et al 2006).

Figure 1.14 shows the case of a two-component blend with a lower critical solution temperature (LCST). The figure shows (upper panel) the free energy of the system as a function of composition at a chosen temperature T_0, and (lower panel) the phase diagram for the system. Between the inflexion points S' and S" the curve is concave upwards, and the mixture will spontaneously separate into components of compositions B' and B". Plotted against temperature on the phase diagram, S' and S" form a curve called the **spinodal** with B' and B" form the **binodal**. Inside the spinodal, the mixture is unstable, but in the metastable region between the two curves, nucleation is required for droplets of the minority phase to grow.

The two types of decomposition generate completely different micro-domain structures before they reach the ultimate equilibrium state. In the nucleation and growth process, the new phase corresponding to B' or B" takes the form of discrete small droplets in a matrix progressing towards B" or B'. When separation follows the spinodal decomposition process, an interconnected bicontinuous morphology develops, which possesses a characteristic length scale, determined by the fluctuation wavelength of fastest growth. The morphology gradually coarsens, and eventually the minority phase will separate into droplets. In practice, the types of morphology are easily distinguished, except at the most mature stages.

Where one of the components of a polymer blend is crystallizable, the process of crystallization will alter the liquid composition, and phase separation can in some circumstances be brought about (Tanaka and Nishi 1985). Regarding which

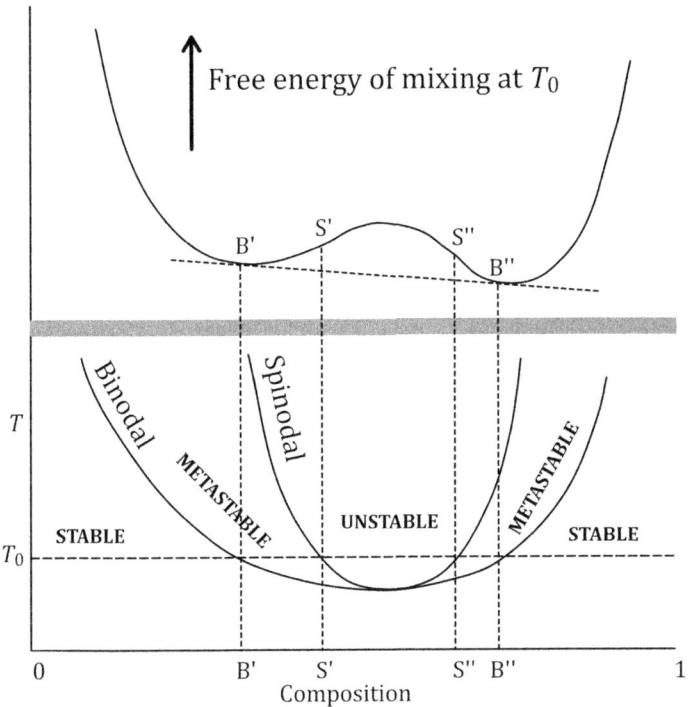

Fig. 1.14 Phase diagram of binary polymer blend with LCST (adapted from Higgins et al. 2010)

polymer blends display LCST and which display upper critical solution temperature (UCST) behaviour, a recent theoretical treatment incorporating the segment–segment nonbonded interaction energy (White and Lipson 2014) appears to approach the experimental situation much more closely than previous theories. Figure 1.15 shows thermodynamic parameters for UCST (left) and LCST (right) behaviour. The $T\Delta S$ (ideal mix) corresponds to what might be calculated from the traditional Flory–Huggins approximation.

When a polymer blend is sheared, and the molecules are stretched, there will be reduction in conformation entropy of the system. This in itself might apply equally to the mixed and unmixed states. However, in terms of enthalpy there is also the energy stored by the polymer molecules during flow, and this can have a significant influence on the magnitude and direction of the change in the phase boundary caused by shear. Computer simulations (Soontaranun et al. 1996) have indicated that flow induces mixing when the excess stored energy is negative and phase separation when the excess stored energy is positive.

Even in a phase separated mixture, each phase will contain a proportion of both components. Where the majority component of one of the phases is crystallizable, crystallization can occur rapidly in that phase, but extend into the phase where it is the minority component. This can be observed in blends of polypropylene with

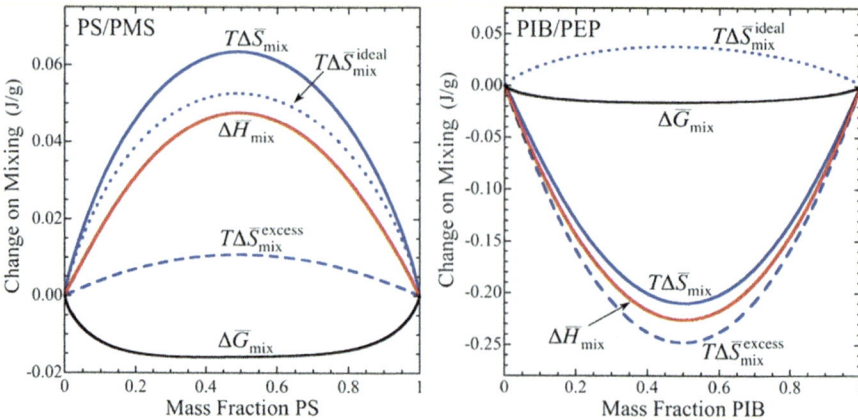

Fig. 1.15 Plots of thermodynamic parameters for mixtures exhibiting UCST (*left-hand figure*) and LCST (*right-hand figure*) from Lipson and White (2014)

linear-low density polyethylene, held for a period above the melting point of the LLDPE (Dong et al. 1998) and of polypropylene with non-crystallizable ethylene-octene copolymer (Svoboda et al. 2009). Where one component starts to crystallize in the compatible region of the phase diagram, the change in composition can cause the composition of the liquid to move into the binodal region, causing phase separation as in polycaprolactone/oligo-styrene blends (Tanaka and Nishi 1985)

1.6.4 Phase Separation on Reaction

Reaction Injection Moulding (RIM) is a technology that enables the rapid production of complex plastic parts directly from a mixture of two reactive materials of low viscosity. The materials are kept in separate containers until required, mixed in specific quantities and injected into a mould. This process allows large complex parts to be produced without the need for high clamping pressures. A large number of materials can be processed through the RIM process, such as polyurethane, nylon, epoxy resins and polyester systems. A typical RIM machine to produce polyurethane components is shown schematically in Fig. 1.16.

In comparison to injection moulding, the RIM process has many scales of structuring, chemical reactions at the smallest scale, products phase separating in to a morphology defined by the RIM operating parameters (Mateus et al. 2013). Multiscale evaluation of the process reveals that the scale morphology is defined at an early stage of the process, long before complete reaction has occurred (Mateus et al. 2016).

1 Scales of Structure in Polymers

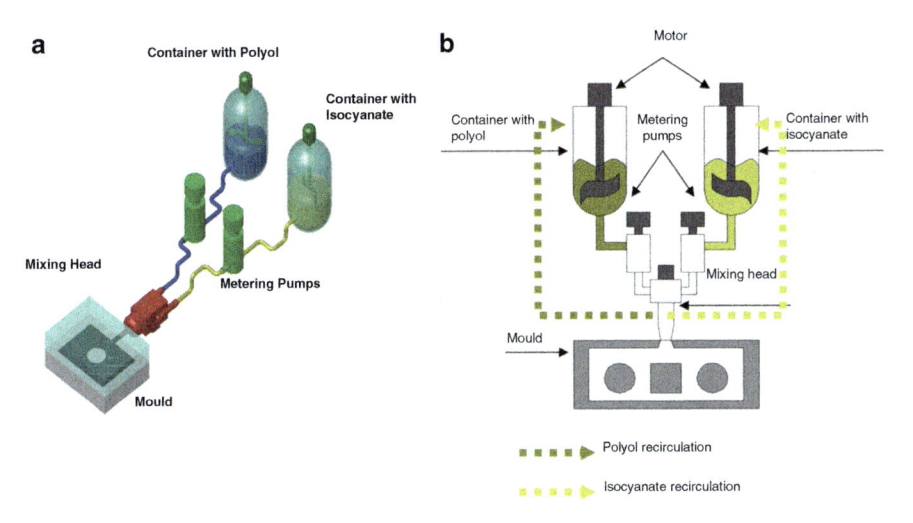

Fig. 1.16 A schematic of the RIM process for polyurethanes

1.7 Large-Scale Processes

The processes outlined above are driven by the forces which exist at the molecular level, a situation which is often described as bottom-up. Over the years, a number of techniques have developed which allows structures to be generated using a top-down approach.

Spin coating is a technique which is used to deposit a thin film of polymer on to a substrate, by spinning that substrate at high speed and introducing a droplet of a dilute solution on to the substrate. The spinning spreads out the solution and leads to high solvent loss leaving a thin film on the surface. This is widely used in the semiconductor industry in the patterning of silicon wafers to produce Large-Scale Integrated Circuits. Cardoso et al. (2012) have used multilayer spin coating to prepare thin films (300 nm to 4.5 μm) of PVDF.

It is generally found for a given solvent evaporation scenario that the film thickness h is related to the rotation speed ω by a power law of the form

$$h \propto \omega^{-b}$$

where b is in the range 0.4–0.9 (Yimsiri and Mackley 2006). The thickness of the films is also strongly dependent on the solvent evaporation rate.

Dip coating is another process for preparing thin films. A substrate is lowered into a solution and then pulled out at a constant and uniform speed. The film thickness is determined by the concentration and the removal speed. Film thickness for dip-coated films also follows a power law dependence on the withdrawal speed. The effect of the solvent evaporation is smaller than in the case of spin coating (Yimsiri and Mackley 2006).

Multilayer extrusion is a technique developed at Case Western University by Professor Eric Baer. The technique involves the co-extrusion of different polymers and can achieve multilayer films without the use of solvents (Mueller et al. 1997).

Electrospinning is a technique for producing nanoscale and microscale fibres from solution or the melt (Mitchell 2015).

Additive manufacturing/3D printing is a family of processes for building a component layer by layer from a digital definition file (Gibson et al. 2009). There are a variety of processes ranging from stereolithography to 3D printing. The latter is the subject of Chap. 7 and involves the deposition a liquid strand of polymer on to a moving stage which then solidifies to build up a 3D structure. The process can involve more than one type of polymer.

1.8 Summary

Polymers are rich in molecular structure variation at the smallest scale of structure, i.e., the scale of covalent bonds and these can give rise to structure and morphology changes at much larger scales which have a major influence on the final properties. The following chapters develop these ideas in specific and important areas of research and technology. The next chapter provides an overview of the common experimental techniques used to study these multiple scales of structures.

References

Abo El Maaty MI, Bassett DC (2001) On fold surface ordering and re-ordering during the crystallization of polyethylene from the melt. Polymer 42(11):4957–4963

Abo El Maaty MI, Bassett DC (2006) On the time for fold surfaces to order during the crystallization of polyethylene from the melt and its dependence on molecular parameters. Polymer 47(21):7469–7476

Abo El Maaty MIA, Olley RH, Bassett DC (1999) On the internal morphologies of high-modulus polyethylene and polypropylene fibres. J Mater Sci 34(9):1975–1989

Alakhova DY, Kabanov AV (2014) Pluronics and MDR reversal: an update. Mol Pharm 11 (8):2566–2578

An Y, Holt JJ, Mitchell GR, Vaughan AS (2006) Influence of molecular composition on the development of microstructure from sheared polyethylene melts: Molecular and lamellar templating. Polymer 47(15):5643–5656

Armistead JP, Hoffman JD (2002) Direct evidence of regimes I, II, and III in linear polyethylene fractions as revealed by spherulite growth rates. Macromolecules 35(10):3895–3913

Armstrong SR, Offord GT, Paul DR, Freeman BD, Hiltner A, Baer E (2014) Co-extruded polymeric films for gas separation membranes. J Appl Polym Sci 131(2):39765

Azzurri F, Alfonso GC (2008) Insights into formation and relaxation of shear-induced nucleation precursors in isotactic polystyrene. Macromolecules 41(4):1377–1383

Baekeland LH (1909) Method of making insoluble products of phenol and formaldehyde. US Patent 942,699

Balsamo V, Muller AJ, Stadler R (1998) Antinucleation effect of the polyethylene block on the polycaprolactone block in ABC triblock copolymers. Macromolecules 31(22):7756–7763

Balsamo V, de Navarro CU, Gil G (2003) Microphase separation vs crystallization in polystyrene-b-polybutadiene-b-poly(epsilon-caprolactone) ABC triblock copolymers. Macromolecules 36(12):4507–4514

Bassett DC, Vaughan AS (1985) On the lamellar morphology of melt-crystallized isotactic polystyrene. Polymer 26(5):717–725

Bassett DC, Hodge AM, Olley RH (1979) Lamellar morphologies in melt-crystallized polyethylene. Faraday Discuss Chem Soc 68:218–224

Bunn CW, Alcock TC (1945) The texture of polyethylene. Trans Faraday Soc 41:317–325

Cardoso VF, Minas G, Lanceros-Mendez S (2012) Multilayer spin-coating deposition of poly (vinylidene fluoride) films for controlling thickness and piezoelectric response. Sens Actuators A Phys 192:76–80

Chen F, Shanks RA, Amarasinghe G (2004) Molecular distribution analysis of melt-crystallized ethylene copolymers. Polym Int 53(11):1795–1805

Cheng SZD, Lotz B (2005) Enthalpic and entropic origins of nucleation barriers during polymer crystallization: the Hoffman–Lauritzen theory and beyond. Polymer 46(20):8662–8681

Cheng SZD, Cao MY, Wunderlich B (1986) Glass transition and melting behavior of poly (oxy-1,4-phenyleneoxy-1,4-phenylenecarbonyl-1,4-phenylene) (PEEK). Macromolecules 19(7):1868–1876

Coleman JN, Khan U, Gun'ko YK (2006) Mechanical reinforcement of polymers using carbon nanotubes. Adv Mater 18(6):689–706

Crist B (2013) Chapter 3: structure of polycrystalline aggregates. In: Piorkowska E, Rutledge G (eds) Handbook of polymer crystallization. Wiley, Hoboken

de Rosa C, Auriemma F (2013) Crystals and crystallinity in polymers: diffraction analysis of ordered and disordered crystals. Wiley, Hoboken

Dettenmaier M, Fischer EW, Stamm M (1980) Calculation of small-angle neutron scattering by macromolecules in the semicrystalline state. Colloid Polym Sci 258(3):343–349

Donald AM, Windle AH, Hanna S (2006) Liquid crystal polymers, 2nd edn. Cambridge University Press, Cambridge

Dong LS, Bassett DC, Olley RH (1998) On nucleation and isothermal changes in supercooling in a partially miscible polypropylene/polyethylene blend. J Macromol Sci B Phys 37(4):527–542

Douy A, Gallot B (1972) Polybutadiene/polystyrene/polybutadiene block copolymers—study of organized structures by X-ray-diffraction, electron-microscopy and differential scanning calorimetry. Die Makromolekulare Chemie 156(1):81–115

Elabd YA, Hickner MA (2011) Block copolymers for fuel cells. Macromolecules 44(1):1–11

Fillon B, Lotz B, Thierry A, Wittmann JC (1993) Self-nucleation and enhanced nucleation of polymers. Definition of a convenient calorimetric "efficiency scale" and evaluation of nucleating additives in isotactic polypropylene (α phase). J Polym Sci B 31(10):1395–1405

Fischer EW (1957) Stufen- und spiralförmiges Kristallwachstum bei Hochpolymeren. Z Naturforsch 12a: 753–754

Gaucher G, Dufresne MH, Sant VP, Kang N, Maysinger D, Leroux J-C (2005) Block copolymer micelles: preparation, characterization and application in drug delivery. J Control Release 109(1–3):169–188

Geil PH (1963) Polymer single crystals. Interscience (WIley) New York

Gibson I, Rosen DW, Stucker B (2009) Additive manufacturing technologies. Springer, New York

Hadjichristidis N, Iatrou H, Pispas S, Pitsikalis M (2000) Anionic polymerization: high vacuum techniques. J Polym Sci A Polym Chem 38(18):3211–3234

Hajduk DA, Harper PE, Gruner SM, Honeker CC, Kim G, Thomas EL, Fetters LJ (1994) The gyroid—a new equilibrium morphology in weakly segregated diblock copolymers. Macromolecules 27(15):4063–4075

Hashimoto T, Fujimura M, Kawai H (1980) Domain-boundary structure of styrene-isoprene block co-polymer films cast from solutions. 5. Molecular-weight dependence of spherical microdomains. Macromolecules 13(6):1660–1669

Herzog RO, Jancke W, Polanyi M (1920) Röntgenspektrographische Beobachtungen an Zellulose. II. Zeitschrift für Physik 3(5):343–348

Higgins JS, Lipson JEG, White RP (2010) A simple approach to polymer mixture miscibility. Philos Trans R Soc A 368:1009–1025

IUPAC (1997) Compendium of chemical terminology, 2nd ed. (the "Gold Book"). Compiled by A. D. McNaught and A. Wilkinson, Blackwell Scientific Publications, Oxford. XML on-line corrected version: http://goldbook.iupac.org (2006) created by M. Nic, J. Jirat, B. Kosata; updates compiled by A. Jenkins. ISBN 0-9678550-9-8. doi:10.1351/goldbook. Stefano V. Meille, Giuseppe Allegra, Phillip H. Geil, Jiasong He, Michael Hess, Jung-Il Jin, Pavel Kratochvíl, Werner Mormann, and Robert Stepto Pure Appl Chem 83(10): 1831–1871, 2011. doi:10.1351/PAC-REC-10-11-13

Jeong S-J, Kim JY, Kim BH, Moon H-S, Kim SO (2013) Directed self-assembly of block copolymers for next generation nanolithography. Mater Today 16(12):468–476

Katz JR (1925) Was sind die Ursachen der eigentümlichen Dehnbarkeit des Kautschuks? Kolloid-Zeitschrift 36(5):300–307

Keller A (1957) A note on single crystal in polymers—evidence for a folded chain configuration. Philos Mag 2(21):1171

Keller A, Hikosaka M, Rastogi S, Toda A, Barham PJ, Goldbeck-Wood G (1994) An approach to the formation and growth of new phases with application to polymer crystallization: effect of finite size, metastability, and Ostwald's rule of stages. J Mater Sci 29(10):2579–2604

Kim G, Han CC, Libera M, Jackson CL (2001) Crystallization within melt ordered semicrystalline block copolymers: exploring the coexistence of microphase-separated and spherulitic morphologies. Macromolecules 34(21):7336–7342

Kitamaru R, Horii F, Murayama K (1986) Phase structure of lamellar crystalline polyethylene by solid-state high-resolution 13C NMR: detection of the crystalline-amorphous interphase. Macromolecules 19(3):636–643

Kumaraswamy G (2005) Crystallization of polymers from stressed melts. J Macromol Sci C Polym Rev 45(4):375–397

Lauritzen JI, Hoffman JD (1973) Extension of theory of growth of chain-folded polymer crystals to large undercoolings. J Appl Phys 44:4340–4352

Lee J-H, Koh CY, Singer JP, Jeon S-J, Maldovan M, Stein O, Thomas EL (2014) 25th anniversary article: ordered polymer structures for the engineering of photons and phonons. Adv Mater 26 (4):532–568

Letchford K, Burt H (2007) A review of the formation and classification of amphiphilic block copolymer nanoparticulate structures: micelles, nanospheres, nanocapsules and polymersomes. Eur J Pharm Biopharm 65(3):259–269

Lewis PR, Price C (1969) Morphology of ABA block polymers. Nature 223:494–495

Lipson JEG, White RP (2014) Connecting theory and experiment to understand miscibility in polymer and small molecule mixtures. J Chem Eng Data 59(10):3289–3300

Mateus A, Bartolo P, Mitchell GR (2013) Chapter 4: rheology and reaction injection moulding. In: Mitchell GR (ed) Rheology: theory, properties and practical applications. Nova Publishers, New York. ISBN 978-1-62618-999-7978-1-62618-999-7

Mateus A, Bartolo P, Mitchell GR (2016) An predictive finite element model for simulating Reaction Injection Moulding submitted to Macromolecular Materials and Engineering

Matsen MW (2002) The standard Gaussian model for block copolymer melts. J Phys Condens Matter 14(2):R21–R47

Matsen MW (2012) Effect of architecture on the phase behavior of AB-type block copolymer melts. Macromolecules 45(4):2161–2165

Matsen MW, Bates FS (1996) Origins of complex self-assembly in block copolymers. Macromolecules 29(23):7641–7644

Mayer R (1974) Organized structures in amorphous styrene-cis-1,4-isoprene block copolymers—low-angle X-ray-scattering and electron-microscopy. Polymer 15(3):137–145

Mitchell GR (2015) Electrospinning: principles, practice and possibilities. Royal Society of Chemistry, London

Miyake GM, Piunova VA, Weitekamp RA, Grubbs RH (2012) Precisely tunable photonic crystals from rapidly self-assembling brush block copolymer blends. Angew Chem Int Ed 51(45):11246–11248

Morawetz H (1995) Polymers: the origins and growth of a science. Dover Publications, New York

Mueller CD, Nazarenko S, Ebeling T, Schuman TL, Hiltner A, Baer E (1997) Novel structures by microlayer coextrusion - Talc filled PP, PC/SAN and HDPE/LLDPE. Polymer Eng. Sci 37:355–362. DOI 10.1002/pen.11678

Nakabayashi K, Mori H (2014) Donor-acceptor block copolymers: synthesis and solar cell applications. Materials 7(4):3274–3290

Olley RH, Bassett DC (1989) On the development of polypropylene spherulites. Polymer 30(3):399–409

Olley RH, Mitchell GR, Moghaddam Y (2014) On row-structures in sheared polypropylene and a propylene-ethylene copolymer. Eur Polym J 53:37–49

Organ SJ, Keller A (1985) Solution Crystallisation of Polyethylene at high temperature Part 1: lateral crystal habits. J Mater. Sci. 20(5):1571–1585

Perret R, Skoulios A (1972a) Study of crystallization of block copolymers having sequence poly(epsilon-caprolactone)-poly(oxyethylene). 1. Copolymers with sequences of very different lengths. Makromolekulare Chemie 162:147

Perret R, Skoulios A (1972b) Study of crystallization of block copolymers having sequence poly(epsilon-caprolactone)-poly(oxyethylene)-poly(epsilon-caprolactone). 2. Copolymers with sequences of similar lengths. Makromolekulare Chemie 162:163

Point JJ, Villers D (1992) Crystals with curved edges: a unified model that mediates between the theories of nucleation-controlled and rough surface growth. Polymer 33(11):2263–2272

Price C, Singleton R, Woods D (1974) Microphase separation in a graft copolymer. Polymer 15(2):117–118

Roche EJ, Thomas EL (1981) Defocus electron-microscopy of multiphase polymers—use and misuse. Polymer 22(3):333–341

Rutledge G (2013) Chapter 6: computer modeling of polymer crystallization. In: Piorkowska E, Rutledge GC (eds) Handbook of polymer crystallization. Wiley, Hoboken

Sadler DM (1984) Chapter 4: structure of crystalline polymers. In: Hall IH (ed) Elsevier Applied Science Publishers, London, p 125

Segalman RA, McCulloch B, Kirmayer S, Urban JJ (2009) Block copolymers for organic optoelectronics. Macromolecules 42(23):9205–9216

Shinohara Y, Yamazoe K, Sakurai T, Kimata S, Maruyama T, Amemiya Y (2012) Effect of structural inhomogeneity on mechanical behavior of injection molded polypropylene investigated with microbeam X-ray scattering. Macromolecules 45(3):1398–1407

Soontaranun W, Higgins JS, Papathanasiou TD (1996) Shear flow and the phase behaviour of polymer blends. Fluid Phase Equilib 121(1–2):273–292

Stevens MP (1990) Polymer chemistry: an introduction, 2nd edn. Oxford University Press, New York

Storks KH (1938) An electron diffraction examination of some linear high polymers. J Am Chem Soc 60(8):1753–1761

Strobl G (2005) A thermodynamic multiphase scheme treating polymer crystallization and melting. Eur Phys J E 18(3):295–309

Svoboda P, Svobodova D, Slobodian P, Ougizawa T, Inoue T (2009) Transmission electron microscopy study of phase morphology in polypropylene/ethylene-octene copolymer blends. Eur Polym J 45(5):1485–1492

Szwarc M (1956) 'Living' polymers. Nature 178:1165

Tambasco M, Lipson JEG, Higgins JS (2006) Blend miscibility and the Flory Huggins interaction parameter: a critical examination. Macromolecules 39(14):4860–4868

Tanaka H, Nishi T (1985) New types of phase-separation behavior during the crystallization process in polymer blends with phase-diagram. Phys Rev Lett 55:1102–1105

Thomas EL, Alward DB, Kinning DJ, Martin DC, Handlin D Jr, Fetters LJ (1986) Ordered bicontinuous double-diamond structure of star block copolymers—a new equilibrium microdomain morphology. Macromolecules 19(8):2197–2202

Thomas EL, Kinning DJ, Alward DB, Henke CS (1987) Ordered packing arrangements of spherical micelles of diblock copolymers in two and tri dimensions. Macromolecules 20(11):2934–2939

Toki S, Sics I, Ran S, Liu L, Hsiao BS, Murakami S, Tosaka M, Kohjiya S, Poompradub S, Ikeda Y, Tsou AH (2004) Strain-induced molecular orientation and crystallization in natural and synthetic rubbers under uniaxial deformation by in-situ synchrotron X-ray study. Rubber Chem Technol 77(2):317–335

Topham PD, Parnell AJ, Hiorns RC (2011) Block copolymer strategies for solar cell technology. J Polym Sci B 49(16):1131–1156

Urbas AM, Fink Y, Thomas EL (1999) Self assembled one dimensionally periodic dielectric reflectors from block copolymer/homopolymer blends. Macromolecules 32(14):4748–4750

Ward IM, Hine PJ (2004) The science and technology of hot compaction. Polymer 45(5):1413–1427

White RP, Lipson JEG (2014) Free volume, cohesive energy density, and internal pressure as predictors of polymer miscibility. Macromolecules 47(12):3959–3968

Wood LA, Bekkedahl N (1946) Crystallization of unvulcanized rubber at different temperatures. J Res Natl Bur Stand 36(6):489–510

Xu H, Xie L, Chen Y-H, Huang H-D, Xu JZ, Zhong G-J, Hsiao BS, Li Z-M (2013) Strong shear flow-driven simultaneous formation of classic shish-kebab, hybrid shish-kebab, and transcrystallinity in poly(lactic acid)/natural fiber biocomposites. ACS Sustain Chem Eng 1(12):1619–1629

Yagpharov M (1986) Thermal-analysis of secondary crystallization in polymers. J Therm Anal 31(5):1073–1082

Yimsiri P, Mackley MR (2006) Spin and dip coating of light-emitting polymer solutions: matching experiment with modeling. Chem Eng Sci 6(11):3496–3505

Zachmann HG, Gehrke R (1986) Interaction between crystallization and orientation. In: Kleintjens LA, Lemstra PJ (eds) Integration of fundamental polymer science and technology. Springer, New York, pp 551–562

Zhang R, Min M, Gao Y, Lu A, Yu X, Huang Y, Lu Z (2008) Row nucleation phenomenon of poly (phenylene sulfide) under shear condition. Mater Lett 62(8–9):1414–1417

Chapter 2
Evaluating Scales of Structures

Saeed Mohan, Robert H. Olley, Alun S. Vaughan, and Geoffrey R. Mitchell

2.1 Introduction

The study of polymeric materials at several scales of structure will almost certainly require the use of a number of complementary techniques. This chapter provides an overview of the experimental techniques available for visualizing and evaluating polymer morphology. The chapter sets out to provide sufficient information to enable the reader to appreciate the research work described in the following chapters. Imaging techniques can provide invaluable information of many different scales. For the nanoscale, transmission electron microscopy is the technique of choice but at the expense of a much reduced sampling volume and more involved sample preparation. Recent developments in scanning electron microscopes, particularly in terms of field emission sources mean that nanometre resolution is readily available with the attendant advantages of the ease of screening large areas. However, as a surface topology technique we need to develop a contrast, which can be readily achieved using, for example, the etching techniques described in this chapter or through examination of a fractured surface. Scattering techniques probe a volume and in some cases information can be obtained in a time-resolving manner, so that the development of structure can be followed in real time. Neutron scattering provides a different contrast to X-ray scattering which can be useful with

---

S. Mohan (✉) • G.R. Mitchell
Centre for Rapid and Sustainable Product Development, Institute Polytechnic of Leiria, Marinha Grande, Portugal
e-mail: s.d.mohan48@gmail.com

R.H. Olley
EMLab, University of Reading, Whiteknights, Reading RG6 6AF, UK

A.S. Vaughan
ECS, Faculty of Physical Sciences and Engineering, University of Southampton, Southampton SO17 1BJ, UK

halogenated polymers or with metal fillers. Recent developments in pulsed neutron sources have resulted in instruments which can provide data in a single calibrated file over length scales from 0.1 to 10 nm. We also show how indirect methods such as thermal analysis can be used to obtain morphological data.

2.2 Indirect Methods

2.2.1 *Differential Scanning Calorimetry*

Differential scanning calorimetry (DSC) has two particular historical applications to polymers. One is the study of the glass transition (McKenna 1989) and the other is the study of crystallization and melting. Compared to simple molecular materials, polymer crystallization always involves considerable supercooling and a range of melting points depending on thermal history and relation to the thickness of the polymer lamellae, ideally as described by the Gibbs–Thomson equation. DSC is most often the technique of choice for observing both crystallization and melting and, ideally, there is a relationship between the two described by the Hoffman–Weeks equation (Hoffman and Weeks 1962).

According to Hoffman nucleation theory (see Sect. 1.6.1), if chains in a polymer crystal stayed as they were laid down, the crystal would melt almost immediately above its crystallization temperature (solid black line at 45° in Fig. 2.1); however, following subsequent processes the lamellar thickness increases by a thickening factor β (2 in the case of polyethylene) giving a slope with its reciprocal γ (here 1/2) in a Hoffman–Weeks plot. The theory of this is quite complicated, and the simple linear plot has been called into question (Marand et al. 1998). Nevertheless, it suffices as a very practical guide to the use of DSC in relating thermal history of a polymer specimen to lamellar morphology.

Testing this relationship was often limited by the fact that for lower crystallization temperatures or at very fast cooling rates (such as found in many industrial processes), thin crystals were formed which on melting could rapidly recrystallize and remelt at higher temperatures (a process generally referred to as reorganization). With recent advances in technology with fast scanning instruments, such crystals can simply melt once, and lower crystallization temperatures are obtainable in the instrument itself. The Hoffman–Weeks relationship has recently been thus tested and found valid for polyethylene (Toda et al. 2014); in Fig. 2.1 the blue diamonds represent conditions attainable in a traditional DSC and the red circles were obtained in a fast scanning calorimeter.

Many of the morphological phenomena in polymers with their concomitant calorimetric behaviour are well displayed in one material, namely isotactic polystyrene.

Figure 2.2 shows the melting endotherm of an isotactic polystyrene specimen which has been fully crystallized at 175 °C (Plans et al. 1984). The first peak I

2 Evaluating Scales of Structures

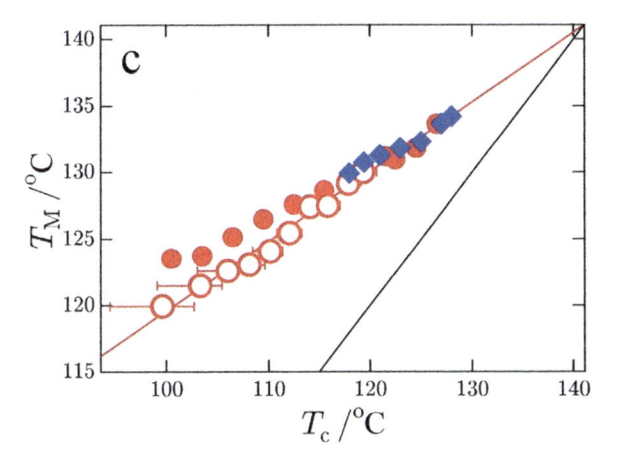

Fig. 2.1 Hoffman–Weeks plot for polyethylene crystallized in a fast DSC. The symbol *open circle* represents melting by fast scan DSC after isothermal crystallization, *filled circle* represents melting after a constant cooling rate and *filled diamond* represents the melting of a conventional DSC after isothermal crystallization. The *solid black line* plots $T_M = T_C$ (Part of Figure 6 reprinted from Polymer, 55(14), 3186–3194 Toda et al. Copyright 2014 with permission from Elsevier)

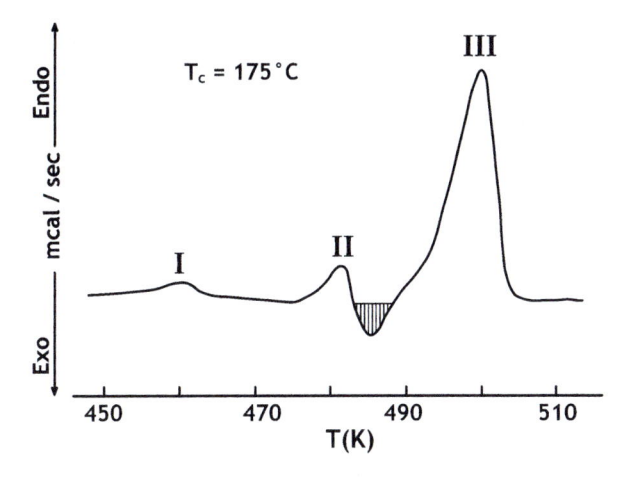

Fig. 2.2 DSC thermogram of pure isotactic polystyrene crystallized at 175 °C for 1 h. An exotherm feature appears between peak II and peak III (Scanning rate: 1.25 °C/min). Adapted with permission from figure 3 in Plans J, MacKnight WJ and Karasz FE, Equilibrium Melting Point Depression for Blends of Isotactic Polystyrene with Poly(2,6-dimethylphenylene oxide), Macromolecules 17:810–814. Copyright (1984) American Chemical Society

shows the melting of secondary crystals formed after the main spherulitic growth has taken place. At peak II the main crystal population starts to melt, but the melt is able to recrystallize rapidly on the still partly solid material, giving the exotherm (shaded). The crystals subsequently formed are more stable, and finally they melt giving peak III, at a temperature where any potential recrystallization is too slow to

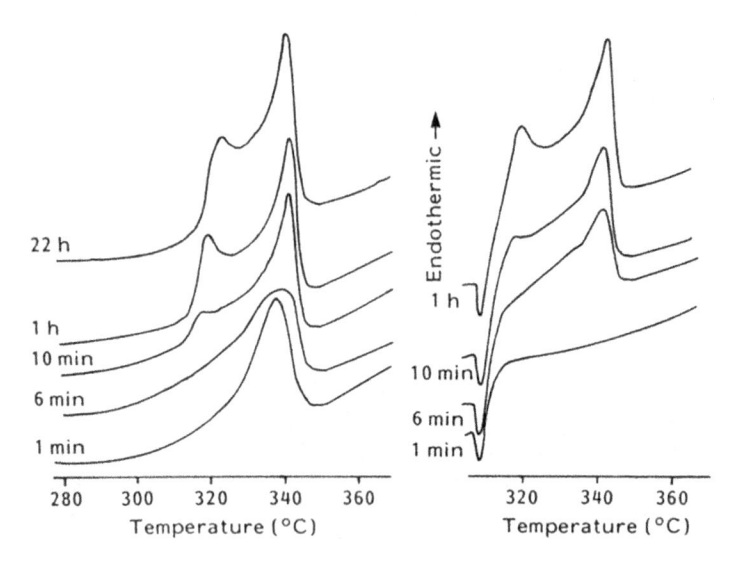

Fig. 2.3 PEEK crystallized from times stated (**a**) cooled after crystallization (**b**) reheated directly from the crystallization temperature (Reproduced with permission from Al-Raheil, Thesis, 1987)

take place over the timescale of the experiment (Plans et al. 1984). Lemstra et al. (1972) previously studied this behaviour and used the DSC not only as an analytical technique, but to prepare small specimens with precise thermal history.

Some early authors, particularly those working on poly ether ethyl ketone (PEEK), interpreted the secondary crystallization peak, which occurs just above the crystallization temperature, as being a primary crystallization peak which was subject to melting and reorganization during the scan. The phenomenon was interpreted and a second explanation presented clearly for other polymers by Yagpharov (1986). By reheating directly from the crystallization temperature, it could be shown that the lower melting peak developed subsequently, deserving the term 'secondary crystallization', and that the higher melting peak was a primary peak, possibly but not necessarily subject to reorganization. Figure 2.3 below is from work conducted at the University of Reading, UK using this technique, taken from the thesis of Al Raheil (1987).

Any interpretation is, however, dependent upon the effect of a crystal's environment on the kinetics of reorganization; in the case of PEEK, irradiation can be used to suppress reorganization kinetics (Vaughan and Sutton 1995) whereupon it appears more likely that the first-formed lamellar population is best thought of in terms of a continuous spectrum that spans the range contained between the apparently distinct primary and secondary DSC peaks. Something similar was proposed earlier (Lee and Porter 1987; Lee et al. 1989). But Lee and Porter's range of experimental observation was far too limited to categorically reject the occurrence of distinct secondary crystallization.

Secondary crystallization is widely accepted today. For example Lee et al. (2004) write:

> ... we propose that secondary crystallization involves the formation of short-range molecular order in the amorphous layers of the lamellar stacks as well as in the amorphous regions between the lamellar stacks. This short-range ordered structure, which is likely a type of single thin lamella, thin lamellae, or fringed micelle-like order, has a lower electron density than the lamellar crystal formed by primary crystallization.

but in ionomers at least Loo et al. (2005) have ruled out the fringed micelle suggestion.

2.2.1.1 Reorganization and 'Morphological Melting'

Reorganization is a very common phenomenon and can often cloud the interpretation of DSC data. Partial melting and recrystallization of a polymer specimen, generally known as annealing, gives rise to thicker crystals with a higher melting point. For specimens crystallized rapidly, this process can take place within the time frame of a typical DSC scan, often 10 min at 10 K min^{-1} in industrial use, typically up to eight times faster in a research laboratory.

A particular case in point is polypropylene. When isotactic polypropylene is crystallized at 145 °C and above, all the material is sufficiently stable not to undergo reorganization at usual scan rates. Typically an endotherm such as in Fig. 2.4 is seen, curve 1 (Weng et al. 2003).

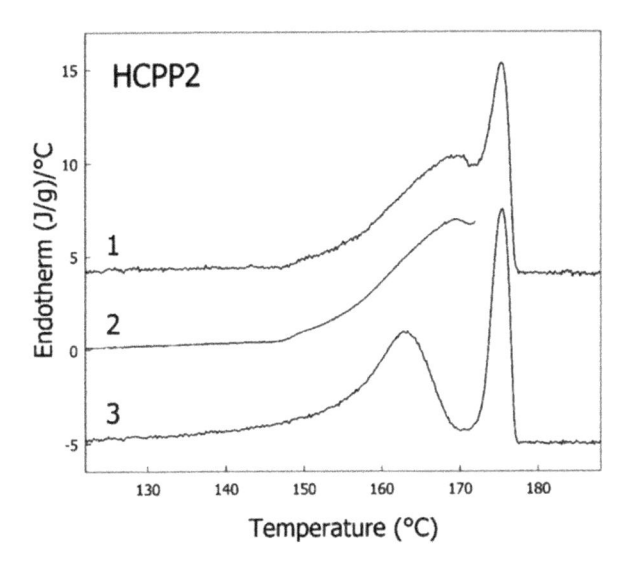

Fig. 2.4 DSC melting endotherms of high crystallinity polypropylene (*1*) crystallized from the melt for 3 h at 145 °C, (*2*) partially melting scan (*3*) fully remelted after the partial melting: From figure 3, Weng J, Olley RH, Bassett DC and Jääskeläinen P, Changes in the Melting Behavior with the Radial Distance in Isotactic Polypropylene Spherulites. J. Polym. Sci. Polym. Phys. 41: 2342–2354. Copyright © 2003 by John Wiley Sons, Inc. Reprinted by permission of John Wiley & Sons, Inc

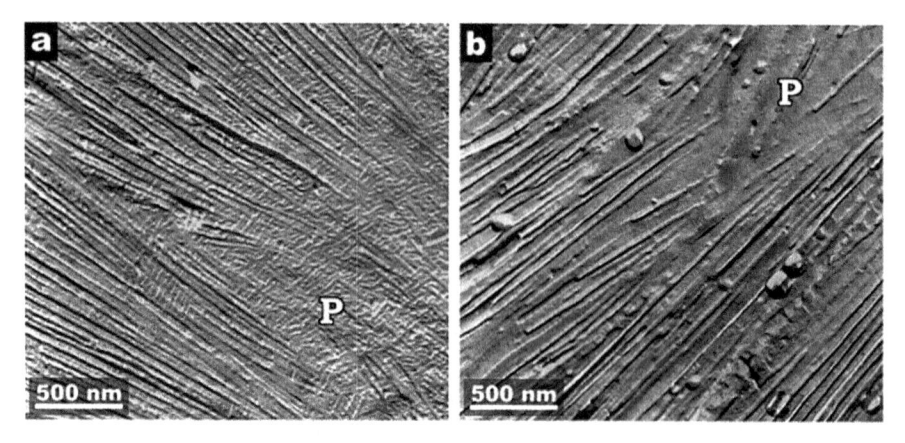

Fig. 2.5 Morphological effects of the recrystallization of isotactic polypropylene at 145 °C (**a**) radial growth enclosing a leaf-shaped area of cross-hatched material, (**b**) a similar area after partial melting (From figure 4 in. Weng J, Olley RH, Bassett DC and Jääskeläinen P, Changes in the Melting Behavior with the Radial Distance in Isotactic Polypropylene Spherulites. J. Polym. Sci. Polym. Phys. 41: 2342–2354. Copyright © 2003 by John Wiley Sons, Inc. Reprinted by permission of John Wiley & Sons, Inc.)

This can then be partially melted (following curve 2) and the specimen quickly removed and quenched, so that the material has recrystallized in a much lower melting state (curve 3). The morphological effect can be seen in Fig. 2.5. The lower melting peak is seen to relate to daughter lamellae formed between the dominant ones, and especially to highly cross-hatched regions which melt out entirely.

If a typical isotactic polypropylene is crystallized to completion at a lower temperature, say 130 °C, and heated at 10 °C/min, only one peak appears, but if heated at 40 or 80 °C/min, a double peak is observed. Even so, for polypropylene crystallized at 130 °C, a traditional DSC cannot heat fast enough, but in unpublished work we have seen that if the specimen is melted in a heating bath at temperature between the two peaks and quenched, then similar morphological effects are observed. Yamada et al. (2003b) also give an example of partial melting to reveal morphological heterogeneity.

However, the advent of superfast DSC shows that at lower crystallization temperatures, the melting difference between the two populations becomes less significant, leading to a one-peak situation (Toda et al. 2014). This technique is ideal for investigating the equilibrium melting point of isotactic polypropylene, but careful work even with traditional DSC (Yamada et al. 2003a) can be used to deduce reorganizational effect and thereby determine this temperature.

2.2.1.2 Thermal Fractionation

Polypropylene is a particularly marked example of how even a relatively homogeneous homopolymer can crystallize isothermally giving a variety of crystal populations with different melting points. A phenomenon generally observed is that the first-formed lamellae, called dominant because they form the basis of the spherulitic architecture, tend to be higher in melting temperature than the subsidiary ones which form later. The spread of melting points in the DSC (Gedde and Jansson 1983) can be accompanied by solvent extraction and gel permeation chromatography (Gedde et al. 1983) to show that in a homopolymer (in this case polyethylene) the crystallization process is accompanied by molecular fractionation, with the lower molecular weights crystallizing more slowly and being concentrated in subsidiary lamellae (Gedde and Jansson 1984).

Fractionation during crystallization occurs in copolymers also, thinking particularly of those where one comonomer constitutes 90 % or more of the monomer mixture. In these there is generally molecular heterogeneity with some molecules having greater comonomer content than others. One example is the commercially available propylene-ethylene 'random' copolymers (Weng et al. 2004) but the most significant by industrial volume are the medium and linear low-density polyethylenes. Different groups have developed their own particular thermal fractionation techniques using the DSC to investigate the heterogeneity of these materials (Arnal 2001; Müller and Arnal 2005; Shanks and Amarasinghe 2000; Chen et al. 2001).

It also gives indications as to how much blends of two different grades of polyethylene co-crystallize in blends. Figure 2.6 below is a comparison of a blend of two branched polyethylenes: at left is the melting endotherm after simple cooling, at right after a thermal fractionation technique.

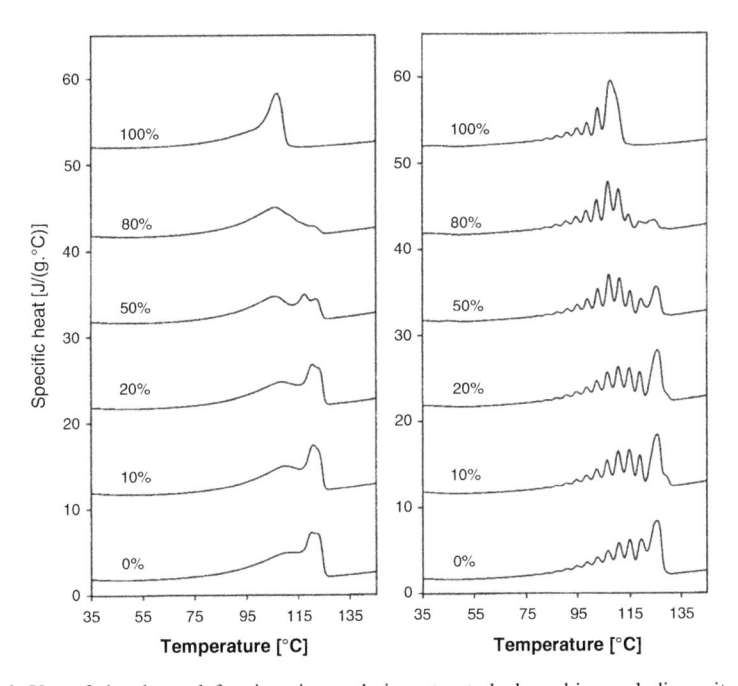

Fig. 2.6 Use of the thermal fractionation technique to study branching polydispersity. (*Left*) Specific heat curves of LLDPE–VLDPE2 blends after cooling at 10 °C/min; (*right*) the same after thermal fractionation (Reprinted from Polymer 41: 4579–4587, Shanks RA and Amarasinghe G. Crystallisation of blends of LLDPE with branched VLDPE Copyright (2000), with permission from Elsevier)

The Vaughan and Sutton (1995) paper is an example of how DSC can also be used to probe molecular degradation following irradiation. This can occur without gross apparent changes to morphology as revealed by TEM/etching. The paper immediately following Vaughan and Stevens (1995) does the same for PET, which is similar to the above-mentioned isotactic polystyrene in crystallization behaviour, although the processes occur considerably faster. However, the most salient feature of that work was that PET is easily degraded by relatively small doses of irradiation, especially compared to the highly radiation-resistant PEEK.

2.2.2 Dynamics

Dynamic Mechanical Thermal Analysis (DMTA) records the temperature-dependence of the viscoelastic properties and determines the modulus of elasticity and the damping values by applying an oscillating force to the sample. Changes in the morphology such as crystallization and phase separation change the viscoelastic properties. For example, Zheng et al. have used DMTA techniques in conjunction with infrared spectroscopy (see Sect. 2.2.3) to study the interactions in semi-interpenetrating polymer networks (Zheng et al. 2013). Dielectric spectroscopy measures the dielectric properties of a sample as a function of frequency. It is based on the interaction of an external oscillating field with the electric dipole moment of the sample. It is a technique for polar molecular systems and has been widely used to study molecular relaxations. Nanocomposites have a significant interfacial area and this has a substantial impact on the dielectric properties and hence dielectric spectroscopy is a very useful tool. A recent example of this concerns the study of a nanosilica/polyethylene system, where due to the lack of strong coupling between the applied field and the materials themselves, water was used as a dielectric probe. Since the solubility of water in polyethylene is extremely low, water molecules tend to associate with residual hydroxyl groups on the nanosilica surface, such that changes in surface structure and interfacial interactions manifest themselves as modified relaxation peaks within the so-called water shell (Lau et al. 2013).

2.2.3 Spectroscopy

Many different spectroscopy techniques can be used to provide structural information concerning polymers. In terms of vibrational spectroscopy, Strobl and Hagedorn (1978) showed how the Raman spectrum of polyethylene could be considered in terms of characteristic bands associated with the crystalline phase, a melt-like amorphous phase and a disordered phase of anisotropic nature. Comparison of structural data obtained in this way with equivalent measures obtained by densitometry and X-ray scattering revealed agreement. Similar approaches have been used to evaluate the crystallinity of isotactic polypropylene (Nielsen et al. 2002).

Nuclear magnetic resonance spectroscopy is a well-established chemical characterization technique widely exploited with polymer solutions to quantify the chemical microstructure of polymer chains (Bovey 2007). Developments in NMR technology and computing led to the development of solid state NMR. This enables site-specific information to be obtained from the solid state. This uses interactions like the dipolar coupling and chemical shift anisotropy to extract information from the solid state. More recently, multidimensional spectroscopy (2D and 3D techniques) correlate a structural parameter such as the chemical shift with parameters that carry information about dynamics and the level of order. Solid state NMR is often seen as strongly connected with crystallography as it provides similar precision information albeit of site-specific interactions and their distances (Schmidt-Rohr and Spiess 1994). As introduced in connection with dielectric spectroscopy, the nature of the interface between polymers and nanoparticles is an area that is of considerable interest and many different spectroscopies have been used to explore this. For example, Miwa et al. (2008) used electron spin resonance (ESR) spectroscopy to examine nanocomposites of poly(methyl acrylate) (PMA) and synthetic fluoromica, in which the PMA had been modified to include a nitroxide spin label. That is, a stable free radical introduced into a material that does not have an intrinsic paramagnetic response. This work demonstrated that, in exfoliated systems, the close proximity of the polymer chains and exfoliated fluoromica reduces the PMA chain mobility within a rigid interface region of thickness 5–15 nm.

2.3 Imaging Methods

Imaging techniques provide invaluable information on the morphology of polymers and nanocomposites. Of course an imaging technique is able to provide information on the spatial distribution of the structure and morphology. Each of the techniques considered here has particular advantages as well as specific sample preparation requirements. It is probable that the techniques employed in a particular project are selected as a consequence of the resolution matching the requirements, otherwise all techniques should be considered useful and effective especially when used in combination.

2.3.1 Light Microscopy

The optical microscope is typically used to reveal structures between 1 µm and 1 mm in size. Although the most intricate details of polymer morphology are revealed by the transmission electron microscope, and to some extent the scanning electron microscope and atomic force microscope, these tiny structures are usually

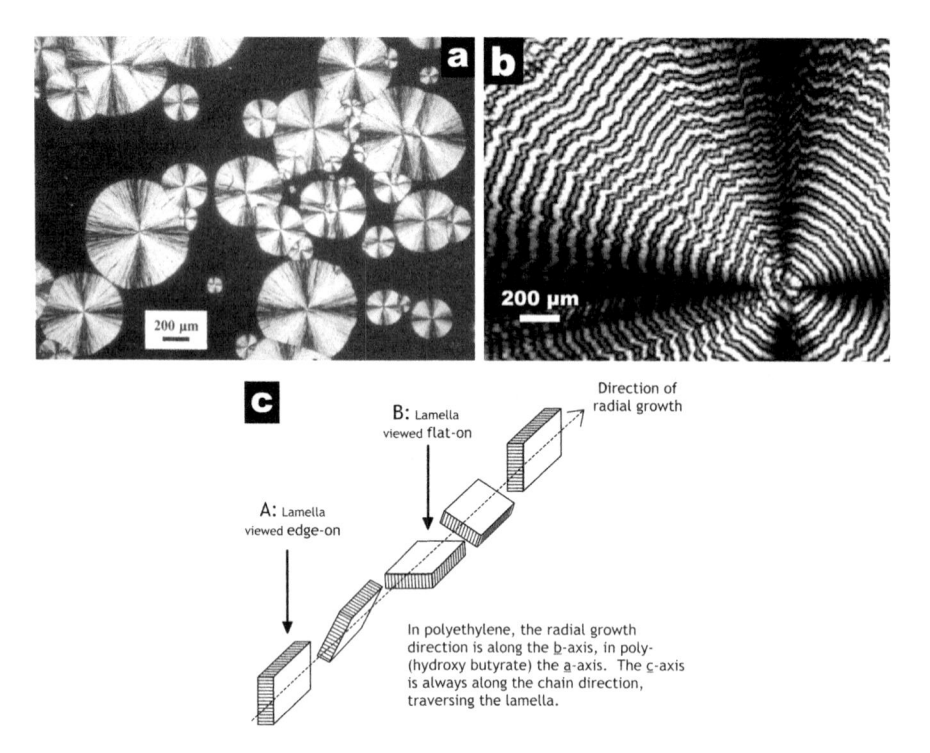

Fig. 2.7 (a) Spherulites in a thin film of polypropylene (Elaine Ann Perkins, University of Reading, UK, unpublished work, 1985), (b) spherulites in polyhydroxybutyrate-co-valerate (Reprinted with permission from Macromolecules, 2010, 43: 4441–4444 Wang Z, Li Y, Yang J, Gou Q, Wu Y, Wu X, Liu P and Gu Q (2010) Twisting of Lamellar Crystals in Poly (3-hydroxybutyrate-co-3-hydroxyvalerate) Ring-Banded Spherulites Copyright (2010) American Chemical Society, (c) Schematic of spherulite twisted orientations. Redrawn from Hsieh et al. (2014)

aggregated into larger structures; for example, the spherulites in Fig. 2.7a (left) are composed of lamellae which require electron microscopy for their resolution, but the variation in size and spatial distribution of the spherulites is much more easily studied by optical microscopy.

Optical microscopy involves a variety of different techniques, and a survey of these will illustrate the breadth of applications to polymer morphology.

The methods of specimen preparation for optical microscopy are various. In fundamental polymer studies, thin films of polymers can be crystallized between a slide and cover slip. In an apparatus such as a hot stage attached to a microscope, the crystallization process can be followed as it occurs. However, two features must be taken into account, firstly that in specimens where large spherulites form, their centres are confined to a narrow plane and are not distributed in depth as they would in a bulk specimen. Also, nucleation of crystal growth may be enhanced on the surface of the slide and cover slip, and transcrystalline layers, rather than the spherulitic structure of the bulk material, may form.

Most often, sectioning is required to examine structure of the material as a whole. For example, there will be great variation in morphology of an injection moulded item between the rapidly chilled surfaces where the polymer came in contact with the cold mould, and the interior where heat was lost much more slowly. However, sections do not have perfectly flat parallel surfaces; the most common form of damage is periodic compression giving rise to 'chatter', though with an inferior blade knife marks parallel to the cutting direction also occur. Interference from these can be greatly reduced by immersing the specimen in a liquid with a refractive index close to that of the polymer. For aliphatic polymers, glycerol ($n = 1.47$) is often a good general purpose liquid, since it is not taken up except by the most polar of polymers such as hydroxyethyl acrylate. For aromatic polymers such as polystyrene, Traylor (1961) dissolved potassium mercuric iodide in the glycerol to increase its refractive index, and there are now many commercial index matching fluids. Sections can also be compressed in such a way that originally circular features such as the bands in spherulites are distorted into ellipses; such distortion is accompanied by an overall background birefringence.

2.3.1.1 Polarized Optical Microscopy (POM)

One of the earliest and most commonly used microscopical methods of examining polymers is between crossed polarizers. Some of the earliest work was determining the birefringence of fibres, then came the study of spherulites in semi-crystalline polymers. Often the spherulites show a simple 'Maltese Cross' pattern where the dark areas show zero-amplitude birefringence, which simply arises from the orientation of the crossed polars. In the polypropylene spherulites (Fig. 2.7a), radial growth has occurred along the a*-axis which is the fastest crystal growth direction, while the b- and c-axis are effectively randomly oriented.

Often, however, the spherulites develop with a twisting orientation (though the individual lamellae are not necessarily twisted) producing a banded pattern when observed under POM. This is illustrated in Fig. 2.7c. For example, in polyethylene the b-axis forms the radial growth direction, but under certain growth conditions the a- and c-axis rotate about the radius with increasing distance from the centre. In polyethylene, the crystals are effectively uniaxial, with the higher refractive index along the chains, namely the c-axis. Looking down at A, the two different refractive indices along b and c cause the crystal to show birefringence. Looking down the c-axis at B, the refractive indices along a and b are effectively equal. This gives a simple alternation of light and dark bands, with the dark representing zero-birefringence extinction. Many other polymers, including many polyesters, are biaxial, and often show double banding (Wang et al. 2010) as observed in Fig. 2.7b.

Figure 2.8 (Kumaraswamy 1999) shows how POM can display orientation and structure gradients in injection moulded materials. These images show polypropylene crystallized in a cell designed to probe the different variables (time, temperature, shear rate) found in this process, and the two parts of the figure illustrate the formation of the characteristic skin-core morphology. The oriented near-skin region

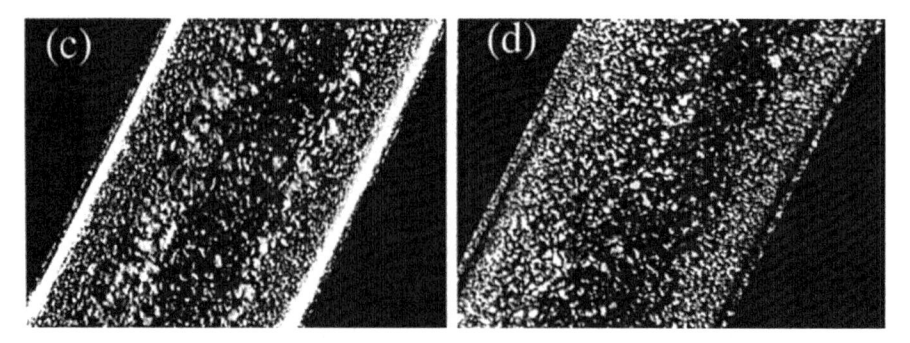

Fig. 2.8 Optical micrographs of sections through injection moulded polypropylene from figure 4 in Kumaraswamy, 1999 (**c**) viewed down the neutral direction (**d**) viewed down the flow direction Reprinted with permission from (G. Kumaraswamy, A.M. Issaian, and J.A. Kornfield Shear-Enhanced Crystallization in Isotactic Polypropylene. 1. Correspondence between in Situ Rheo-Optics and ex Situ Structure Determination, Macromolecules, 32, 7537–7547). Copyright (1999) American Chemical Society

appears bright when viewed down the neutral direction (c) and dark when viewed down the flow direction (d) indicating that the skin has crystallites with the cylindrical symmetry characteristic of polypropylene row structures.

2.3.1.2 Use of Tint Plates

A bright appearance between vertically and horizontally cross polars indicates that, in the diagonal directions, one refractive index is greater than the other, but does not indicate which is which. The ambiguity can be resolved by use of tint plates, of which the first-order compensation plate (quartz sensitive tint) is the most popular. This shifts one component of the beam by one green wavelength (550–580 nm), so that without a specimen, green light destructively interferes, and mainly red and blue are observed, giving rise to the characteristic magenta colour. One direction of the plate will be marked as slow or positive, and if the higher refractive index is parallel to this, a higher order blue interference colour will be produced, while if it is perpendicular, a lower order yellow colour will be seen.

Spherulites of aliphatic hydrocarbon polymers such as polyethylene generally have their molecular chains with the highest refractive index oriented perpendicular to the radial growth direction. This is the case with the polypropylene spherulites shown in Fig. 2.9a, which were crystallized at 140 °C and consequently contain very few cross-hatched lamellae, so in this respect they are like polyethylene spherulites. The lower refractive index in this view is radial in direction, giving a yellow colour where the radius is parallel to the slow direction of the tint plate; these are therefore termed negative spherulites.

In Fig. 2.9b, the specimen was crystallized at 120 °C, so that there is a preponderance of cross-hatching lamellae located roughly perpendicular to the dominant radially growing ones. This reverses the birefringence, giving positive spherulites

2 Evaluating Scales of Structures 41

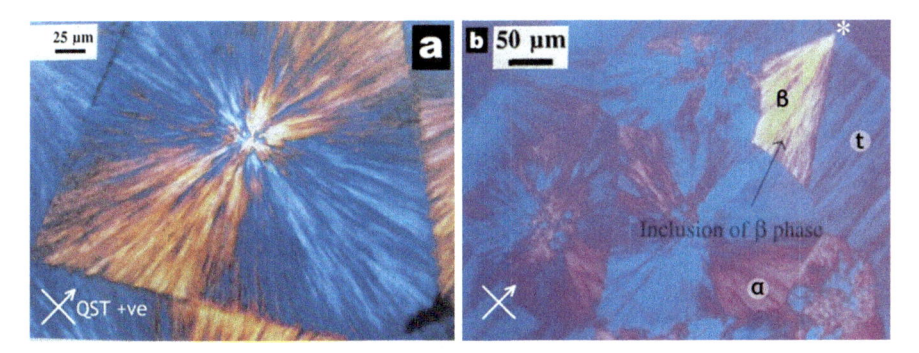

Fig. 2.9 Showing use of Quartz sensitive tint. (*Left*) polypropylene crystallized at 140 °C (−ve spherulite); (*right*) various orientations in film transcrystallized at 120 °C (+ve spherulites with −ve beta inclusion) (Reproduced with permission from White HM, University of Reading thesis, 1995)

in the majority α-phase. However, the isolated inclusion of the β-phase in the transcrystalline layer (t) does not contain any cross-hatching, and it displays strong birefringence characteristic of a negative spherulite.

One hydrocarbon polymer that gives positive spherulites is isotactic polystyrene; here the effect of aromatic rings with strong polarizability perpendicular to the molecular chains dominates over the polarizability of the chain.

Another form of tint plate commonly used is the mica quarter-wave plate with a retardation of 140–155 nm. This gives a grey background becoming brighter or darker with the birefringence of the specimen, and is suitable for black-and-white imaging.

2.3.1.3 Circularly Polarized Light

Dark areas observed in the use of crossed polarizers by themselves can be due to zero-amplitude or zero-birefringence extinction, and sometimes it is advantageous to remove the zero-amplitude extinction. This can be achieved by including a pair of crossed quarter-wave plates in the optical path (Olley and Bassett 1989). When a spherulitic sample is observed in the way, the dark Maltese cross is no longer part of the picture, and one can see the spherulites as bright objects with their mutual boundaries much more clearly displayed. This is particularly helpful when looking at the early stages of spherulitic growth, where the growing object takes a form often called a 'hedrite' or 'axialite', or in the special case of isotactic polypropylene, a 'quadrite'. Figure 2.10 shows a film of polypropylene with quadrites in various orientations, with the top picture displayed in circularly polarized light. The object at top left is seen in one of the 'cardinal' views, with the b-axis of the earliest formed crystals pointing towards the observer. Without the Maltese cross, the fourfold rosette form of the object is clearly displayed. Moreover, the dark 'leaf-like' structure at the centre is a characteristic feature of the development. However,

Fig. 2.10 Use of circularly polarized light (top) and phase contrast (bottom) to show development of polypropylene spherulites at 150 °C (Reprinted from Polymer 30: 399–409, Olley RH and Bassett DC. On the Development of Polypropylene Spherulites, Copyright (1989), with permission from Elsevier)

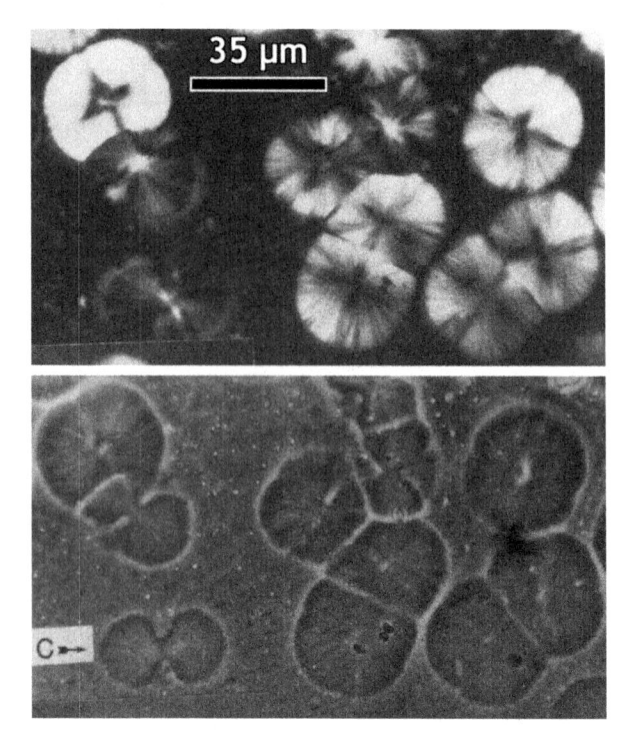

the general brightness indicates that over most of the object, the optic axis lies in the plane of the film, perpendicular to the b-axis. The two objects beneath appear very dark and sheaf-like in form, but were they developing freely in a bulk specimen they would be the same in form as the one above, but with their b-axis located in the plane of the film, and their optic axis (effectively the **c**-axis) pointing towards the observer.

Where spherulites of more than one crystal phase are growing together, use of circularly polarized light makes delineating and distinguishing the different types of object much easier (He and Olley 2000).

2.3.1.4 Phase Contrast Microscopy

This refers to the technique invented by Zernike, which makes visible the phase shifts produced in a transparent specimen by local variations in refractive index. For the technique as applied to polymers, see Hemsley (1989), Chapter 2; for a description of how the technique works, see Murphy and Davidson (2012), Chapter 7.

An ideal phase contrast specimen is either a well-prepared section or a thin film melted between slide and cover slip. Much of the problem of knife marks or other surface structure in sections and thin films can be alleviated by 'oiling out' with a suitable liquid (Traylor 1961).

The method is very suitable for examining particle dispersions and inclusions, especially the rubbery inclusions in toughened plastics. The method is sensitive, and does not involve polarized light, so can reveal features in specimens with overall or localized orientation which might be lost where there is strong birefringence contrast. An example is the clear revelation of the outlines and some internal structure in the early development of polypropylene spherulites, as in the lower part of Fig. 2.10. However, a halo is produced around the objects picked out, and where there is a large concentration of particles, or in thicker specimens, the overlapping haloes can confuse the picture.

An early application was the examination of polyethylene crystals formed from dilute solution, which are 'phase objects par excellence' (Keller 1968). As is generally the case, the technique proved an ideal precursor to electron microscopy (Keller and Bassett 1960). Quite recently, the technique has been used to observe the development of the earliest stages of polypropylene crystals directly in the melt (Yamada et al. 2011).

2.3.1.5 Interference Microscopy (Nomarski)

There is a variety of techniques which come under the heading of interference microscopy, and one early example is the determination of the thickness of single crystals with a two beam microscope (Wunderlich and Sullivan 1962).

However, the most widely used type of interference microscopy is that giving Differential Interference Contrast (DIC), widely known as Nomarski contrast.

The Nomarski technique is directional, because the very small separation of the two beams, about 1 µm, samples small differences in height along the splitting direction only. So, for example, if one is examining a piece of polypropylene tape whose surface is covered with ridges parallel to the tape length, then if the specimen is rotated so that the ridges are perpendicular to the beam separation, they will appear heavily contrasted, but if they lie parallel, then they will almost disappear.

In the system used by the author, the two beams were separated along a diagonal in the field of view. Figure 2.11 shows how they could be used to determine whether a blemish in an otherwise flat surface was a pit or a dome.

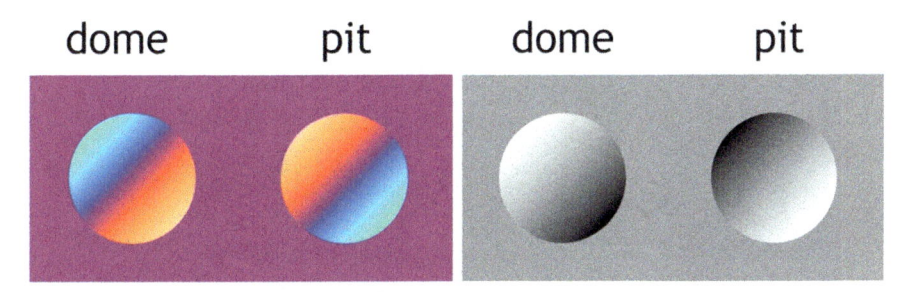

Fig. 2.11 Appearance of domes and pits in an otherwise flat surface under Nomarski: for description see text

Normally the microscope can be adjusted so that, on a flat surface, the two beams were out of phase by 1 wavelength of green light, so that the surface appeared first order purple (the same colour as given by a quartz sensitive tint). Higher or lower order colours appeared according to whether the slope on the element of surface being sampled increased or decreased the phase difference. At right, the microscope is adjusted to give a smaller path difference, grey on the Newton colour scale, and this configuration is suitable for black-and-white photography.

This technique works very well for solution-grown crystals, but much better contrast is achieved if a reflective gold (or carbon) is applied. If the crystals are deposited on mica and carbon coated, they can be floated off for electron microscopy.

Where the surface itself is of interest, this is the technique par excellence. For example, blown polypropylene films used in food packaging sometimes show submillimetre 'haze rings', which degrade their appearance. These were studied by, among other techniques, Nomarski reflection, and related to the presence of defects in the polypropylene tube prior to blowing (Olley and Bassett 1994).

One very productive use of Nomarski contrast is examination of etched surfaces prior to electron microscopy. Here one does not want to coat the specimen, so in looking at an etched disc prior to replication, it is better to put it on some black card (surface to be replicated uppermost) to eliminate much of the scattered and reflected light from below.

Figure 2.12 shows (a, left) a micrograph (crossed polars) of section from a specimen of poly ether ether ketone (PEEK) taken as it was being prepared for etching, and (b, right) the etched specimen examined under Nomarski contrast. First

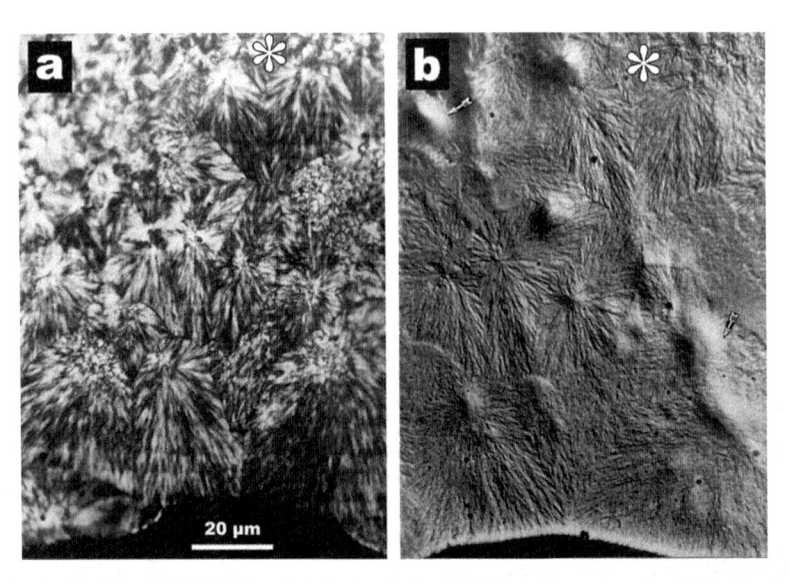

Fig. 2.12 Section of PEEK film and etched stub from which section was taken (Reprinted with permissions from Al-Raheil IAM, University of Reading thesis, 1987)

to notice is the strong correspondence between the morphology of the section and the etched morphology, though it is not exact because of a small depth difference in the specimen. The asterisks at the top mark a pair of 'comet-like' spherulites which are easily discerned in both views.

This pair was taken during the development of the etching procedure, and besides the spherulitic morphology displays some small artefacts (arrowed) which were eliminated with the reagent as eventually published.

This technique is also very useful for examining replicas of etched specimens after they have been extracted and placed on number grids, enabling one to locate specific features under the transmission electron microscope.

2.3.1.6 Ancillary Techniques

One advantage of optical microscopy is that it is easily used in conjunction with ancillary techniques, where heating, stretching and other processes are applied to the specimen. Various manufacturers supply hot stages in which the specimen can be studied while being heated or cooled, allowing direct observation of crystallization and melting processes, and phase separation in blends (Shabana et al. 2000).

As well as simple mechanical testing, it is also possible to study the evolution of morphology under shear conditions, again both in terms of crystallization (Shen et al. 2013; Chan and Gao 2005) and phase separation and ordering in blends (Lin et al. 2012; Zou et al. 2012; Trindade et al. 2004; Kielhorn et al. 2000).

In the authors' experience, the optical microscope is an invaluable first port of call, even when the target is the finest level of morphology only resolvable by the transmission electron microscope. It is best to start with a very low-power objective, and to move to higher ones. This practice can show up variations in structure throughout a specimen which might totally be missed when going straight to the higher power technique. In addition, sometimes it becomes apparent that what one is looking at is not what one is looking for; it may reveal films or deposits of foreign material, or even that the specimens may have been mixed up. To quote from Hemsley (1989): 'In no sense is the light microscope the poor relation of the electron microscope, neither is the latter to be considered by the light specialist only when all else fails . . .. One hesitates to insist that any examination with an electron microscope should be preceded by light microscopy, but polymer science and technology would benefit if this were a more frequent procedure'.

2.3.2 Scanning Electron Microscopy

Scanning electron microscopy (SEM) is a widely available technique in which a fine electron beam is rastered across the sample surface. The electrons interact with atoms in the surface component of the sample, producing various signals (Fig. 2.13) that can be detected and that contain information about the sample's surface

Fig. 2.13 Interaction of the electron beam in a SEM with the surface layers of the sample. The *grey area* shows the interaction volume and the location of the origin of the signals used in a SEM to construct an image

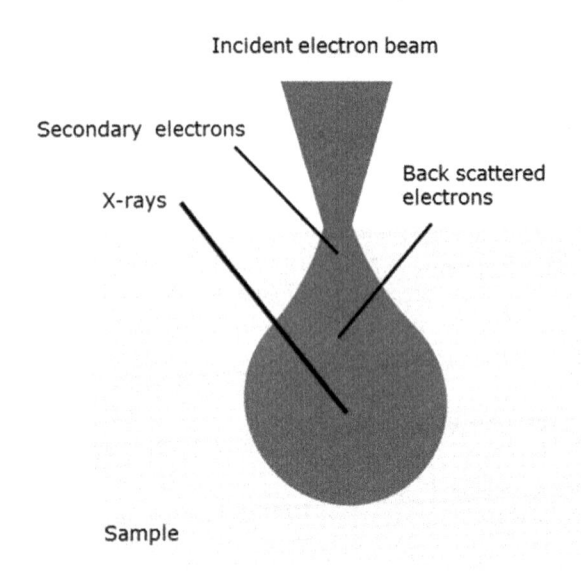

topography and composition. The electron beam is generally scanned in a raster scan pattern, and the beam's position is combined with the detected signal to produce an image. The so-called secondary electrons are the most commonly used signal. These are produced by excitation of atoms close to the sample surface or in the gold coating often applied to insulating samples to dissipate the build-up of charge. These electrons are of low energy and hence originate in the surface layers. In contrast, the back scattered electrons, electrons from the incident beam that have been scattered by the atoms in the sample, are of high energy and may come from much deeper in the sample as shown in Fig. 2.13. In this case, the efficiency of scattering depends on the atomic number, the larger the value of Z the greater the scattering. As a consequence, an image formed from back scattered electrons may reveal elemental composition variations which can be helpful in revealing the structure of composites. However, due to the larger interaction volume, the resolution of the image will be lower than that formed using secondary electrons. The electrons in the probe beam can excite characteristic X-rays from the atoms in the sample. Which X-rays are excited will depend upon the accelerating voltage. With a suitable energy dispersive X-ray detector images can be formed from the distribution of particular elements in the sample. Again because the interaction volume is large, the resolution of the image will be lower. Recently techniques have been developed which provide for elemental imaging at the nanoscale.

2.3.2.1 Contrast

As the SEM is a topological technique, we need to reveal the morphology of the sample. If there are compositional variations, then the images formed by characteristic X-rays or by back scattered electrons may suffice. However, if this is insufficient, another very common approach is to prepare a fractured surface. Here the samples are broken under brittle conditions, often by first freezing in liquid nitrogen, then fracturing to reveal the variation in morphology. For polymer blends and mixtures, the staining technique often used in TEM (discussed under TEM) may be helpful in providing a contrast to discern between materials. Another approach is to etch the surface, preferentially removing material and revealing the underlying morphology. The etching technique is described in the following section.

2.3.2.2 Sandwiching and Embedding Techniques

While often simply exposing a microtomed surface is sufficient for the purpose, sometimes specimens require support, especially thin films which cannot stand by themselves and where the top and bottom layers would be removed by the etchant. Sandwiching techniques have been developed, whereby the top and bottom surfaces are supported and protected by a material which can be etched along with the polymer specimen. An example is a work on woven polypropylene tapes, both before and after consolidation (Jordan et al. 2003).

Fibre and weaves can be embedded in a mounting medium which can be etched along with the polymer composing the fibre (Abo El Maaty et al. 1999). Details of such a technique are shown in the Fig. 2.14 below.

In the micrographs below, an original weave is shown in a sandwich. At left, A is the high-impact polystyrene and B is the epoxy resin (Fig. 2.15).

At right, in the regions marked C, one tape is emerging, so an etch cross-section of the tape is seen. At D, another tape is running parallel to the surface.

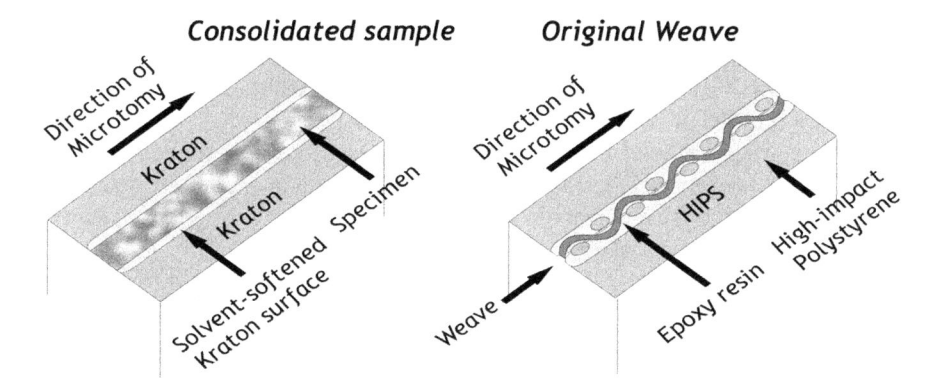

Fig. 2.14 A comparison of two sandwiching techniques, applicable to solid specimens (*left*) and those with open texture (*right*). Adapted from Jordan et al. (2003)

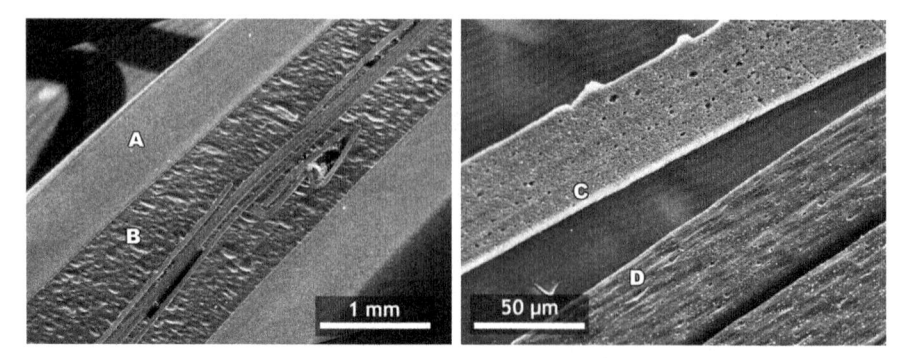

Fig. 2.15 SEM of etched polypropylene weave specimens, showing (*left*) sandwiching technique (*right*) different structures of polypropylene tapes in transverse and longitudinal directions (Image taken by Robert Olley, 2003, University of Reading, UK. The specimen was provided by the University of Leeds)

2.3.3 Etching

Both SEM and TEM (to follow) generally require some preparation to make the morphology visible. The etching technique is equally applicable to both, and so is included here between the two sections.

Etching of polymers is very similar to the process in metallography, where a surface is treated with a chemical agent to reveal the underlying phase or crystalline structure, which are then studied under the electron or optical microscopes. A variety of etching procedures are reviewed in (Bassett et al. 2003).

Most etching techniques involve a liquid etchant. Most commonly used are permanganic reagents, generally consisting of a dilute solution of potassium permanganate in a mixture of sulfuric and orthophosphoric acids, either dry or with a certain amount of water. Different compositions are suitable for different classes of polyethylene (Shahin et al. 1999), other polyolefins (Patel and Bassett 1994; Weng et al. 2003, 2004), isotactic polystyrene (Bassett and Vaughan 1985), PEEK (Olley et al. 1986), etc. This is the most common of the oxidizing etchants, though chromic reagents are also used. Polyesters are generally treated with an alkaline reagent, typically a solution of potassium hydroxide in an alcohol.

Etching of a semi-crystalline polymer can be visualized by a simple model of plate-like crystals embedded in a rubbery matrix. Three etching rates are observed: the fastest etching is the removal of the amorphous polymer constituting the matrix; the broad basal surfaces of the plate-like crystals are almost totally resistant to etching, but the thin side-surfaces are eroded at an intermediate rate (Fig. 2.16). This is described more fully in Olley (1986).

The morphology of banded spherulitic polyethylene as in Fig. 2.17 (left) illustrates this well. The orientation of crystals rotate periodically along the spherulite radius; at A the resistant basal surfaces of the crystals are seen standing proud of the general etched plane; at B the lamellae are displayed edge-on, so are somewhat less

2 Evaluating Scales of Structures

Fig. 2.16 Schematic of etching of a polymer crystal: (**a**) free molecular end (cilium); (**b**) tight chain fold; (**c**) loose chain fold; (**d**) short stem freed from side surface by chemical attack (Reproduced with Permission from Olley 1986)

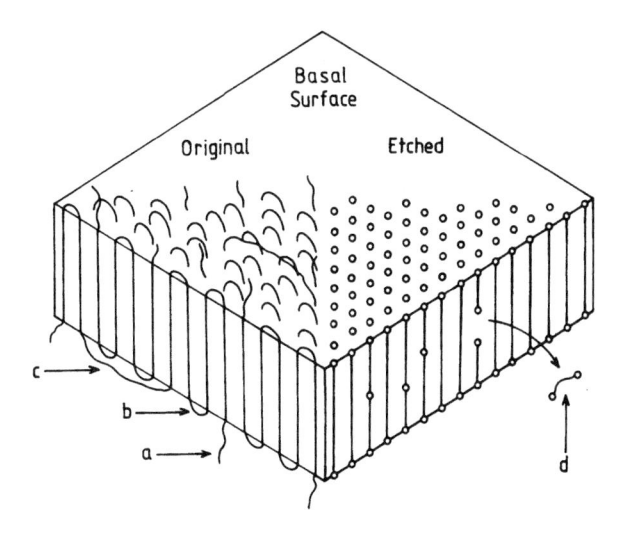

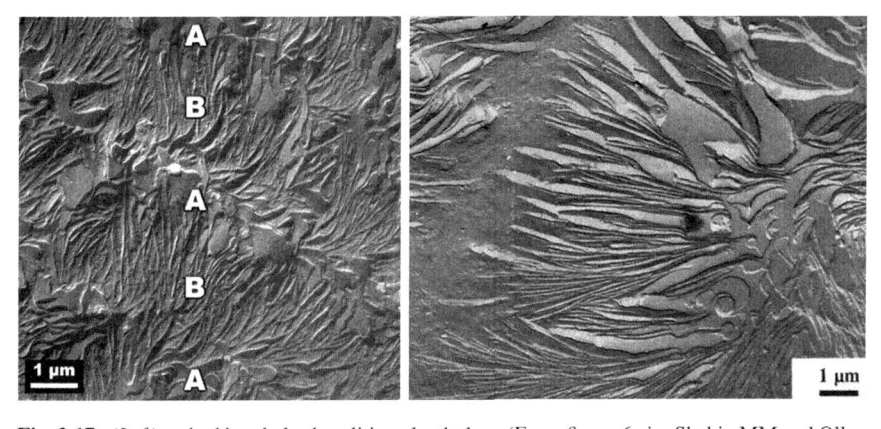

Fig. 2.17 (*Left*) etched banded spherulitic polyethylene (From figure 6c in, Shahin MM and Olley RH. Novel Etching Phenomena in Poly(3-hydroxy butyrate) and Poly(oxymethylene) Spherulites. J Polym Sci Part B: Polym Phys 40: 124–133. Copyright © 2002 by John Wiley Sons, Inc. Reprinted by permission of John Wiley & Sons, Inc.); (*right*) Original caption: TEM micrographs showing the morphology of sample BPA 124/20 (20 % LPE). It is clear that distinct boundary regions exist between the lamellar aggregates that grow at 124 °C. J. Mater. Sci.,1997, 32, (17), 4523–31, Structure-property relationships in polyethylene blends: the effect of morphology on electrical breakdown strength. Hosier, I. L., Vaughan, A. S. and Swingler, S. G, Figure 8 Reprinted with kind permission from Springer Science and Business Media

resistant to the etch. The amorphous regions are more highly etched, and lie below the protruding crystalline parts (Shahin and Olley 2002).

The specimen at right in Fig. 2.17 is an example of crystallizing a polyethylene blend as a proposed cable material under a controlled thermal regime in order to optimize the morphology in regard to electrical breakdown behaviour (Hosier et al. 1997). We have included a TEM image here as the higher resolution is useful in identifying the features. Large spherulitic objects made largely of high-density PE are revealed in a matrix where low-density PE predominates.

Etching is used to display not only semi-crystalline morphology, but also multiphase structure in polymer blends which may be totally non-crystalline. An early example is work on high-impact polystyrene and similar materials by Bucknall et al. (1972). From the abstract: 'These techniques reveal details of orientation in injection mouldings, of internal structure in composite rubber particles, and of crazing and shear band formation. The etch method avoids the specimen distortion inherent in sectioning'. A more recent example is one following the spinodal decomposition of blends of tetramethyl polycarbonate and polystyrene (Shabana et al. 1993). During blend decomposition, one or more of the components may crystallize, and the crystal morphology as well as the phase structure can be observed, for example, in a phase-separating copolymer of polyethylene naphthoate with polyhydroxybenzoate (Shabana et al. 1996) or a blend of polycaprolactone with oligo-styrene (Shabana et al. 2000). The technique may also be used to reveal texture in liquid crystal polymers (Ford et al. 1990).

2.3.4 Transmission Electron Microscopy

In transmission electron microscopy (TEM), a beam of electrons is passed through a thin sample, such that an image is formed as a result of absorption or diffraction contrast. In the case of polymers, a combination of disorder and radiation sensitivity means that, of these, absorption contrast is most important, in which case, high resolution images can be generated where image contrast is based on the spatial variation in electron density. In the case of materials such as nanocomposites, the distribution of the nanoparticles can therefore easily be imaged, as a result of the difference in the atomic number between the nanoparticles and the matrix polymer, as shown in Fig. 2.18.

However, obtaining useful structural information concerning the lamellar texture of semi-crystalline polymers or the phase distribution in polymer blends or block copolymers is often complicated by the fact that the variation in electron density between the structural components of interest is insufficient to provide meaningful image contrast. In such circumstances, additional sample preparation procedures, beyond ensuring that the sample is sufficiently thin, are required.

2.3.4.1 Inducing Contrast

If etching of polymers, as described above, is equivalent to similar process in metallography, then staining has analogues in biology, where the very same TEM contrast problems exist. Essentially, staining involves exposing a polymeric sample to a chemical reagent, which diffuses into the structure and becomes preferentially located at certain sites. The classic example of this concerns studies of the structure

Fig. 2.18 Colloidal silver embedded in an epoxy matrix (Image provided courtesy of Dr Suvi Virtanen/Alun Vaughan, University of Southampton, UK)

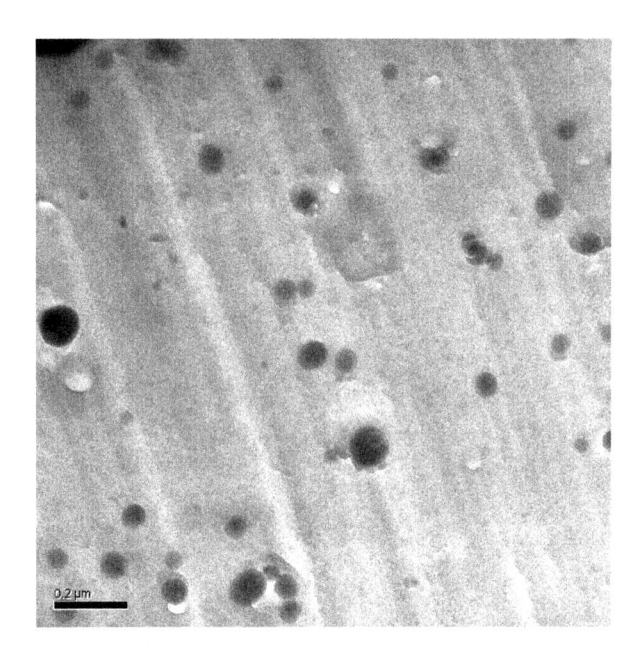

of rubber toughened polymer blends in which osmium tetroxide (OsO_4) reacts with and preferentially stains unsaturated rubbery phases. The more aggressive analogue, which consequently has greater utility, is ruthenium tetroxide (RuO_4); Trent et al. (1983) provide an excellent overview of the staining methodologies and polymer types where RuO_4 has been applied with success. While many different stains have been developed to enhance image contrast in the TEM, a major issue with such reagents concerns their aggressive nature, which raises safety concerns during use and in disposal. In the case of staining reagents such as OsO_4 and RuO_4 that can be used to stain polymers in the vapour phase, the staining device described by Owen and Vesely (1985) has been found to be extremely convenient. A related issue concerns 'overstaining', whereby exposure to the stain introduces artefacts into the specimen (Chou et al. 2002). Figure 2.19 shows images of two different materials stained with RuO_4; (a) from Trent et al. (1983) is a spherulitic morphology in a sample of high density polyethylene while (b) is a section from a block copolymer of polyethylene and atactic polypropylene (Hong et al. 2001); the amorphous atactic polypropylene is much more heavily stained overall than the semi-crystalline polyethylene in which the stain has also revealed the fine semi-crystalline lamellar morphology.

The etching techniques described above in connection with SEM can also be used for TEM, in conjunction with so-called replication. In this, an accurate facsimile of the sample's surface topography is produced, frequently in the form of a thin carbon film, which is then coated at an oblique with an electron dense metal (shadowed). As a consequence, the distribution of metal on the surface is non-uniform and directly related to the surface topography of the specimen and, hence, the morphological features present within it. Figure 2.20a shows an equivalent structure, as revealed by etching followed by replication, to the lamellar

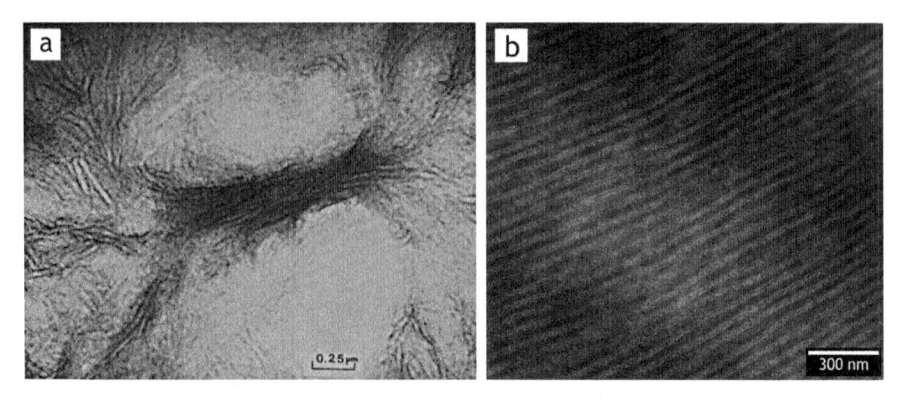

Fig. 2.19 TEM micrographs of (**a**) a spherulite centre in a thin film of high density polyethylene stained with RuO$_4$, (**b**) section of polyethylene-atactic polypropylene block copolymer stained with RuO$_4$. Reproduced with permission (**a**) from Figure 15 in Trent, Scheinbeim and Couchman, Macromolecules, 1983, 16: 589–598 (Copyright 1983) American Chemical Society. (**b**) from Figure 3 in Hong, Copyright (2001) with permission from Elsevier

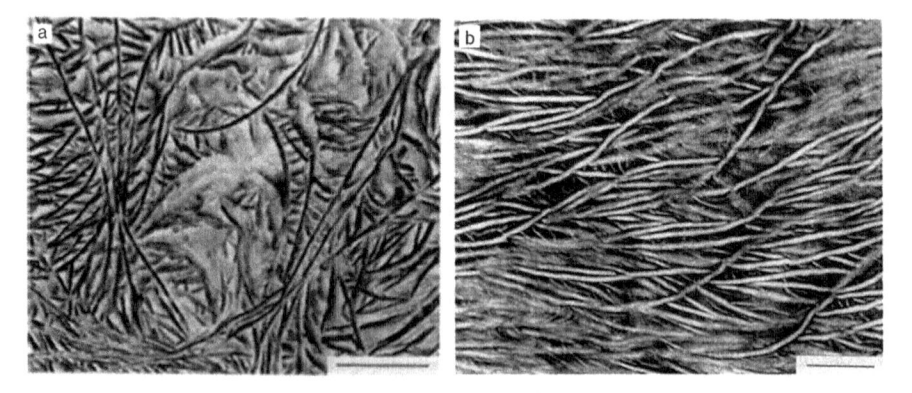

Fig. 2.20 TEM micrographs of 60 % LPE/LDPE blends cooled at 1 °C/min. (**a**) Sample prepared through etching and replication Scale bar is 1 μm (**b**) Stained sections, cut from material fixed with chlorosulfonic acid. Scale bar is 200 nm. Reprinted from Polymer, 40:337–348, Morgan RL, Hill MJ and Barham PJ, Morphology, melting behaviour and co-crystallization in polyethylene blends: the effect of cooling rate on two homogeneously mixed blends, Copyright (1999), with permission from Elsevier

texture shown in Fig. 2.20b which was prepared by staining with chlorosulfonic acid and sectioning. Fig. 2.20a represents a case of a direct replica, where the carbon and shadowing metal are applied directly to the etched polyethylene. Removal of the replica is typically performed using a polyacrylic glue (Morgan et al. 1999) or polyacrylic acid. Two stage (indirect) replicas are more commonly used; a thin sheet of cellulose acetate is applied to the etched surface, peeled off and coated with shadowing metal and carbon. The replica is then placed on top of a TEM grid and freed by dissolving the cellulose acetate away with acetone. Weng et al. (2003, 2004) showed examples of replicas prepared by this technique.

2.3.5 Atomic Force Microscopy

Scanning probe microscopy involves a family of techniques in which atomic force microscopy (AFM) is the most popular which is relevant to this work. Scanning tunnelling microscopy can yield atomistic level images under good conditions. In AFM, the so-called contact mode provides an image of the surface topology. The resolution can approach the nanometre scale under ideal conditions but tip adhesion may limit the usefulness with soft materials such as polymers. The non-contact method involves an oscillating cantilever and under appropriate conditions, images of the spatial distribution of different phases or material can be formed where the contrast essentially arises from the differences in the stiffness of the phases.

Although for the highest resolution care has to be taken to isolate vibrations and screen electric fields, AFM is essentially performed in the laboratory environment. As such, it is possible to perform experiments such as heating and cooling in the AFM and the sample does not require treatment to reveal the morphology as in the case of SEM. Time-resolving studies are now possible with the latest developments and this has been used to good effect by Hobbs et al. to observe the growth of polyethylene crystals (Hobbs 2007). Further developments to the technology have allowed for the use of higher frame rate AFM, termed a video AFM, which has been utilized to good effect to show the growth of polyethylene oxide crystals (Hobbs et al. 2005).

2.4 Scattering Methods

Scattering methods are a collection of powerful techniques for evaluation the molecular organization of materials including polymers. One of the earliest applications of X-ray diffraction at the start of the twentieth century was with polymers. There is considerable diversity in the instrumentation available for use with polymers using both laboratory sources and synchrotron national facilities. The type of instrumentation used relates to the type and scale of the information required. Figure 2.21 shows a schematic of the scales of structures which can be studied using readily available equipment.

In a typical scattering experiment, the incident beam on a sample is elastically scattered by the sample through an angle, θ, described by the equation $|\mathbf{Q}| = Q = 4\pi\sin(\theta)/\lambda$, where 2θ is the scattering angle, λ is the wavelength of the incident beam and Q is the scattering vector. The angle through which the beam is scattered will determine the structural information that can be discerned as shown Fig. 2.21. For example, small angle scattering can be utilized to determine features on the length scales of several to a hundred nanometres, such as the long period and lamellae thickness, whilst wide angle examines the features from several nanometres down to angstrom levels where it can be used in determination of the crystalline unit cell, crystal size and the polymer chain conformation.

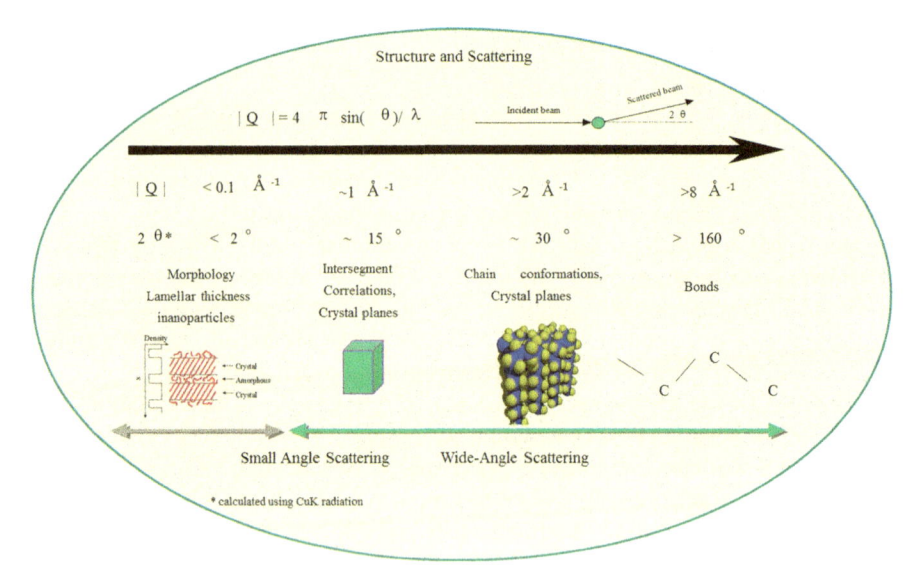

Fig. 2.21 Various length scales examinable by scattering methods (Reproduced with the permission of the author of the chapter)

2.4.1 X-Ray Scattering

X-ray scattering arises from electrons where the observed intensity of scattering is dependant on the number of electrons present on an atomic species contained within the sample ($I \sim Z^2$), hence heavier elements scatter more intensely than lighter ones.

2.4.1.1 Small Angle X-Ray Scattering

Numerous small angle X-ray scattering (SAXS) experiments have been conducted on extracting information regarding the long period which, with knowledge of the degree of crystallinity, can be used to calculate the crystalline and amorphous thicknesses from SAXS data. This is done by considering the material as a two-phase system of periodic alternatively stacked lamellar crystals and amorphous regions. The methodology utilizes a Fourier transform of the data to extract an interface distribution function (Ruland 1977) or the correlation function (Strobl and Schneider 1980) containing the information on the materials long period. Strobl used the correlation function to extract information from polyethylene SAXS data with further consideration of how inhomogeneities and lamellar curvature would impact the results (Strobl and Schneider 1980). Some knowledge of the crystallinity is required in order to determine the values for the crystal and amorphous thicknesses, which would make the combination of wide angle X-ray scattering and SAXS a useful tool in a thorough analysis of a material. An example of a 2D SAXS pattern, and the corresponding data reduction to a one-dimensional plot of intensity

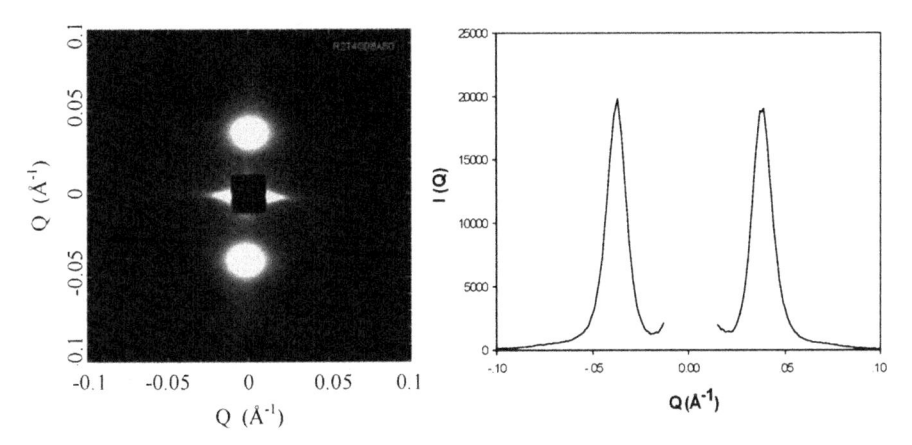

Fig. 2.22 SAXS curve of a crystallizable copolymers of l-lactide, ε-caprolactone and glycolide. Wavelength of X-rays are 1.4 Å. Reprinted from figure 9a in Polymer 46 (17), 6411–6428 W. Channuan, J. Siripitayananon, R. Molloy, M. Sriyai, F.J. Davis and G.R. Mitchell, The structure of crystallisable copolymers of l-lactide, ε-caprolactone and glycolide, Copyright (2005) with permission from Elsevier

as a function of Q, is provided in Fig. 2.22. The scattering pattern was obtained from an annealed fibre of a crystallizable copolymer of l-lactide, ε-caprolactone and glycolide. The 2D scattering pattern exhibits highly anisotropic features (Channuan et al. 2005).

Another instance where SAXS has been a useful tool for polymer characterization relates to the examination of block copolymers. Block copolymers can form structural domains that are on length scales detectable by the small angle scattering technique. A series of scattering peaks occur with each peak position having a separation indicative of the structure formed. Examples of this would be if a peak occurred at $Q = 1$ Å^{-1}, reflections would occur at 2, 3, 4, 5,... for lamellar domains, whilst body centred cubic will have peak positions $\sqrt{2}$, $\sqrt{3}$, $\sqrt{4}$, $\sqrt{5}$,... (Hamley and Castelletto 2004).

The use of SAXS as a characterization technique for polymeric materials has been extended to include studies of liquid crystal polymers, polymer blends, nanocomposite materials, polymer gels and solutions (Chu and Hsiao 2001).

2.4.1.2 Wide Angle X-Ray Scattering

Wide angle X-ray scattering (WAXS) provides information on the unit cells where the data obtained can be utilized in determination of the lattice parameters, *hkl* reflection planes and hence the lattice structure i.e. face centred cubic, hexagonal, etc. In addition to the unit cell information, the level of crystallinity can be determined by considering the polymer as a two-phase material, amorphous and crystalline. The ratio of the second moment of the data corresponding to the sharp

Fig. 2.23 WAXS Curve of Polyethylene oxide (Mv—100,000 Da). Data was obtained on a powder X-ray diffractometer with the sample in a powdered form at room temperature. X-ray wavelength used was 1.54 Å

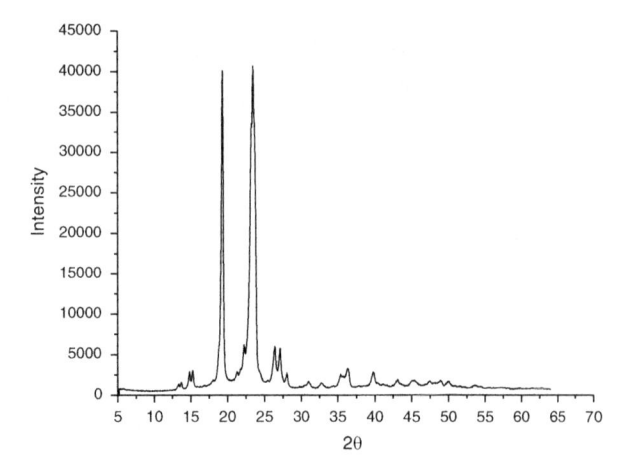

crystalline peaks to the total second moment of the scattering curve provides the degree of crystallinity of the sample from the scattering data. However, suitable corrections for incoherent backgrounds, air scattering, thermal motion and lattice imperfections need to be accounted for (Ruland 1961, 1964). Ruland first applied this method to determine the degree of crystallinity in nylon 6 and 7 (Ruland 1964). An example of a wide angle scattering curve is observed below in Fig. 2.23 for a sample of polyethylene oxide.

2D WAXS patterns can be used to determine the direction of preferred orientation and the orientation distribution function or a set of orientation parameters, P_{2n}, which describe the orientation distribution function $D(\alpha)$.

$$D(\alpha) = \sum_{2n=0}^{2n=\infty} D(\alpha) P_{2n}(\cos\alpha)$$

where $P_2 = (3\cos^2\alpha - 1)/2$

In many works, attention is focused on $\langle P_2 \rangle$, although X-ray scattering is a technique able to provide quantitative estimates of all of the orientation parameters. In contrast, birefringence is only related to $\langle P_2 \rangle$ as it is a second rank tensor property. Similar restrictions relate to all other second rank properties. For a perfectly aligned sample, the value of $\langle P_2 \rangle$ is 1, whereas a randomly orientated distribution with no preferential alignment $\langle P_2 \rangle$ has a value of 0. Methods are available for examination of preferential alignment in semi-crystalline and amorphous materials as described by Mitchell (2016). The methodology developed by Mitchell and co-workers in which the scattering is described in terms of a series of spherical harmonics provides a powerful tool box for the analysis of 2D scattering patterns. These techniques were successfully applied to isotactic polypropylene containing a thermotropic liquid crystal (Mitchell et al. 2005). This same methodology was used by Edwards et al. to examine the orientation of polymer lamellae in PCL electrospun fibres collected onto a rotating collector at differing take-up speeds. An increasing level of orientation was observed in the scattering patterns

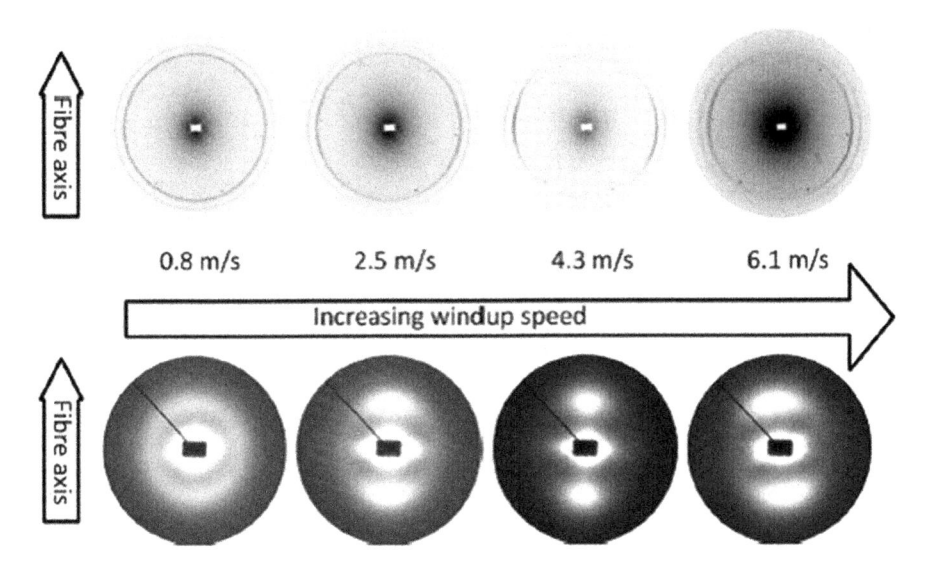

Fig. 2.24 PCL electrospun fibre scattering for samples collected at differing take-up speeds. Wide angle X-ray scattering patterns (*top*) for electrospun fibres of PCL prepared using the indicating tangential velocities and the small angle X-ray scattering patterns (*bottom*) for the same samples. The outer edge of the SAXS patterns corresponds to $|Q| = 0.08$ Å^{-1}. Reprinted from European Polymer Journal, 46 (6). pp. 1175–1183, Edwards, M. D., Mitchell, G. R., Mohan, S. D. and Olley, R. H. Copyright (2010) Development of orientation during electrospinning of fibres of poly (ε-caprolactone) with permission from Elsevier

(Fig. 2.24), shown by the azimuthal arcing of the intensity distribution with increasing collection speed until a critical speed was reached that induced fibre breakage (Edwards et al. 2010). The obtained scattering patterns were described as a convolution of the fibre orientation and crystal orientation, for which fibre misalignment can be corrected for to establish the orientation of the crystals in a perfectly aligned sample. Similar methodologies can be applied to small-angle scattering data.

The use of WAXS is not limited to the study of semi-crystalline materials but can also be applied to purely amorphous materials. WAXS patterns obtained can contain information on the inter- and intra-chain segments and the trans-gauche conformation of the chains. For example, Lovell et al. performed WAXS on atactic polystyrene showing the scattering contributions from the inter- and intra-chain segments, and estimated the trans-gauche conformational arrangement of the chain by comparison to the scattering one would expect for various trans-gauche configurations. An example of this is given below in Fig. 2.25 for atactic polystyrene (Lovell et al. 1979).

2.4.1.3 SAXS/WAXS Instruments

WAXS techniques have become commonly used in characterization of polymer samples with lab scale equipment available for use. Although small angle X-ray lab

Fig. 2.25 (a) experimental X-ray data (b) model of scattering for a tttt chain, (c) model of scattering for ttgg (where *t* trans, *g* gauche) (Reproduced from Lovell et al. 1979 with permission of The Royal Society of Chemistry)

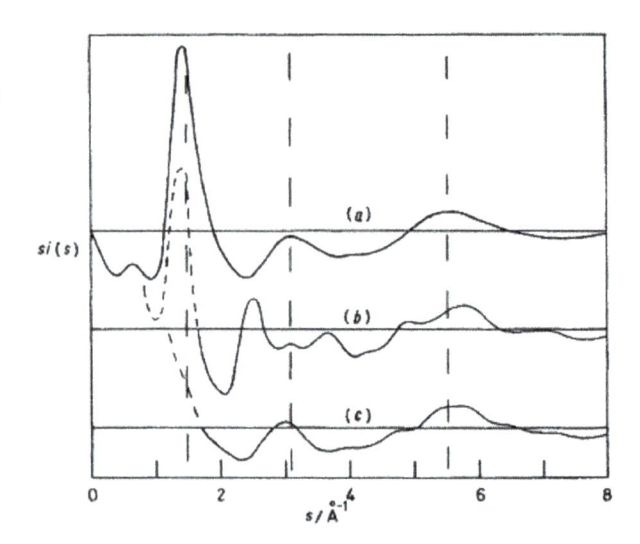

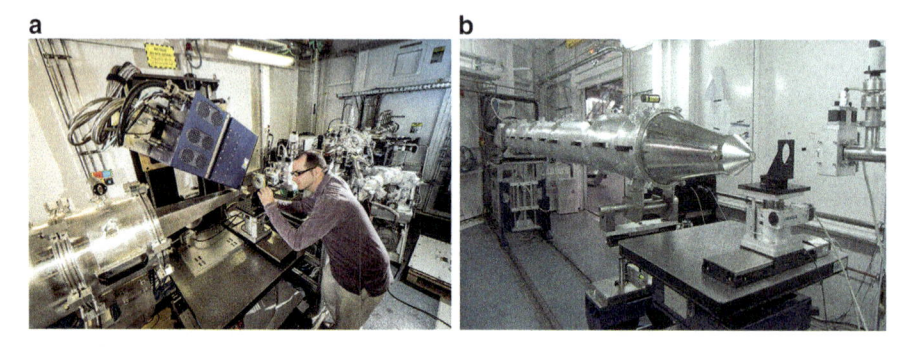

Fig. 2.26 NCD Beamline at ALBA. (a) Image of the WAXS detector (*Blue box* in the top centre of image) (Picture taken by Sergio Ruiz) and (b) Image of part of the SAXS beamline (Picture taken by Juan Carlos Martínez). Images provided courtesy of Alba Synchrotron

sources are available, there is significantly higher flux available at synchrotron sources allowing for rapid and in situ studies of samples (Fig. 2.26).

2.4.2 Neutron Scattering

Neutron scattering occurs from the nucleus of the molecule as opposed to the X-rays which scatter from electrons. As the neutrons scatter from the atomic nucleus, which occupies a significantly smaller volume than that of the electron clouds surrounding atoms, neutron scattering is considered a point scattering technique, hence the observed pattern does not fall off with increasing scattering angle as observed with X-ray scattering. Whilst the observed X-ray intensity is proportional to the atomic number of the scattering atom, neutron scattering lengths exhibit no particular order, with scattering lengths appearing random with atomic species. However, there is a difference in scattering lengths between isotopic species, hence a contrast can be provided by utilizing an isotopic labelling technique.

2.4.2.1 Small Angle Neutron Scattering

Small angle neutron scattering (SANS) can be utilized to study structural development in a polymer; similar to the SAXS technique, however, the unique isotopic labelling techniques available in a SANS experiment allow for study of the polymer chain dimensions in the bulk and solution state. For example, a hydrogenous polymer matrix that contains perdeuterated polymer chains can provide a contrast which allows the study of the macroscopic chain conformation in the bulk state. The perdeuterated polymer chain is considered to be chemically identical to that of the hydrogenous polymer, hence a study of the polymer in its natural state can be conducted. If a similar experiment were to be conducted with SAXS, it would require the substitution of a heavier element on the polymer chains to provide a scattering contrast, meaning a different chemical species is being studied compared to the normal material. This unique isotopic scattering length difference provided for neutron scattering experiments makes it a powerful tool for the investigation of polymer systems. Various examples are present in the literature where the study of the polymer chain dimensions has been conducted on a variety of polymers. Comparisons of the literature results for these SANS measurements show the theta, glassy and molten polymer state have similar conformation reflecting an ideal random walk as shown in a review collated by Wignall and Melnichenko (2005). For example, both atactic polymers, polystyrene and poly methyl methacrylate were shown to have radii of gyration that varied with the molecular weight of the sample by the relation $R_g = 0.275 M_W^{0.5}$, similar to that measured in the theta state (Wignall and Melnichenko 2005). In addition to bulk materials, thin films and fibres can be studied by SANS to examine the chain conformation. Studies of isotopically labelled electrospun fibres were conducted by Mohan et al. with results showing a small level of preferential alignment of the polymer chain along the fibre axis induced by the electrospinning process, with further extension possible due to mechanical deformation from a rotating collector (Fig. 2.27) (Mohan et al. 2011).

2.4.2.2 Broad Q Neutron Diffraction

Wide angle neutron scattering, commonly referred to as broad Q neutron diffraction, has been performed on various crystalline and amorphous polymers. Exploiting the contrast variation between isotopes as was done in SANS, the broad Q neutron scattering can be used to examine various scattering features. For example, neutron diffraction was performed on amorphous atactic polystyrene, which had selectively deuterated components. This allowed for the examination of

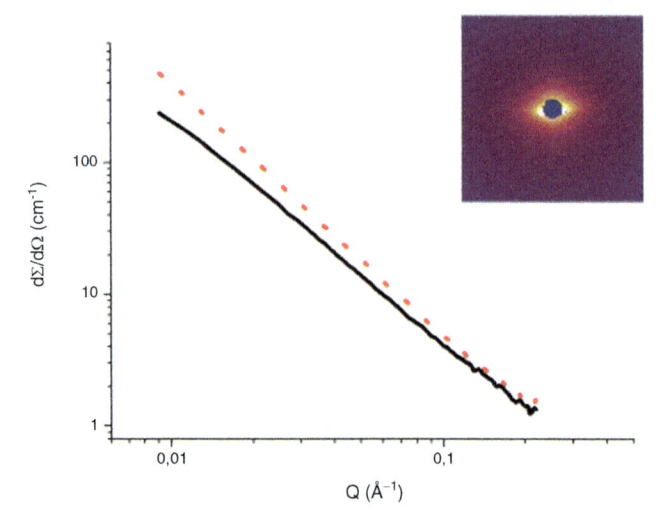

Fig. 2.27 Scattering curves for an electrospun fibre sample of (50/50) hydrogenous/perdueterated polystyrene collected onto a rotating collector. The data has been reduced to a 1D plot of the differential scattering cross section as a function of Q, where the curves represent the scattering both parallel (*solid line*) and perpendicular (*dotted line*) to the fibre axis (vertical on page) obtained from the 2D scattering pattern (*inset*). *Inset* runs from -0.05 Å$^{-1} < Q < 0.05$ Å^{-1} (Adapted from Mohan et al. 2011 with permission from The Royal Society of Chemistry)

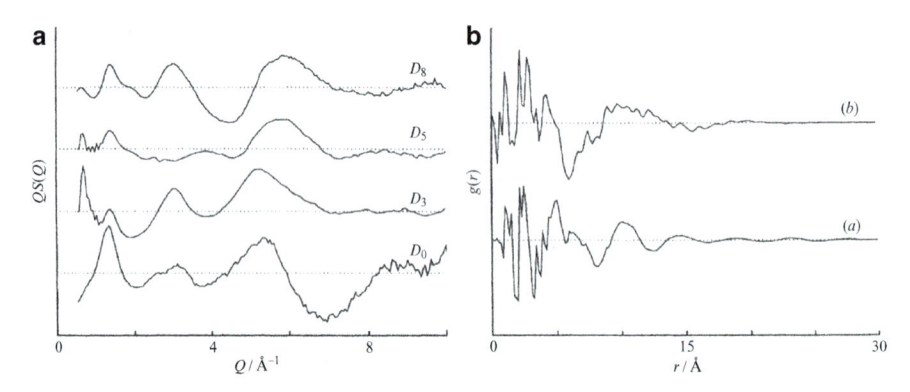

Fig. 2.28 (**a**) Structure factors for selectively deuterated polystyrene. (**b**) Fourier transform of (**a**) fully deuterated monomer and (**b**) partial correlation function relating to backbone atoms (Reproduced from Mitchell et al. 1994, with permission of The Royal Society of Chemistry)

the contribution to the scattering from various components on the polymer chain (Fig. 2.28). The structure factor for these scattering patterns is shown below in Fig. 2.28a where it can be seen that the various deuteration levels influence the scattering obtained. Here the scattering at approximately $Q \sim 1$–1.5 Å^{-1} arises from scattering between molecules from neighbouring chains, whilst $Q > 1.5$ Å^{-1} relates to the correlations from molecules on the same chain. The Fourier transform into a real space correlation function is provided in Fig. 2.28b. The lower values relate to

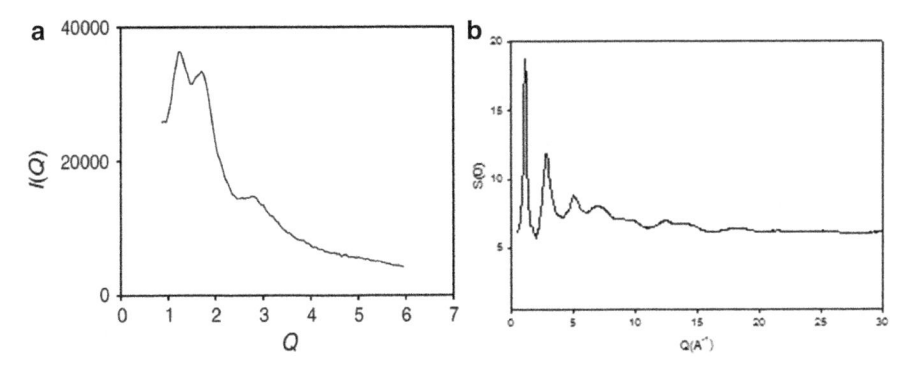

Fig. 2.29 (a) Wide-angle X-ray scattering curve for a sample of perdeuterated poly (vinylchloride) (b) Neutron scattering function for the sample subjected to X-ray scattering in the *left hand image* (Reproduced from Mitchell 2011)

bond lengths present, C–D and C–C, whilst the higher length terms reflect the inter- and intra-chain segment contributions. Figure 2.28b shows a repeating oscillation at approximately 10 Å, 15 Å, 20 Å, 25 Å which relates to the scattering from the phenyl rings along the polymer chain.

The technique can also be applied to the examination of semi-crystalline polymer systems, for example, Fig. 2.29 shows the WAXS and broad Q neutron diffraction from the same sample of d-PVC. In the X-ray scattering data (Fig. 2.29a) the material appears very disordered, whilst the neutron scattering from the same perdeuterated sample of PVC shows a typical melt structure factor similar to polyethylene although the first peak is somewhat sharper reflecting a higher level of order which perhaps is a consequence of the more polar structure (Mitchell 2011).

As the intensities of scattered X-rays depend on the atomic number, $I \sim Z^2$, the scattering for a sample of PVC will be dominated by the scattering from the Cl Atoms ($Z = 17$), compared with the scattering from C ($Z = 6$) or H ($Z = 1$). In contrast, neutron scattering arises from changes in the neutron scattering length b which is determined by the atomic nucleus. For d-PVC $b_{cl} = 9.577$, $b_C = 6.646$, $b_D = 6.671$, hence the scattering is not dominated over by the presence of the chlorine allowing for a more detailed study of the structure.

2.4.2.3 Neutron Scattering Instruments

Neutrons are produced either in a nuclear reactor (e.g. the Institut Laue Langevin) or at a spallation source (i.e. ISIS Neutron Facility). In the latter most scattering experiments are performed using angular dispersion as in a typical X-ray scattering experiment, whereas at a spallation source, experiments are performed using time of flight techniques which opens up new possibilities as described below. Typically neutron scattering fluxes are lower and the interaction with the sample weaker and

Fig. 2.30 The NIMROD instrument at ISIS (Image provided courtesy of ISIS Pulsed Neutron Source, STFC, Didcot, UK)

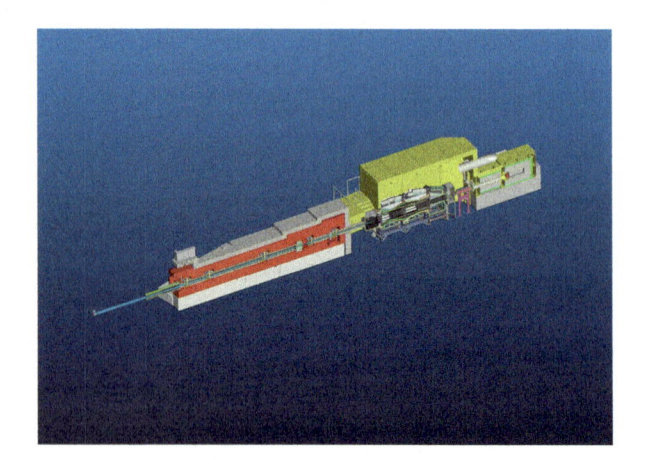

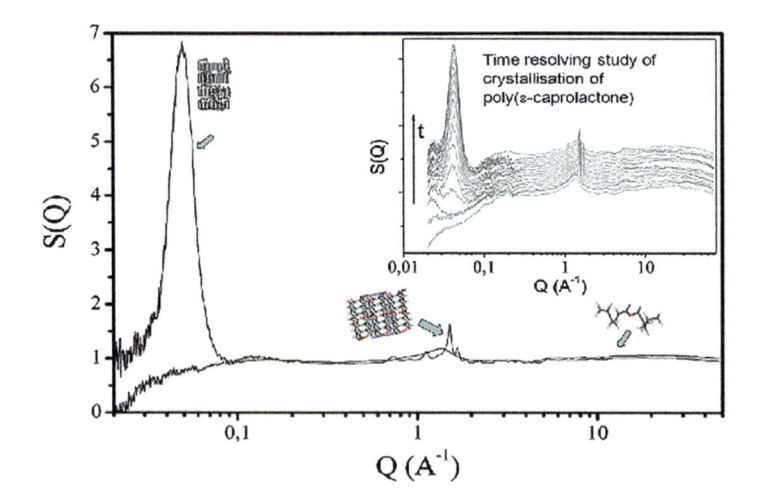

Fig. 2.31 Time-resolved data obtained on NIMROD for PCL in both the semi-crystalline and melt phases. *Inset* shows time-resolved data captured during crystallization

hence the neutron scattering signals are much diminished. Much effort has been made to reduce the data collection cycle time, for example, using an area detector (LOQ ISIS first target station Heenan et al. (1997) or many detectors (e.g. NIMROD (Fig. 2.30) second target station ISIS Bowron et al. (2010)).

Figure 2.31 shows such data obtainable for a time-resolved study of PCL on the instrument NIMROD at ISIS, STFC, Didcot, UK with a cycle time of 400 s. The scattering data shown ranges from $Q = 0.02$ to 50 Å^{-1} and is on a single vertical scale in absolute units.

2.5 Summary

We have covered a wide range of techniques in this chapter. We can see a continual improvement in the facilities for the study of structure at multiple scales. Simultaneous SAXS and WAXS experiments are now routinely performed at synchrotron-based beam lines, while the NIMROD instruments take this a further step by routinely providing multiple scale data on a single intensity scale. Despite these developments, however exciting, there remains a need to utilize a range of complementary techniques involving both imaging and scattering both to cover the structural scales and to probe the scientific issues.

References

Abo El Maaty MI, Olley RH, Bassett DC (1999) On the internal morphologies of high-modulus polyethylene and polypropylene fibres. J Mater Sci 34:1975–1989

Al Raheil IAM (1987) Development of crystalline morphology in polyethylene and PEEK. PhD Thesis, University of Reading

Arnal ML, Sanchez JJ, Müller AJ (2001) Miscibility of linear and branched polyethylene blends by thermal fractionation: use of the successive selfnucleation and annealing (SSA) technique. Polymer 42:6877–6890

Bassett DC, Vaughan AS (1985) On the lamellar morphology of melt crystallized isotactic polystyrene. Polymer 26(717):725

Bassett DC, Olley RH, Vaughan AS (2003) Specimen preparation for TEM of polymers. In: Pethrick RA, Viney C (eds) Techniques in polymer organisation and morphology characterisation. Experimental methods in polymer characterisation. Wiley, Chichester

Bovey FA (2007) NMR of polymers. eMagRes. Wiley Online

Bowron DT, Soper AK, Jones K, Ansell S, Birch S, Norris J, Perrott L, Riedel D, Rhodes NJ, Wakefield SR, Botti A, Ricci MA, Grazzi F, Zoppi M (2010) NIMROD: the Near and InterMediate Range Order Diffractometer of the ISIS second target station. Rev Sci Instrum 81:033905

Bucknall CB, Drinkwater IC, Keast WE (1972) An etch method for microscopy of rubber-toughened plastics. Polymer 13:115–118

Chan C-K, Gao P (2005) Shear-induced interactions in blends of HMMPE containing a small amount of thermotropic copolyester HBA/HQ/SA. Polymer 46:10890–10896

Channuan W, Siripitayananon J, Molloy R, Sriyai M, Davis FJ, Mitchell GR (2005) The structure of crystallisable copolymers of l-lactide, ε-caprolactone and glycolide. Polymer 46:6411–6428

Chen F, Shanks RA, Amarasinghe G (2001) Crystallisation of single-site polyethylene blends investigated by thermal fractionation techniques. Polymer 42:4579–4587

Chou TM, Prayoonthong P, Aitouchen A, Libera M (2002) Nanoscale artifacts in RuO4-stained poly(styrene). Polymer 43:2085–2088

Chu B, Hsiao BS (2001) Small-angle X-ray scattering of polymers. Chem Rev 101:1727–1762

Edwards MD, Mitchell GR, Mohan SD, Olley RH (2010) Development of orientation during electrospinning of fibres of poly(ε-caprolactone). Eur Polym J 46:1175–1183

Ford JR, Bassett DC, Mitchell DR, Ryan TG (1990) Morphology of a main chain liquid crystal polymer containing semi-flexible coupling chain. Mol Cryst Liq Cryst 180:233–243

Gedde UW, Jansson JF (1983) Molecular fractionation in melt-crystallized polyethylene 1. Differential scanning calorimetry. Polymer 24:1521–1531

Gedde UW, Jansson JF (1984) Molecular fractionation in melt-crystallized polyethylene 3. Microscopy of solvent-treated samples. Polymer 25:1263–1267

Gedde UW, Eklund S, Jansson JF (1983) Molecular fractionation in melt-crystallized polyethylene 2. Effect of solvent-extraction on the structure as studied by differential scanning calorimetry and gel-permeation chromatography. Polymer 24:1532–1540

Hamley IW, Castelletto V (2004) Small-angle scattering of block copolymers in the melt, solution and crystal states. Prog Polym Sci 29:909–948

He Z, Olley RH (2000) On spherulitic forms in an aromatic polyesteramide. Polymer 41:1157–1165

Heenan RK, Penfold J, King SM (1997) SANS at pulsed neutron sources: present and future prospects. J Appl Cryst 30:1140–1147

Hemsley DA (ed) (1989) Applied polymer light microscopy. Springer, Berlin

Hobbs JK (2007) Insights into polymer crystallization from in-situ atomic force microscopy. In: Progress in understanding of polymer crystallization, vol 714. Lecture notes in physics. Springer, Berlin, pp 373–389

Hobbs JK, Vasileva C, Humphris ADL (2005) Real time observation of crystallization in polyethylene oxide with video rate atomic force microscopy. Polymer 46:10226–10236

Hoffman JD, Weeks JJ (1962) Melting process and the equilibrium melting temperature of polychlorotrifluoroethylene. J Res Natl Bur Stand (US) A66: 13

Hong S, Bushelman AA, MacKnight WJ, Gido SP, Lohse DJ, Fetters LJ (2001) Morphology of semicrystalline block copolymers: polyethylene-b-atactic-polypropylene. Polymer 42:5909–5914

Hosier IL, Vaughan AS, Swingler SG (1997) Structure-property relationships in polyethylene blends: the effect of morphology on electrical breakdown strength. J Mater Sci 32(17):4523–4531

Hsieh YT, Ishige R, Higaki Y, Woo EM, Takahara A (2014) Microscopy and microbeam X-ray analyses in poly(3-hydroxybutyrate-co-3-hydroxyvalerate) with amorphous poly(vinyl acetate). Polymer 55:6906–6914

Jordan ND, Bassett DC, Olley RH, Hine PJ, Ward IM (2003) The hot compaction behaviour of woven oriented polypropylene fibres and tapes. II. Morphology of cloths before and after compaction. Polymer 44:1133–1143

Keller A (1968) Polymer crystals. Rep Prog Phys 31:623–704

Keller A, Bassett DC (1960) Complementary light and electron microscope investigations on the habit and structure of crystals, with particular reference to long chain compounds. Proc R Microsc Soc 79:243–261

Kielhorn L, Colby RH, Han CC (2000) Relaxation behavior of polymer blends after the cessation of shear. Macromolecules 33:2486–2496

Kumaraswamy G, Issaian AM, Kornfield JA (1999) Shear-enhanced crystallization in isotactic polypropylene. 1. Correspondence between in situ rheo-optics and ex situ structure determination. Macromolecules 32:7537–7547

Lau KY, Vaughan AS, Chen G, Hosier IL, Holt AF (2013) Absorption current behaviour of polyethylene/silica nanocomposites. Dielectrics 2013(10–12):24–25

Lee Y, Porter RS (1987) Double-melting behavior of poly(ether ether ketone). Macromolecules 20:1336–1341

Lee Y, Porter RS, Lint JS (1989) On the double-melting behavior of poly(ether ether ketone). Macromolecules 22:1756–1760

Lee B, Shin TJ, Lee SW, Yoon J, Kim K, Ree M (2004) Secondary crystallization behavior of poly(ethylene isophthalate-co-terephthalate): time-resolved small-angle X-ray scattering and calorimetry studies. Macromolecules 37:4174–4184

Lemstra PJ, Kooistra T, Challa G (1972) Melting behavior of isotactic polystyrene. J Polym Sci A-2 Polym Phys 10:823–833

Lin D, Cheng H, Zou F, Ning W, Han CC (2012) Morphology evolution of a bisphenol A polycarbonate/poly (styrene-co-acrylonitrile) blend under shear and after shear cessation. Polymer 53:1298–1305

Loo Y-L, Wakabayashi K, Huang YE, Register RA, Hsiao BS (2005) Thin crystal melting produces the low-temperature endotherm in ethylene/methacrylic acid ionomers. Polymer 46:5118–5124

Lovell R, Mitchell GR, Windle A (1979) Wide-angle X-ray scattering study of structural parameters in non-crystalline polymers. Faraday Discuss Chem Soc 68:46–57

Marand H, Xu JN, Srinivas S (1998) Determination of the equilibrium melting temperature of polymer crystals: Linear and nonlinear Hoffman-Weeks extrapolations. Macromolecules 31:8219–8229

McKenna GB (1989) Glass formation and glassy behavior. In: Booth C, Price C (eds) Comprehensive polymer science, vol 2. Polymer properties. Pergamon, Oxford, pp 311–363

Mitchell GR (2011) Chapter 47: neutron diffraction from polymers and other soft matter. In: Imae T, Kanaya T, Furusaka M, Torikai N (eds) Neutrons in soft matter. Wiley, Hoboken

Mitchell GR (2016) Scattering methods for polymer orientation characterisation. Springer, Berlin

Mitchell GR, Rosi-Schwartz B, Ward DJ, Warner M (1994) Local order in polymer glasses and melts. Phil Trans R Soc London A 348:97–115

Mitchell GR, Saengsuwan S, Bualek-Limcharoen S (2005) Evaluation of preferred orientation in multi-component polymer systems using x-ray scattering procedures. Progr Coll Polym Sci 130:149–158

Miwa Y, Drews AR, Schlick S (2008) Unique structure and dynamics of poly(ethylene oxide) in layered silicate nanocomposites: accelerated segmental mobility revealed by simulating ESR spectra of spin-labels, XRD, FTIR, and DSC. Macromolecules 41:4701–4708

Mohan SD, Mitchell GR, Davis FJ (2011) Chain extension in electrospun polystyrene fibres: a SANS study. Soft Matter 7:4397–4404

Morgan RL, Hill MJ, Barham PJ (1999) Morphology, melting behaviour and co-crystallization in polyethylene blends: the effect of cooling rate on two homogeneously mixed blends. Polymer 40:337–348

Müller AJ, Arnal ML (2005) Thermal fractionation of polymers. Prog Polym Sci 30:559–603

Murphy DB, Davidson MW (2012) Fundamentals of light microscopy and electronic imaging, 2nd edn. Wiley-Blackwell, New York

Nielsen AS, Batchelder DN, Pyrz R (2002) Estimation of crystallinity of isotactic polypropylene using Raman spectroscopy. Polymer 43:2671–2676

Olley RH (1986) Selective etching of polymeric materials. Sci Prog 70:17–43

Olley RH, Bassett DC (1989) On the development of polypropylene spherulites. Polymer 30:399–409

Olley RH, Bassett DC (1994) On Surface-Morphology and Drawing of Polypropylene Films. J Macromol Sci Phys B33:209–227

Olley RH, Bassett DC, Blundell DJ (1986) Permanganic etching of PEEK. Polymer 27:344–348

Owen G, Vesely D (1985) An easy and economical method for staining electron microscopy specimens with osmium tetroxide. Proc R Microsc Soc 20:297

Patel D, Bassett DC (1994) On spherulitic crystallization and the morphology of melt-crystallized isotactic poly(4-methylpentene-1). Proc R Soc A 445:577–595

Plans J, MacKnight WJ, Karasz FE (1984) Equilibrium Melting Point Depression for Blends of Isotactic Polystyrene with poly(2,6-dimethylphenylene oxide). Macromolecules 17:810–814

Ruland W (1961) X-ray determination of crystallinity and diffuse disorder scattering. Acta Crystallogr 14:1180–1185

Ruland W (1964) Crystallinity and disorder parameters in nylon 6 and nylon 7. Polymer 5:89–102

Ruland W (1977) Determination of the interface distribution function of lamellar two-phase systems. Coll Polym Sci 255:417

Schmidt-Rohr K, Spiess HW (1994) Multidimensional solid-state NMR and polymers. Elsevier, Amsterdam

Shabana HM, Guo W, Olley RH, Bassett DC (1993) Electron-microscopic observation of spinodal decomposition in blends of tetramethyl polycarbonate and polystyrene. Polymer 34:1313–1315

Shabana HM, Olley RH, Bassett DC, Zachmann H-G (1996) On crystallization and phase separation phenomena in PEN/PHBA copolyesters. J Macromol Sci Phys B35:691–708

Shabana HM, Olley RH, Bassett DC, Jungnickel BJ (2000) Phase separation induced by crystallization in blends of polycaprolactone and polystyrene: an investigation by etching and electron microscopy. Polymer 41:5513–5523

Shahin MM, Olley RH (2002) Novel etching phenomena in poly(3-hydroxy butyrate) and poly (oxymethylene) spherulites. J Polym Sci B Polym Phys 40:124–133

Shahin MM, Olley RH, Blissett MJ (1999) Refinement of etching techniques to reveal lamellar profiles in polyethylene banded spherulites. J Polym Sci Polym Polym Phys 37:2279–2286

Shanks RA, Amarasinghe G (2000) Crystallisation of blends of LLDPE with branched VLDPE. Polymer 41:4579–4587

Shen B, Liang Y, Kornfield JA, Han CC (2013) Mechanism for shish formation under shear flow: an interpretation from an in situ morphological study. Macromolecules 46:1528–1542

Strobl GR, Hagedorn V (1978) Raman spectroscopic method for determining the crystallinity of polyethylene. J Polym Sci Polym Phys 16:1181–1193

Strobl GR, Schneider M (1980) Direct evaluation of the electron density correlation function of partially crystalline polymers. J Polym Sci Polym Phys 18:1343–1359

Toda A, Taguchi K, Nozaki K, Konishi M (2014) Melting behaviors of polyethylene crystals: an application of fast-scan DSC. Polymer 55:3186–3194

Traylor A (1961) A sample preparation technique for the application of phase contrast microscopy to polystyrene-type polymers. Anal Chem 33:1629–1630

Trent JS, Scheinbeim JI, Couchman PR (1983) Ruthenium tetraoxide staining of polymers for electron microscopy. Macromolecules 16(4):589–598

Trindade AC, Godinhoa MH, Figueirinhas JL (2004) Shear induced finite orientational order in urethane/urea elastomers. Polymer 45:5551–5555

Vaughan AS, Stevens GC (1995) On radiation effects in poly(ethylene terephthalate): a comparison with PEEK. Polymer 36:1541–1547

Vaughan AS, Sutton SJ (1995) On radiation effects in oriented PEEK. Polymer 36:1549–1554

Wang Z, Li Y, Yang J, Gou Q, Wu Y, Wu X, Liu P, Gu Q (2010) Twisting of lamellar crystals in poly(3-hydroxybutyrate-co-3-hydroxyvalerate) ring-banded spherulites. Macromolecules 43:4441–4444

Weng J, Olley RH, Bassett DC, Jääskeläinen P (2003) Changes in the melting behavior with the radial distance in isotactic polypropylene spherulites. J Polym Sci Polym Phys 41:2342–2354

Weng J, Olley RH, Bassett DC, Jääskeläinen P (2004) Crystallization of propylene–ethylene random copolymers. J Polym Sci Polym Phys 42:3318–3332

Wignall GD, Melnichenko YB (2005) Recent applications of small-angle neutron scattering in strongly interacting soft condensed matter. Rep Prog Phys 68:1761–1810

Wunderlich B, Sullivan P (1962) Interference microscopy of crystalline high polymers. Determinations of the thickness of single crystals. J Polym Sci 56:19–25

Yagpharov M (1986) Thermal analysis of secondary crystallization in polymers. J Therm Anal Calorim 31:1073–1082

Yamada K, Hikosaka M, Akihiko Toda A, Yamazaki S, Tagashira K (2003a) Equilibrium melting temperature of isotactic polypropylene with high tacticity: 1. Determination by differential scanning calorimetry. Macromolecules 36:4790–4801

Yamada K, Hikosaka M, Toda A, Yamazaki S, Tagashira K (2003b) Equilibrium melting temperature of isotactic polypropylene with high tacticity. 2. Determination by optical microscopy. Macromolecules 36:4802–4812

Yamada K, Kajioka H, Nozaki K, Toda A (2011) Morphology and growth of single crystals of isotactic polypropylene from the melt. J Macromol Sci Phys 50:236–247

Zheng K, Zhang J, Cheng J (2013) Morphology, structure, miscibility, and properties of wholly soy-based semi-interpenetrating polymer networks from soy–oil–polyol-based polyurethane and modified soy protein isolate. Ind Eng Chem Res 52:14335–14341

Zou F, Dong X, Lin D, Liu W, Wang D, Han CC (2012) Morphological and rheological responses to the transient and steady shear flow for a phase-separated polybutadiene/polyisoprene blend. Polymer 53:4818–4826

Chapter 3
Crystallization in Nanocomposites

Geoffrey R. Mitchell, Donatella Duraccio, Imran Khan, Aurora Nogales, and Robert Olley

3.1 Introduction

Nanocomposites based on polymer matrices have been studied extensively (Paul and Robeson 2008). In the majority of these studies, the focus has been the enhancements in properties through the addition of relatively small quantities of nanoparticles. In this chapter, we centre our attention on the effects, if any, of the nanoparticles on the behaviour of the polymer matrix and in particular the nucleation and growth of crystal phases in the case of crystallizable polymers. We first consider the influence of macroscopic particles on this behaviour by examining the influence of polymer fibres on the matrix behaviour. We then consider the situation in polydisperse melts where extended objects can be formed from the high molecular weight fraction in the melt, and then finally we review work reported in this field in the context of nanoscale fillers.

G.R. Mitchell (✉) • I. Khan
Centre for Rapid and Sustainable Product Development, Institute Polytechnic of Leiria, Marinha Grande 2430 028, Portugal
e-mail: geoffrey.mitchell@ipleiria.pt

D. Duraccio
Italian National Research Council Institute for Agricultural and Earthmoving Machines IMAMOTER, Strada delle Cacce, 73, Torino 10135, Italy

A. Nogales
Macromolecular Physics Department at the Instituto de Estructura de la Materia, CSIC, Serrano 121, Madrid 28006, Spain

R. Olley
EMLAB, University of Reading, Whiteknights, Reading RG6 6AF, UK

3.2 Templating

3.2.1 Linear Nucleation

Preferred orientation in polymers (see Sect. 1.6.1) can deliver enhanced mechanical properties, especially when the level of preferred orientation is high. Most familiar to many is the case of nylon rope. Polyethylene is another material where very high strength fibres have been used to make ropes and bullet-proof vests. Two kinds of these are commercially available, namely Dyneema® from DSM and Spectra® from Honeywell.

To incorporate these advantages into bulk items, the making of polymer fibre-based composites has been investigated. Where the matrix is of a different material such as an epoxy resin, compatibility is a problem, both in terms of differential thermal expansion and adhesion. To avoid this, self-composites have been prepared by the technology of hot compaction (Ward and Hine 2004). The initial research focused on exploiting high modulus melt spun polyethylene fibres. These were laid down in parallel in a metal mould, which was heated under a small contact pressure to prevent the fibres contracting, to a temperature where the material was just beginning to melt. A much higher pressure was briefly applied to force the fibres together, with perhaps some deformation, and to eliminate the air in order that the fibres were bathed in molten polyethylene. On cooling the molten polyethylene recrystallized, producing a fibre composite with a semi-crystalline matrix of the same polymer as the fibres.

In these composites, much of the original mechanical strength is retained in the composite (Hine et al. 1993) which is achieved due to the development of a particular characteristic morphology (Olley et al. 1993). This is shown in Fig. 3.1

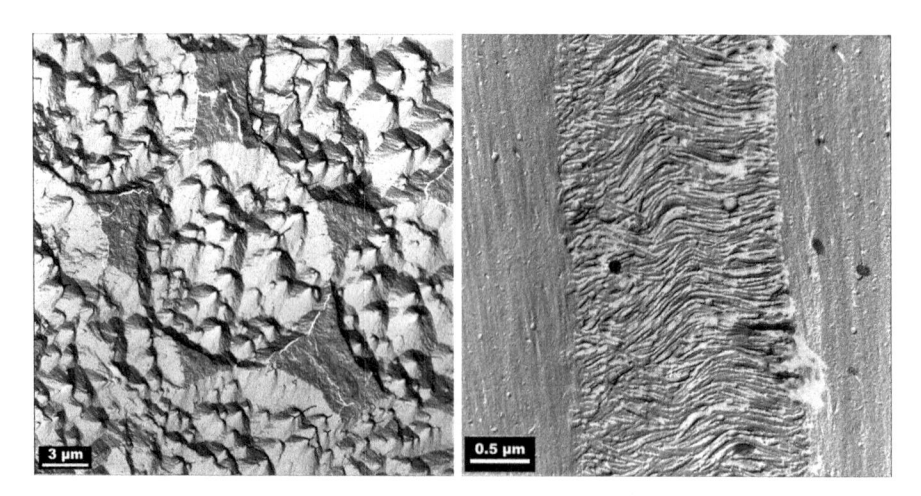

Fig. 3.1 The morphology of fibre compaction from Olley et al. (1993): reproduced with permission from the Royal Society. (**a**) The *left hand image* shows an etched section taken normal to the initial fibre direction, (**b**) the *right hand image* shows an etched section taken parallel to the initial fibre direction

which demonstrates sections transverse and parallel to the initial fibre axis. Figure 3.1a is a transverse section which has been etched (see Chap. 2): this composite was prepared from fibres with a circular cross-section, inside of which are seen pits developed round nanovoids which run along the length of the fibre. Between the fibres the surfaces of lamellae are seen, indicating that molecular axes in the crystals are also parallel to the fibre length. Figure 3.1b is a section parallel to the fibres. To the left and right of this micrograph are parts of two fibres aligned vertically; in the middle lamellae are seen which have grown out from the surface of each fibre. This is the characteristic morphology of compacted fibre composites.

The first examples of self-composites were unidirectional composites based on polyethylene, but later the technique was modified by laying down alternating perpendicular layers of polyethylene fibres producing a two-dimensional composite sheet, which maintained a sizeable fraction of the strength of the original fibres in all directions along the plane of the sheet (Hine et al. 2000). The technique was extended to other materials including polypropylene. This polymer is available as high-modulus tapes woven into cloths, for example geotextiles, which can be compacted (Amornsakchai et al. 2000; Jordan et al. 2003). This is currently the largest commercial application of the technique, which is used in Samsonite Curv® luggage.

Figure 3.2 shows the results of a crystallization experiment, where a high-modulus polyethylene fibre (Tekmilon®) has been inserted into molten polyethylene; the high molecular weight and orientation of the fibre allow it to remain solid a few degrees above the melting point of the surrounding polymer. On cooling the sample to the crystallization temperature, crystals nucleate on the surface of the fibre (Fig. 3.2) as they do in fibre compactions (Fig. 3.1).

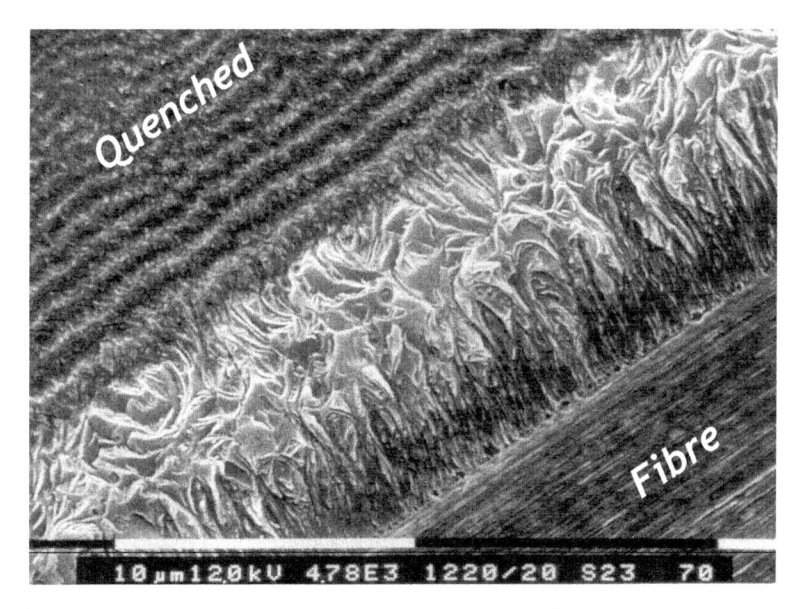

Fig. 3.2 Sclair 2907 polyethylene (high density, crystallized at 123 °C for 70 s) (Abo el Maaty and Bassett 2005)

Fig. 3.3 Polypropylene row structure: adapted from (White and Bassett 1998)

Initially, the crystal growth follows a direction normal to the orientation of the fibre (right hand picture above). With a fibre isolated in the melt, this closely parallel mode of growth is only sustainable for a short distance, owing to the presence of non-crystallized cilia, and after a short distance the closely spaced parallel lamellae give way to a form of cylindrulitic growth whose morphology resembles that of the outer regions of a large spherulite (See Chap. 1). This geometry has been extensively studied with respect to variations in the crystallization processes (Abo el Maaty and Bassett 2005, 2006), in order to give a basis for the formation of banding in polymer spherulites.

Polypropylene can be easily induced to form row structures by extension of the melt. An example of the resultant morphology is shown below. Closely spaced shish-kebab lamellae have grown directly out from the central row nucleus, but as growth proceeds outwards, increasing numbers of cross-hatching lamellae are found. The cross-hatching lamellae are found almost perpendicular to the lamellae growing out from the initial nucleus (Fig. 3.3).

In injection-moulded polypropylene products, densely packed row structures are typically found near the surface of the object. At least one process has been developed which is able to deliver this morphology throughout the moulded object, namely shear-controlled orientation injection moulding (SCORIM) (Kalay and Bevis 1997).

Polyethylene, especially a material containing a proportion of high molecular chain lengths, can be induced to form row structures by applying sufficient shear to the melt (An et al. 2006). If these are closely spaced, the material will be filled with parallel "shish-kebab" growth as in the Fig. 3.4:

3.3 Crystallization and Flow

We have seen in the previous section that objects such as fibres or bundles of polymer chains can act as nucleating surfaces for polymer lamellae. The geometry of the lamellae is normal to the surface as can be seen clearly in Fig. 3.1. As a

Fig. 3.4 Representative TEM morphology seen in quenched linear PE crystallized isothermally at 115°C. Sample had been sheared at 25 s^{-1}. The arrow indicates the shear flow direction in the melt. Y. An et al. (2006) Polymer 47: 5643–5656

consequence, the morphology of the sample is determined by the orientation of the nucleating object. The most common form of a nucleating object is approximately spherical. Although interesting in themselves spherical nucleants do not lead to morphological control other than in the density of nucleating sites. In contrast, if we are able to develop anisotropic nuclei such as rod-shaped particles, then this leads directly to a common orientation of the polymer lamellar crystals. We describe these objects as *directing* the crystallization. As we have seen in the previous section, this directing does not continue forever, as other processes cause the lamellar crystals to diverge and lose their initial preferred orientation. If we wish to define the morphology throughout the material, we need to provide sufficient directing nuclei such that the directed volumes associated with each nucleus fill space (see Chap. 5). Moreover, we will need to provide the nuclei themselves with a common alignment. In principle, this could be achieved using an electric, magnetic, or thermal field but for ease of application we restrict our discussion to the use of flow fields as widely used in conventional polymer processing.

We expect that a rod-like object embedded in an elastic medium will rotate as the matrix is deformed. Anisotropic objects within an elastic medium which retain their integrity on deformation will follow the predictions of the pseudo-affine model with respect to the orientation of the objects as shown in Fig. 3.5.

Polymer processing involves shear flow as well as simple extensional flow. Shear flow is a more complex arrangement with a three-dimensional nature to it (Fig. 3.6) and a weak extensional flow component. Rigid rods in a viscous fluid were first considered theoretically by Jeffery (1922). Jeffery showed that objects exhibited a trajectory often called a Jeffery's orbit which rotated the axis of the

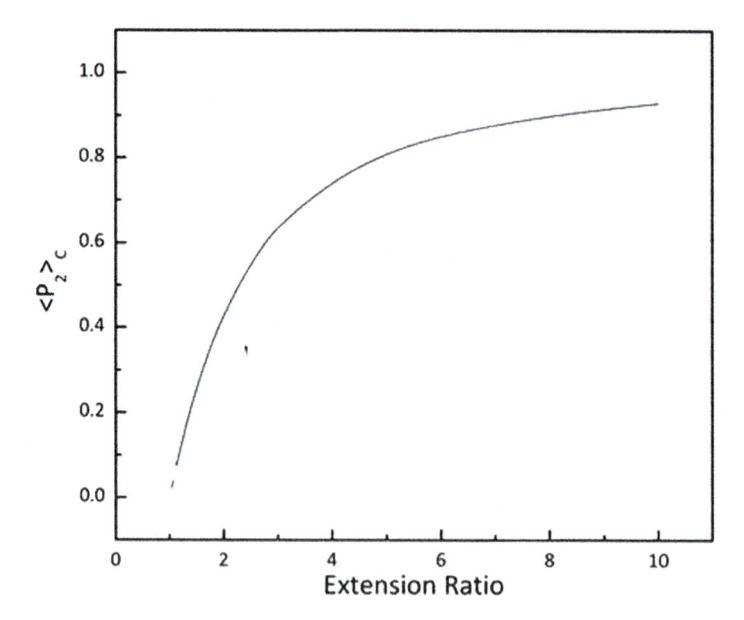

Fig. 3.5 A plot of the orientation parameter $\langle P_2 \rangle_c$ against extension ratio λ calculated on the basis of the pseudo-affine model for non-deformable rods embedded in an elastic medium

Fig. 3.6 The essentials of shear flow geometry

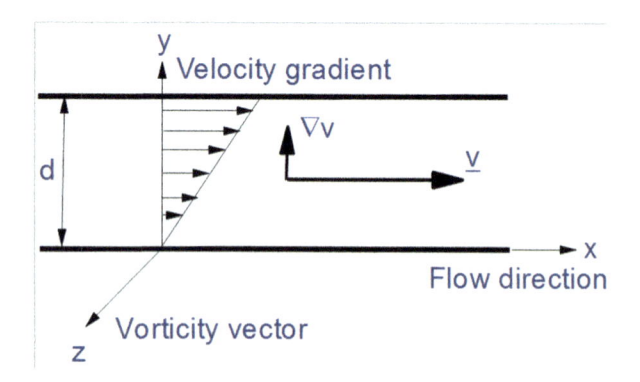

anisotropic unit from parallel to the vorticity vector to parallel to the flow direction. The frequency of that orbit depends on the aspect ratio of the object, Jeffrey considered ellipsoidal particles and in some cases a stable alignment could be obtained which was also related to the aspect ratio and the properties of the fluid matrix.

Hobbie et al. (2003) used optical measurements of semi-dilute dispersions of polymer dispersed multiwalled carbon nanotubes to determine that for a weakly elastic polymer melt, the tubes aligned preferentially parallel to the flow direction at low stresses with a transition to alignment with the vorticity vector above a critical stress, corresponding to a Deborah number of 0.15. The Deborah number is defined as the ratio of the relaxation time characterizing the time it takes for a material to adjust to applied stresses, and the characteristic time scale of an experiment probing the response of the material. At low Deborah numbers, the material is fluid-like but at higher Deborah numbers the material exhibits non-Newtonian behaviour that is typical of solid materials. For a highly elastic polymer solution, the tubes align parallel to the flow field at high shear rates.

In summary, the behaviour in flow of nanoparticles will depend upon their shape size and the nature of the polymer matrix.

3.4 Nanoparticles

3.4.1 Introduction

Eric Drexler introduced the term nanotechnology to describe the act of engineering materials on a very small scale. Since then, the production of nanomaterials and nanoparticles has blossomed. A "Nanoparticle" is a particle with an upper particle size limit of 100 nm. They are a link between bulk materials and atomic or molecular structures. While bulk materials have constant physical properties regardless of its size, among nanoparticles the size often dictates the physical and chemical properties. This indicates that properties of materials change as their size approaches the nanoscale. This feature can be attributed to the change in surface-to-volume ratio. A higher surface-to-volume ratio provides the opportunity that the structure of interest can interact with its environment (e.g. polymer matrix) more efficiently leading to property enhancement. Moreover, the main reason why nanoparticles have different optical, electrical, magnetic, chemical, and mechanical properties from their bulk counterparts are that in this size-range quantum effects start to predominate (Hristozov and Malsch 2009). By means of quantum effects one can intuitively understand that they start to display quantum physics effects.

The major challenges, for nanoparticle applications, are their strong tendency of aggregation, agglomeration, and rapid sedimentation which limit the property enhancement in different matrices. Two approaches have been commonly used in improving nanoparticle stability: electrostatic and steric stabilization. Electrostatic repulsion is achieved by imparting or increasing the surface charge while steric stabilization is typically attained by the adsorption of long-chain organic molecules (e.g. surfactants). Combined electro-steric stabilization is also promising with the use of ionic polymeric molecules (e.g. polyelectrolytes).

Nanoscale particles can be classified in two ways, i.e. based on (i) structure and (ii) nature of nanoparticles. Structurally, they can be further distinguished as 1-D, 2-D and 3-D nanoparticles where 1-D, 2-D and 3-D indicate one-dimensional, two-dimensional and three-dimensional nanoparticles. Practically speaking, a

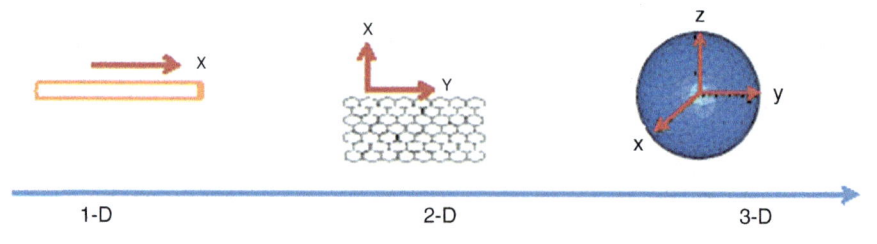

Fig. 3.7 A schematic depiction of 1-D, 2-D and 3-D nanoparticles

one-dimensional nanoparticle does not exist at all but what we mean as 1-D nanoparticle is that the length along two of the sides are relatively much smaller than the other side such that it that looks like a 1-D nanoparticle. For example, a long cylindrical particle having more than couple of hundred nanometres length and a few nanometres in diameter can be considered as a 1-D nanoparticle. 2-D nanoparticles are plate-like or two-dimensional sheets while 3-D nanoparticles can be thought of as spherical or sphere-like particle, e.g. a balloon or football as depicted in Fig. 3.7. Carbon nanotubes, graphene and fullerene are examples of 1-D, 2-D and 3-D nanoparticles, respectively.

The nature of nanoparticles can be categorized as carbonaceous nanoparticles, metallic nanoparticles and clay nanoparticles, all of which are well-studied nanoparticle families. Carbonaceous nanoparticles are a family of nanoparticles including single wall carbon nanotubes (SWCNT), double wall carbon nanotubes (DWCNT), multiwall wall carbon nanotubes (MWCNT), graphene, graphene oxide and fullerenes. The metallic nanoparticle is a large class of nanomaterials including but not limited to silver (Ag) nanoparticles, Gold (Au) nanoparticles, iron (Fe) nanoparticles and copper nanoparticles. Clay nanoparticles are receiving attention not only due to their competitive prices but also due to the tubular-shaped as well as plate-like structures. The size and shape of such nanoparticles depends on the synthesis method. The choice of nanoparticle for use in a composite material will depend on both the structure and nature of the nanoparticle to achieve a particular functionality. Members from the carbonaceous and clay families are further discussed before considering their behaviour in polymeric matrices.

3.4.2 Carbon Nanotubes

Since their discovery over two decades ago, carbon nanotubes have been heralded as a material to initiate a wave of major technological advancements during the twenty-first century. The low density of carbon in comparison to other elements, i.e. metal or clay makes it a versatile family. The incorporation of carbon nanotubes into a matrix material often leads to property enhancement in the material.

3 Crystallization in Nanocomposites

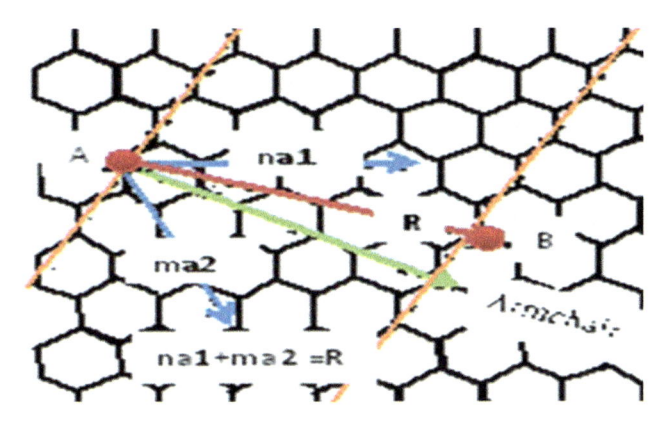

Fig. 3.8 Schematic diagram showing how a hexagonal sheet of graphene is rolled to form a CNT with chiral vector (Wilder et al. 1998) [Reproduced Courtesy of Nature Publishing Group]

Carbon nanotubes (CNTs) are members of the Fullerene structural family. Their name is derived from their long, hollow structure with the walls formed by one-atom-thick sheets of carbon, called graphene (Fig. 3.8). The manner in which these sheets are connected to form tubes at specific and discrete ("chiral") angles and the combination of the angle and radius decides the nanotube properties; for example, whether the individual nanotube exhibits metal or semiconductor like behaviour. The values of n and m are used not only for the chirality or "twist" but also to draw a line between metallic and semiconducting nanotubes. In other words, chirality in turn affects the conductance, density, lattice structure, and certain other properties of the nanotube. A nanotube is considered metallic if the value $n - m$ is divisible by three. Otherwise, the nanotube is semiconducting.

Consequently, when tubes are formed with random values of n and m, we would expect that two-thirds of nanotubes would be semiconducting, while the other third would be metallic, and this has been experimentally confirmed to be the case (Wilder et al. 1998). The diameter of a carbon nanotube is related to the chiral vector (n, m) using the relation

$$d = \left(n^2 + m^2 + nm\right)^{1/2} 0.0783 \, \text{nm} \tag{3.1}$$

where "d" is diameter of the tube and n, m are integers of the vector equation $\mathbf{R} = n\mathbf{a}_1 + m\mathbf{a}_2$ as shown in Fig. 3.8 (Wilder et al. 1998). Metallic and semiconducting single-walled carbon nanotubes on SiO_2 can have obviously different contrast in scanning electron microscopy due to their conductivity difference and thus can be effectively and efficiently identified as shown in Fig. 3.9 (Jie 2012, Li 2012).

The structure of CNTs consists of one, two or more, graphene sheets rolled up to make the tubes leading to variations from single (S), double (D) and multi-wall (MW) carbon nanotubes, respectively. Nanotubes naturally align themselves into "ropes" held together by van der Waals forces, more specifically, pi-stacking. The nature of the dispersion problem for CNTs is rather different from other

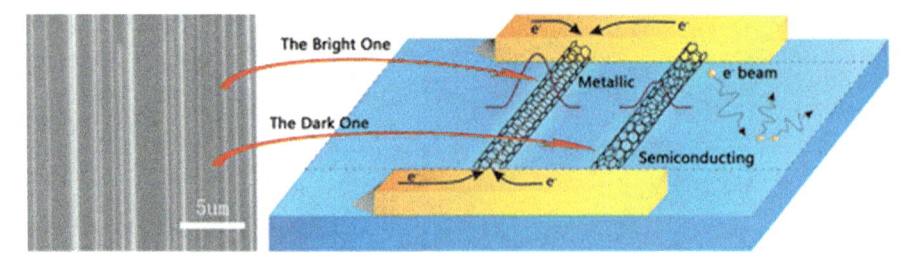

Fig. 3.9 Identification of metallic and semiconducting single-walled carbon nanotubes in Scanning Electron Microscopy. Adapted with permission from Jie (2012). Copyright (2015) American Chemical Society

conventional fillers, such as spherical particles and carbon fibres, because CNTs are characterized by a small tube diameter on the nanometre scale and with a high aspect ratio (>1000) and thus possess an extremely large surface area. A variety of techniques have been used to disperse carbon nanotubes in the matrix material to obtain the desired properties. These include but are not limited to use of polar solvents, mechanical mixing, ultrasonication and functionalization of carbon nanotubes. Functionalization can be achieved in two different ways, i.e. ionic functionalization and covalent functionalization. In ionic functionalization, an ionic salt is used to disperse carbon nanotubes while in covalent functionalization carbon nanotubes are acid treated to get attach the carboxylic group on the surface of tubes. The repulsion between these attached carboxylic groups restricts or hinders the aggregation process which consequently leads towards improved dispersions of the CNTs. Scanning electron microscope (SEM) and transmission electron microscope (TEM) are the most widely used techniques to observe how well the tubes are dispersed. Figure 3.10 shows multiwall carbon nanotubes observed using TEM, revealing that tubes are not only curled but even highly entangled due to van der Waals forces of attraction leading to bundles of tubes.

Because of their elongated cylindrical structure, CNTs are more favourable for enhancing electronic and charge transport properties. Besides their interesting electronic properties CNTs show a set of unique mechanical properties as well owing to an extremely high level of strength due to the bonds between the single carbon atoms. The chemical bonding in CNTs is composed entirely of sp^2 bonds which are stronger than the sp^3 bonds found in diamond making them a unique candidate with the highest tensile strength, elastic modulus and perhaps the stiffest material on earth with extra light weight (Rasmussen and Ebbesen 2014).

Graphene: The development of graphene has a very high pace due to its two-dimensional structure having a variety of impressive properties. The most common properties associated with graphene on which it is selected for different applications range from high electron mobility, Young's modulus, high thermal conductivity and optical absorbance. In other words, graphene is super-strong and stiff, amazingly thin, almost completely transparent, extremely light weight, and an amazing conductor of electricity and heat. The properties associated with the

Fig. 3.10 Transmission electron microscopy (TEM) images of MWCNTs

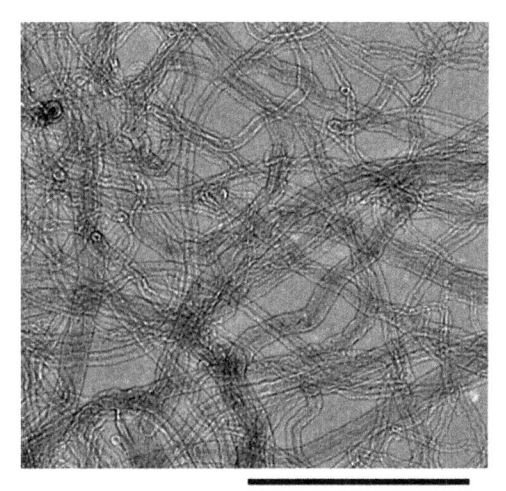

100 nm
HV=200.0kV
Direct Mag: 115000x

graphene structure depends on the synthesis method. Currently, chemical vapour deposition (CVD) is considered to be the best method for fabricating high-quality graphene in large quantities. This method involves depositing gaseous carbon atoms on a copper foil, then transferring the graphene film to a wide range of substrates, such as silicon. The CVD process can be controlled to minimize the amount of defects. Although weak points at grain boundaries and multilayer regions are quite common in CVD graphene, they have been shown to have a minimal effect on performance in most cases. The favourable price/quality ratio makes CVD graphene (Fig. 3.11) suitable for many applications such as flexible/transparent electronics, gas barriers, and anti-corrosion coating. Liquid phase and thermal exfoliation both involve splitting up graphite to form graphene flakes—either using chemical solvents or thermal shock. These methods are very scalable to large-scale production—however, the flakes are usually multilayered, not single layered, and the number of layers can be hard to control. The process can also introduce impurities into the graphene. Silicon carbide can be used to synthesize graphene layers when Si atoms are sublimated; the remaining face of the silicon carbide becomes a thin graphite surface. Nowadays, it is possible to control the number of graphene layers and to obtain a very high quality over a wide area. The high cost of silicon carbide and the requirement for elevated temperatures to achieve the sublimation are the main shortcomings of this process.

An issue in graphene research, which is similar to that which arises with carbon nanotubes, is the use of terms. The term "graphene" is used in a generic manner and not in a precise way to describe many graphene-based materials synthesized and studied (Bianco et al. 2013). Peter (2014) recently proposed a model which considers the number of graphene layers, the average lateral size, and the carbon-

Graphene synthesis and Quality over price scalling

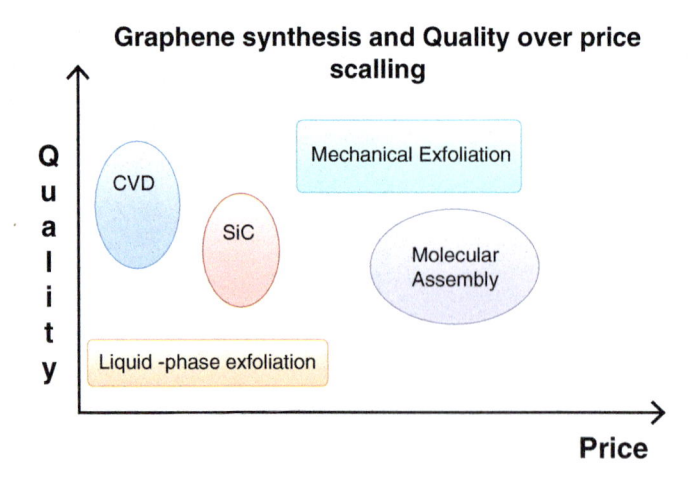

Fig. 3.11 A general trend to describe graphene quality over price for different synthesis methods

to-oxygen (C/O) atomic ratio as the three fundamental properties that cover the largest set of current graphene-based materials (GBMs) encountered in practice. One can distinguish easily the different types of graphene structure from this model which is depicted in Fig. 3.12.

The electronic properties of graphene are sensitive to geometry. For electronic applications, graphene is generally laid on a substrate made of a bulk material. Owing to the presence of dangling chemical bonds, 3D materials tend to be poor substrates for 2D materials such as graphene. One solution is to use hexagonal boron nitride (h-BN), which is both an insulator and a 2D material, and so does not have the problematic bonds. For a complete electronic characterization of the graphene film, down to submicrometre grains, synchrotron-based conventional and nanoresolved photoelectron spectroscopies are used (Razado-Colambo et al. 2015).

There are numerous potential applications for graphene. One particular application may be as a gas barrier, for example, blocking water vapour or oxygen from interacting with sensitive materials. In addition, the strong carbon–carbon bonds within a sheet of graphene open up the possibility for using either sheets of pure graphene in applications where the strength can be exploited directly or adding graphene to other materials to improve their mechanical strength. It is also used as anti-corrosion coating. It is the most promising nanofiller due to its high surface area, aspect ratio, tensile strength, thermal and electrical conductivity, EMI shielding ability, flexibility and transparency (Potts et al. 2011).

3 Crystallization in Nanocomposites

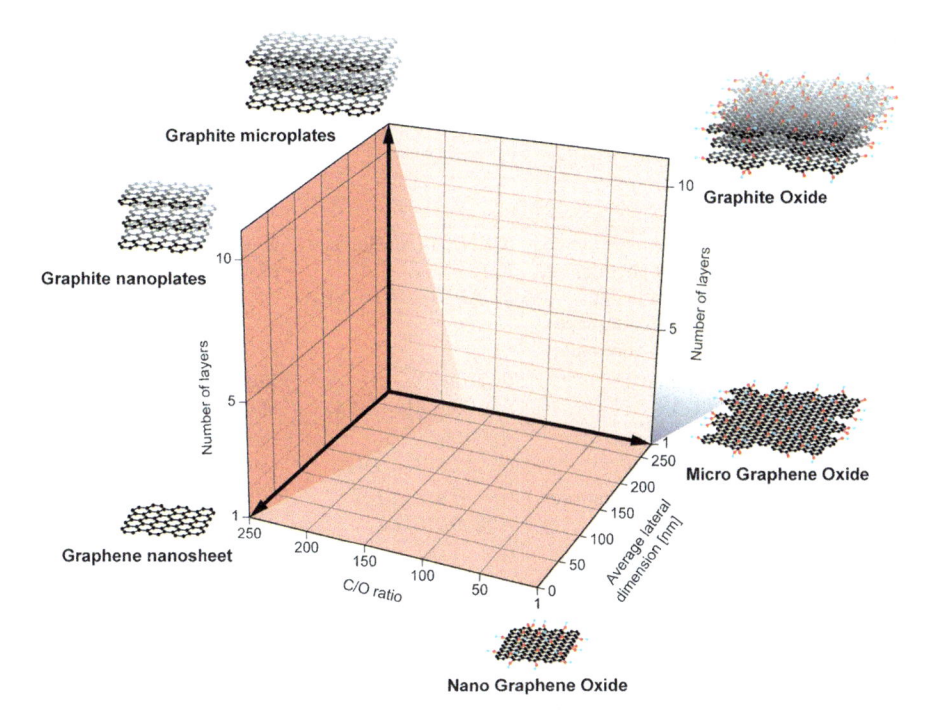

Fig. 3.12 Classification grid for the categorization of different graphene types according to three fundamental GBM properties: number of graphene layers, average lateral dimension and atomic carbon/oxygen ratio. The different materials drawn at the six corners of the box represent the ideal cases according to the lateral dimensions and the number of layers reported in the literature. The values of the three axes are related to the GBMs at the nanoscale, but it is feasible to expand the values to the microscale (Wick et al 2014) Courtesy: Reproduced with permission of John Wiley and Sons

3.4.3 Nanoclay

In terms of chemistry, all clay minerals may simply be described as hydrous silicates. In terms of their natural locations, clay minerals can be divided into two classes, residual clay and transported clay (or sedimentary clay) (Siddiqui and Ahmed 2005). Residual clays could be produced by the chemical decomposition of rocks caused by weathering conditions, e.g. granite containing silica and alumina; by the solution of rocks, i.e. limestone; and by the disintegration and solution of shale. The second type, transported clay, is removed from the original deposit through erosion and deposited to a distant place. Clay minerals may be divided into four major groups, mainly in terms of the variation in the layered structure. These include the Kaolinite group, the Montmorillonite/smectite group, the Illite group, and the Chlorite group and are summarized in Table 3.1 with general formula (Uddin 2008).

Table 3.1 Four major groups of clay minerals, mainly in terms of the variation in the layered structure

	Group name	Member minerals	General formula	Remarks
1	Kaolinite	Kaolinite, dickite, nacrite	$Al_2Si_2O_5(OH)_4$	Members are polymorphs (composed of the same formula and different structure)
2	Montmorillonite or Smectite	Montmorillonite, pyrophyllite, talc, vermiculite, sauconite, saponite, nontronite	$(Ca,Na,H)(Al,Mg,Fe,Zn)_2(Si,Al)_4O_{10}(OH)_2\text{-}XH_2O$	**X** indicates varying level of water in mineral type
3	Illite	Illite	$(K,H)Al_2(Si,Al)_4O_{10}(OH)_2\text{-}XH_2O$	**X** indicates varying level of water in mineral type
4	Chlorite	(i) Amesite	(i) $(Mg,Fe)_4Al_4Si_2O_{10}(OH)_8$	Each member mineral has separate formula; this group has relatively larger member minerals and is sometimes considered as a separate group, not as part of clays
		(ii) Chamosite	(ii) $(Fe,Mg)_3Fe_3AlSi_3O_{10}(OH)_8$	
		(iii) Cookeite	(iii) $LiAl_5Si_3O_{10}(OH)_8$	
		(iv) Nimite, etc.	(iv) $(Ni,Mg,Fe,Al)_6AlSi_3O_{10}(OH)_8$	

(Uddin 2008) Courtesy: With permission from Springer Publications

In general, nanoclays have a large class of categories with different compositions which will alter the properties hence only two members, halloysite and montmorillonite nanoparticles, are thoroughly discussed to give an insight to nanoclays in general.

Halloysites Nanotubes: Halloysite is a naturally occurring eco-friendly clay mineral which is predominantly composed of multi-walled nanotubular-shaped crystals and characterized by a high specific surface area and unique surface properties (Joussein et al. 2005). In terms of chemical composition, Halloysite nanotubes (HNT) are similar to Kaolin and can be considered as rolled Kaolin sheets with an inner diameter of 10–20 nm, outer diameters of 40–70 nm and lengths of 500–1500 nm. The empirical formula for halloysites is $Al_2Si_2O_5(OH)_4$ and its main constituents are aluminium (20.90 %), silicon (21.76 %) and hydrogen (1.56 %). Halloysite typically forms by hydrothermal alteration of alumino-silicate minerals. The internal side of Halloysite is composed of Al_2O_3 while the external is mainly SiO_2. The layers have a tendency to roll into nanotubes to correct the lateral misfit between the silicate and aluminate sheets. Since Halloysite nanotubes belong to an aluminosilicate group, they can be decomposed by both strong inorganic acids and bases, resulting in changes to their morphology and structure. An example HNTs are shown in the electron micrographs in Fig. 3.13 (Wei et al. 2014).

Due to the easy dispensability of HNTs, it is expected that they could be dispersed relatively uniformly in thermoplastics by direct melt blending, especially for polymers with high polarity such as polyamides. Due to the variety of

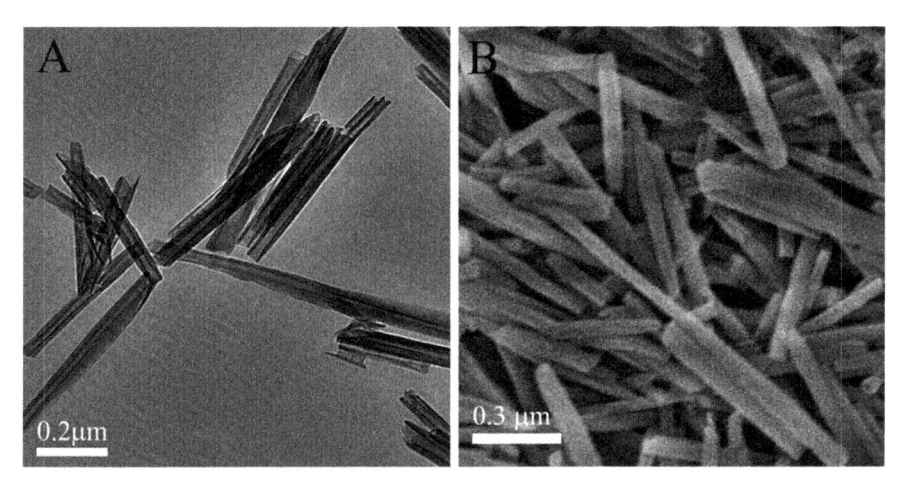

Fig. 3.13 Images of halloysite clay, (**a**) Transmission electron micrograph of tubes dispersed in water and dried on a copper grid; (**b**) Scanning electron micrograph of dry HNT powder (Wei et al. 2014) [Reproduced with permission of "The Royal Society of Chemistry"]

crystallization conditions and geological occurrence, HNTs adopt different morphologies such as tubular, spheroidal and plate-like particles, of which the tubular structure is the most common. Empty tubular structures of halloysite makes it a versatile container for loading with active chemical agents. It has been loaded with organic and inorganic substances including corrosion inhibitors, biocides, metals and salts to prepare functional polymer nanocomposites. Halloysite exhibits higher adsorption capacity for both cationic and anionic dyes because it has negative SiO_2 outermost and positive Al_2O_3 inner tube surface; therefore, these clay nanotubes have efficient bivalent absorbency (Zhao et al. 2013). The flexural properties, tensile strength and impact strength all are noticeably improved with the incorporation of HNTs (Du et al. 2010) (Table 3.2).

HNTs have a wide range of applications such as for use in anticancer therapy, for sustained delivery of certain agents, acting as a template or nanoreactor for biocatalyst, use in personal care and cosmetics and even used as environment protective (Ravindra et al. 2012).

Montmorillonite (MMT) nanoclays have attracted much research interests over the past decade and are widely used for dispersion in polymers due to the high aspect ratio and the large interface of the polymer–nanoclay interaction. MMT is a hydrophilic and inorganic 2-D structure belonging to the nanoclay family (Fig. 3.14). MMT is a subclass of smectite with 2:1 phyllosilicate mineral characterized as having greater than 50 % octahedral charge. It is generally represented as $(Na,Ca)_{0.33}$ $(Al,Mg)_2$ $(Si_4O_{10})(OH)_2 \cdot nH_2O$ with repeating units and monoclinic prismatic structure. MMT is less effective than halloysite when employed with a matrix material in context to electrical, mechanical and thermal properties. Naturally available MMT clays are hydrophilic, and they must be made organophilic (hydrophobic) to have compatibility with most host polymers. Organic treatment is

Table 3.2 Comparison of mechanical property enhancement of three polymer matrices material [polypropylene (PP), polybutylene terephthalate (PBT) and polyamide (PA)] when employed with HNTs

Material	Flexural modulus (GPa)	Flexural strength (MPa)	Tensile strength (MPa)	Impact strength (kJm^{-2})
Neat PP	1.37	44.5	33.5	4.05
PP/HNT/(100/5)	1.75	51.5	35.2	5.51
Neat PBT	2.3	79.7	52.3	5.65
PBT/HNT/(100/5)	2.6	82.7	54.9	4.65
Neat PA	2.71	110	77.0	5.25
PA/HNT/(100/5)	3.23	118.5	82.1	5.75

(Mingliang 2010) Reproduced, courtesy of "John Wiley and Sons"

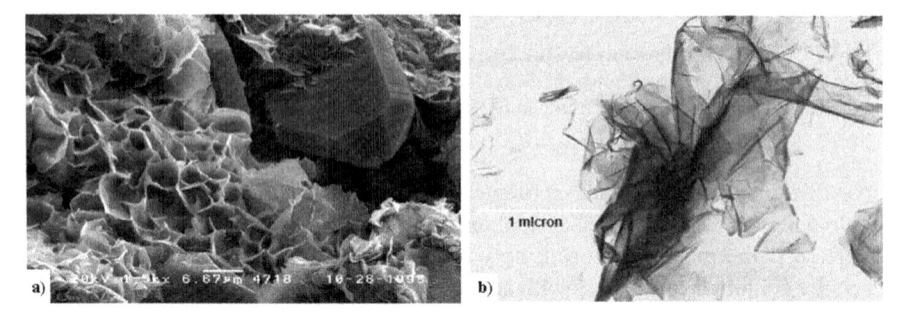

Fig. 3.14 (**a, b**) Montmorillonite (MMT) nanoclay. (Wallace 2015). Source [https://nice.asu.edu/nano/montmorillonite-mmt-nanoclays-ashpalt]

typically accomplished via organic cations, namely onium or quaternary ammonium-based salts in MMT clays. When treated MMT clays are mixed into the host matrix polymer, two types of morphologies evolve, intercalated and delaminated/exfoliated structure. The intercalated structure is a well-ordered multilayered structure of silicates, where the polymer chains gets into the interlayer spaces (inter-gallery region). On the other hand, in the exfoliated/delaminated structure the polymer chains separate individual silicate layers well apart, e.g. 80–100 Å or more, and no longer close enough to interact with each other.

MMT nanoclay has a potential to alter material properties when used as a filler material (Ogata et al 1997, You et al 2011). Epoxy-MMT nanocomposite is one of the key examples indicating an increase in storage modulus (E′) with an increase in MMT loading. It was also reported that volume conductivity and surface conductivity values of the nanocomposites decreased with increasing DC voltage (Rashmi et al. 2011). A more recent study reveals the significance and use of MMT in hydrogels to improve its properties (Ting et al 2015). Li et al. (2015) observed the influence of the MMT on the swelling degree at equilibrium of P(ATC-AAm)/MMT nanocomposite hydrogels was shown in Fig. 3.15. The nanocomposite hydrogels showed normal swelling behaviour at low loading while

3 Crystallization in Nanocomposites

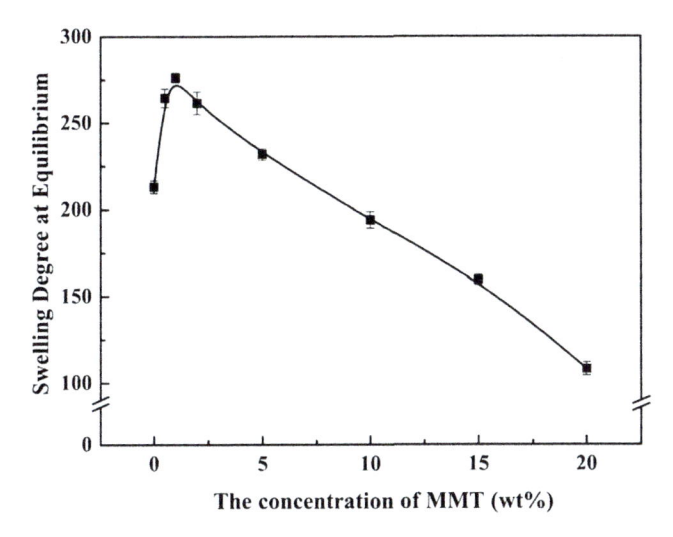

Fig. 3.15 The swelling degree at equilibrium of nanocomposite hydrogels with various MMT contents (Ting et al 2015) [Reproduced with permission of "John Wiley and Sons"]

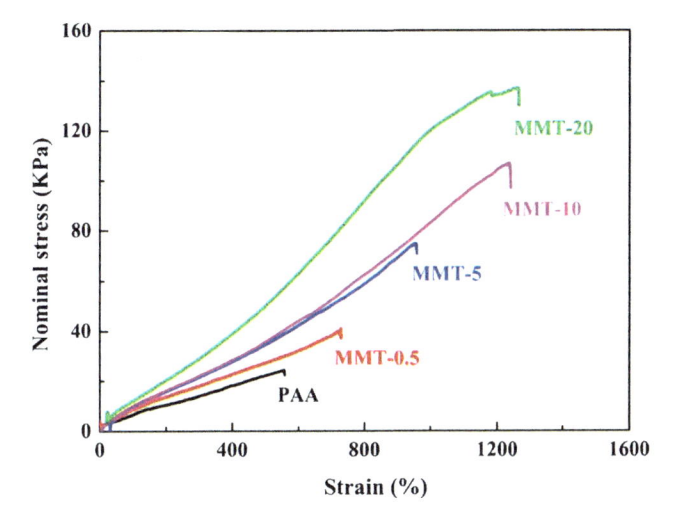

Fig. 3.16 Tensile stress–strain curves of P(ATC-AAm)/MMT nanocomposite hydrogels. (Ting et al 2015) [Reproduced with permission of "John Wiley and Sons"]

increasing the MMT amount in the nanocomposite gels led to decrease the swelling degree at equilibrium. Moreover, the tensile stress–strain curves are shown in Fig. 3.16, obviously, the addition of MMT led to an increase in Young's modulus. The reported work clearly signifies the MMT platelets as cross-linkers in hydrogel resulted in the enhanced mechanical properties.

3.5 Nanocomposites with Carbon Nanotubes

The outstanding properties of individual carbon nanotubes (CNT) combined with their low mass density make them ideal candidates as reinforcing nanoparticles for high performance polymer materials. The idea is that, due to their aspect ratio, they can modify the properties of the polymer matrix even at very low concentrations. However, mechanical properties of CNT nanocomposites, are still very far from the theoretical value of the filler particles. In order to transfer the CNT properties to the composite material one essential and difficult step involves the fine dispersion of the nanotubes within the polymer matrix. However, because of van der Waals attraction among nanotubes and their large surface areas, CNT tend to form agglomerates. CNT arrange themselves into bundles or ropes that may achieve the micron size. Without proper dispersion, the properties of the nanocomposite are more similar to those of a regular composite than to those of individual CNT.

Among the different methods of dispersion of CNT into a polymer matrix, one may distinguish two: mechanical methods and solution methods. Mechanical methods have been extensively employed (Linares et al. 2008; Jin and Park 2011). In order to destroy the bundles, very energetic mechanical methods should be employed. Among this mechanical methods are ball milling (Song et al. 2013), stirring and extrusion (Zhang et al. 2004) and ultrasonication (Bin et al. 2003). These high energy methods are very effective in dispersing nanotubes in liquid matrices, but it shows severe problems when trying to disperse CNTs in high viscosity matrices. This is one of the reasons why solution methods have been developed. CNT are dispersed into a polymer solution, and the mixture is allowed to coprecipitate or it is casted (Chae et al. 2005; Fornes et al. 2006). A variation of this method consists on in situ polymerization of the polymer in a solvent with CNT dispersed (Broza et al. 2005; Nogales et al. 2004a, b; Garcia-Gutiérrez et al. 2006; Hernández et al. 2006).

The mixing method has a strong impact on the properties of the nanocomposites. In general, low percolation thresholds for electrical conductivity are achieved in the case of solution mixed systems, than in direct melt mixing methods. As an example, in situ polymerized nanocomposites with Poly(butylene terephthalate) as the polymer matrix and single wall carbon nanotubes as filler has shown one of the lowest percolation threshold for electrical conductivity in thermoplastic polymer nanocomposites (Nogales et al. 2004a, b). A comparison can be made between the conductivity of the nanocomposite prepared by different methods (Fig. 3.17), and in general, it has been observed that solution mixing is a more efficient method for dispersion.

However, in some cases, the properties of the matrix polymer and those of the carbon nanotubes, results in synergetic effects in the form of new materials with desirable combined properties, like shape recovery materials (Cho et al. 2005) or conducting transparent polymer materials as shown in Fig. 3.18 (Hernández et al. 2009a, b).

3 Crystallization in Nanocomposites

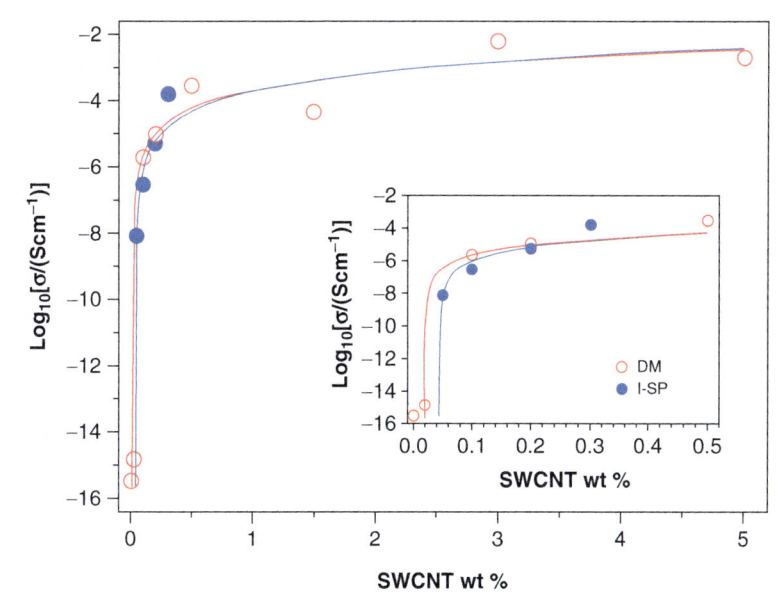

Fig. 3.17 Electrical conductivity versus SWCNT weight concentration for nanocomposites of Poly(Butylene Terephthalate) prepared by direct mixing (DM) and for nanocomposites prepared by in situ polymerization (I-SP). Reprinted with permission from Elsevier (Hernández et al. 2006)

Fig. 3.18 Transparency of conducting nanocomposites of similar thickness with 0.1 wt% of SWCNT prepared by direct mixing (*left side*) and by in situ polymerization (*right side*). Reprinted with permission from Elsevier (Hernández et al. 2006)

The final property of a CNT/polymer nanocomposite cannot be predicted simply by considering the addition of those of each of the components. The presence of CNT in a polymer melt has a strong impact on the subsequent morphology of the crystallized composite. In general, it is accepted that CNT may act as heterogeneous nucleating agents in the polymer matrix (Lim et al. 2015; Bhattacharyya et al. 2003). Polymer processing involves the application of combination of shear and elongational flows coupled with complex temperature profile. Due to the high aspect ratio of CNT, they tend to align with the flow (Garcia-Gutiérrez et al. 2006). In order to disentangle all these effects, emphasis has been focused on the physical aspects of these modifications. For that, scattering experiments to assess the structure have been performed, either in rheometer or in shear cell, so that thermal and flow effects can be separated (García-Gutiérrez et al. 2008; Hernández et al. 2009a, b, Chen et al 2011).

3.5.1 Nanocomposites with Halloysites

Ning (2007) have studied i-PP loaded with 1 and 10 % w/w of Halloysite nanotubes with diameters reported to be in the range of 50–300 nm and a length/diameter ratio of 3–10. As the non-isothermal crystallization curves (Fig. 3.19) show there is a clear but small nucleation effect from the halloysite nanotubes. Despite this the larger scale morphology was reported to remain the same for samples containing halloysite nanotubes as displayed for the i-PP alone.

Mitchell and Duraccio (2011) carried out similar studies using polyethylene oxide and 2 % of halloysite nanotubes using a small shear cell which facilitated the sample being sheared in the melt before cooling to crystallize. The shear cell was mounted to an SAXS beamline at the Elettra Sychrotron Facility and SAXS patterns were recorded throughout the heating-shearing-cooling cycle. They found that when shearing took place at lower temperatures, the system crystallized during shear with an isotropic distribution of lamellar crystals as revealed by the SAXS

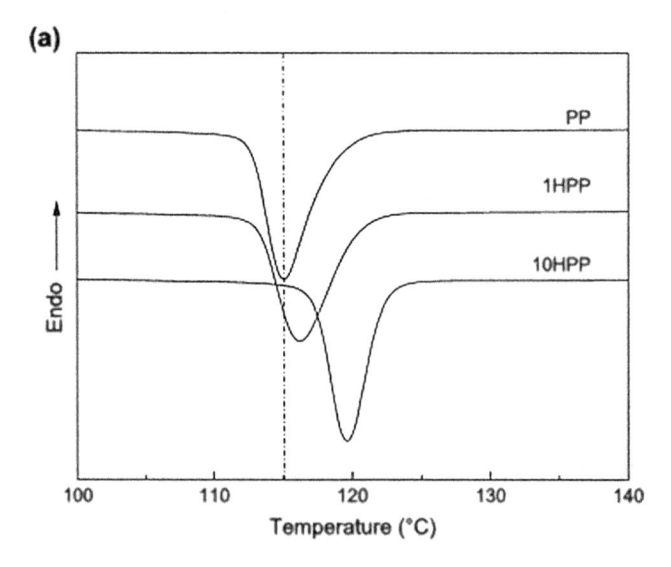

Fig. 3.19 Non-isothermal crystallization curves for i-PP and i-PP/HNT composites. Cooling rate is 10 C/min. Reproduced from Polymer 48 (2007) 7374–7384 with permission from Elsevier— Ning (2007)

data (Mitchell and Duraccio 2011). As the shear temperature increased, the orientation distribution of lamellar crystals became anisotropic presumably as a consequence of crystallization nucleated by the halloysite nanotubes became a more dominant feature of the crystallization. On shearing at even higher temperature, the distribution of crystals became isotropic as the memory of the anisotropy induced in the melt relaxed before the crystallization took place.

3.5.2 Nanocomposites with Clay Platelets

Processing operations such as injection and compression moulding, extrusion, fibre spinning, etc. are likely involved for the preparation and commercialization of polymer/clay nanocomposites. The flow applied during processing play a decisive role in crystallization kinetics and in the morphology obtained. For neat polymers, it has been demonstrated that the flow accelerates crystallization kinetics when compared with the quiescent melts (Keller and Kolnaar 1997; Kumaraswamy et al. 1999) and some crystallization mechanism has been proposed (Somani et al. 2000; Seki et al. 2002), whereas for polymer/clay nanocomposites, studies are still lacking and most of results have been reported for iPP and nylon 6/clay nanocomposites.

As in the static conditions, the incorporation of nanoscale-layered silicates impacts the crystallization under shear in terms of kinetics, structure and degree of crystallinity. In particular, two main effects have to be taken into account during the crystallization of a polymer: (a) the enhancement of nucleating speed and crystallization rate (Sun et al. 2009; Nowacki et al. 2004; Somwangthanaroj et al. 2003), being the clay is a heterogeneous nucleating agent; (b) the decrease of crystallinity and degree of perfection of the crystal (Wang et al. 2005), due to the physical hindrance of nanoclay layers to the motion of polymer chains.

Two other aspects should be considered during crystallization under shear: the clay nanolayer orientation with respect to the flow direction and polymer matrix orientation due to the influence of clay on the viscoelastic behaviour of the melt polymer. Both affect the final polymer/clay nanocomposite morphology and its macroscopic properties. The two-dimensional (2D) clay may align in three primary directions under shear, often referred to as a, b and c orientations (Butler 1999). In the perpendicular (a) orientation, the surface normal aligns parallel to the vorticity direction, and the particles lie in the flow-gradient plane; in the transverse (b) orientation, the surface normal aligns parallel to the flow direction, and the particles lie in the vorticity gradient plane; in the parallel (c) direction, the surface normal aligns parallel to the shear gradient direction, and the particles lie in the flow-vorticity plane. As reported in literature, the orientation of the clay platelets depend on many parameters such as their structure in the polymer matrix (Rozanski et al. 2009), their concentration (Sun et al. 2009) and shear rate (Lele et al. 2002). The general or intuitive response of the clay platelet orientation in a polymer matrix is in the c orientation. This has been described by studies such as for nylon–clay nanocomposites (Krishnamoorti and Vaia 2001; Krishnamoorti and Yurekli 2001; Medellin-Rodriguez et al. 2001; Medellin-Rodríguez et al. 2003). Anyway different orientations have also been found (Medellin-Rodríguez et al. 2003; Okamoto et al. 2001) as for example particular superstructure called "house of cards structure" reported by Okamoto et al. in polypropylene/clay nanocomposite melt under elongational flow (Okamoto et al. 2001).

In a recent work (Dykes et al. 2012), there has been found an unexpected resistance of nanoclay to shear flow-induced orientation in PS/clay nanocomposites. Despite the small magnitude of the estimated rotational diffusivity of a single clay platelet, significant orientation in steady shear flow only develops at shear rates greater than 1 s^{-1}. Similarly, flow-induced orientation in large-amplitude oscillatory shear only develops at frequencies greater than 0.1 Hz. The inability of shear at low rates and frequencies to produce significant alignment suggests the existence of a structural relaxation process that is much faster than that of rotational Brownian motion of the clay particles. Direct evidence of such a fast process is found in measurements of orientation relaxation following cessation of shear flow. In the article, it is hypothesized that the fast nanoclay relaxation arises from either relaxation of shear-induced distortion of partially flexible exfoliated clay sheets in a highly entangled nanoparticle network, or a direct coupling of particle and polymer dynamics through the chemical tethering of particle chains that occurs during the in situ polymerization synthesis of these materials. Analogously (Rozanski et al. 2009), it has been found in iPP/clay exfoliated nanocomposites that the shear-induced orientation of the clay platelets is very weak in spite of the shear rate of 2 s^{-1}.

The effect of shear on the orientation of polymer–clay nanocomposites has been examined by a number of groups (Somani et al. 2000, 2001; Fong et al. 2002; Jimenez et al. 1997), and in some cases, the three-dimensional (3D) orientation of the polymer and clay has been determined (Kojima et al. 1994; Bafna et al. 2003; Varlot et al. 2001).

Different authors (Nowacki et al. 2004; Somwangthanaroj et al. 2003; Wang et al. 2005; Yalcin et al. 2003; Koo et al. 2005), studying early stage of shear-induced crystallization, reported stronger orientation of polymer crystals in nanocomposites with organo-modified MMT (o-MMT) processed by injection moulding and extrusion than that in neat polymers with the same thermo-mechanical history.

A typical "shish-kebab" crystalline structure has been found by Maiti and Okamoto (2003) and Kim et al. (2001) in polyamide/organoclay nanocomposite and by Choi and Kim (2004) in PP/EPR/talc nanocomposite where a preferential orientation of polymer lamellae perpendicular to the surface of organoclay layers was inspected by TEM measurements. The unique observation of lamellar orientation on the clay layers was ascribed to nucleation and epitaxial crystallization at the interface between layered silicate and polymer matrix especially the surfaces of clay platelets acted as heterogeneous nucleation sites. Orientation of iPP crystals was also enhanced in iPP/PP-MA/o-MMT injection-moulded parts, especially manufactured by dynamic packing injection moulding (Wang et al. 2005). MMT platelets aligned along the flow direction contributed to oriented crystallization by constraining the growth of iPP crystals. Koo et al. (2005) found that the clay enhanced iPP orientation in extruded iPP/PP-MA/o-MMT nanocomposites resulting in two iPP crystal fractions with either c or a* axis oriented along the flow direction and with b axis oriented in the film thickness direction. Analogously, the amount of crystal orientation in injection-moulded nylon6/clay nanocomposites were found higher than that of pure nylon (Yalcin et al. 2003), and this was attributed to the so-called shear amplification mechanism that a great enhancement of local stress occurred in the small interparticles region of two adjacent layered tactoids with different velocities. In another work (Medellín-Rodríguez et al. 2003), the complete spherulitic formation was suppressed even in the core region of specimen at high mould temperature due to the nucleation density that was significantly increased through orientation of macromolecules and the foreign particle effect of the nanoclay.

In contrast with the above results, Zhang et al. (2003) reported a decreased orientation of iPP in iPP/PP-MA/o-MMT nanocomposites during dynamic packing injection moulding as compared to that of neat iPP solidified under the same conditions. Similarly, Medellin-Rodriguez et al. (2001) found that the overall molecular orientation of the crystals was found to decrease with the clay content in Nylon 6/clay nanocomposites subject to uniaxial deformation, with some clay platelets oriented perpendicular to the film surface and the molecular axis of the Nylon 6 crystals parallel to the stretching direction. This was explained by the presence of the clay platelets as well as the rotation of the clay platelets during deformation, which hindered the orientation of the nylon crystals.

Sun et al. (2009) have proposed a model for iPP/organomodified clay composites when the content of clay is higher than the percolation threshold (Fig. 3.20).

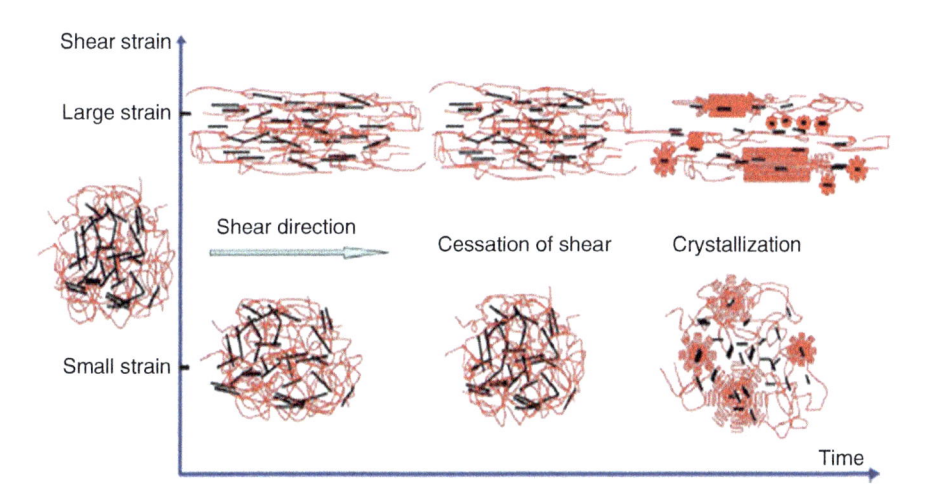

Fig. 3.20 The proposed model for the shear-induced crystallization of iPP/OMMT composite with OMMT content higher than the percolation threshold. Reproduced with permission from Sun et al. (2009)

In particular, they found that when the shear strain was not big enough to destroy the fillers network in the matrix, the crystallization of iPP was similar with that of the unsheared sample. When shear strain was large enough, the fillers network was destroyed and clay layers were aligned along the flow direction. There formed oriented crystals including cylindrites and strings of spherulites, which were small in size because the aligned clay layers acted as heterogeneous nucleation agents to promote crystallization of iPP.

Rozanski et al. (2009) studied iPP/PPMA/OMMT nanocomposites in order to clarify the augmented nucleation. They found that the level of clay dispersion was more important than molecular weight of iPP in the nucleation effect. Better exfoliation and dispersion of the clay in nanocomposites with the same o-MMT content denotes a decreased amount of iPP chains intercalated in o-MMT galleries and a larger interphase area between o-MMT platelets exposed and a surrounding polymer. Therefore, the orientation-induced nucleation is enhanced by the clay dispersion as also observed by Wang et al. (2005). The morphology shown by crystallized iPP/PPMA/OMMT nanocomposites after shearing at $2\ s^{-1}$ is reported in Fig. 3.21a and compared with that obtained in static conditions for the same nanocomposites (Fig. 3.21b). In Fig. 3.21a spherulites are clearly visible, whereas in Fig. 3.22b different polycrystalline aggregates composed by elongated sectors radiating from the same centres are evident. The authors stated that such structure might originate from instantaneous nucleation on neighbouring site or because the secondary growth did not fill sufficiently the space between earlier crystallized lamellae.

3.5.3 Nanocomposites with Graphene

As we have seen some nanoparticles are able to direct the crystallization of the matrix material. Mitchell et al. (2015) have used time-resolving x-ray scattering techniques at a synchrotron beam line to follow the structural evolution in a

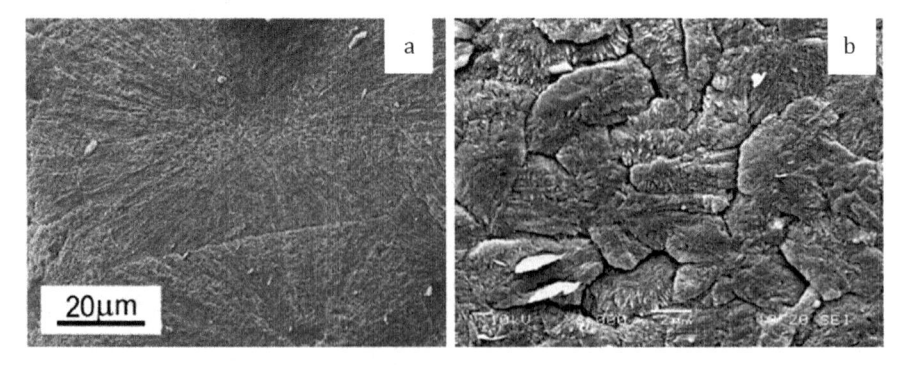

Fig. 3.21 SEM micrographs of etched iPP/clay nanocomposites crystallized (**a**) in static conditions (**b**) crystallized after shearing at $2\ s^{-1}$ in a plane parallel to the shear direction, shear direction—horizontal. Reproduced with permission from Rozanski et al. (2009)

3 Crystallization in Nanocomposites

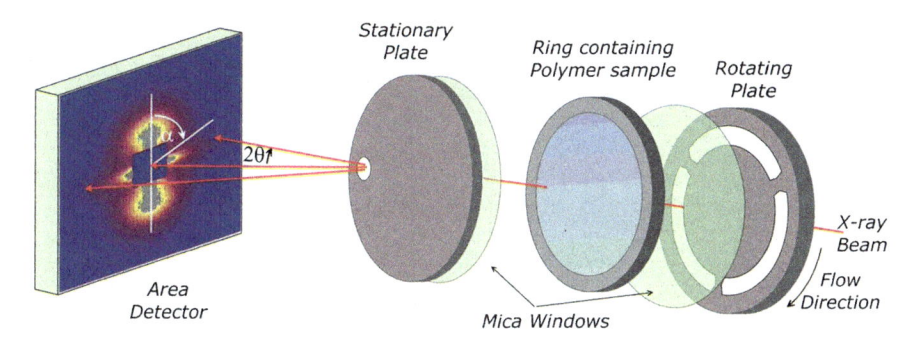

Fig. 3.22 Schematic of the shear cell used with the scattering vector $\underline{Q}$ lies in the plane defined by the flow direction and the vorticity vector (Nogales et al 2004a)

Fig. 3.23 Schematic of the shear cell used in which the scattering vector $\underline{Q}$ lies in the plane defined by the flow direction and the velocity gradient

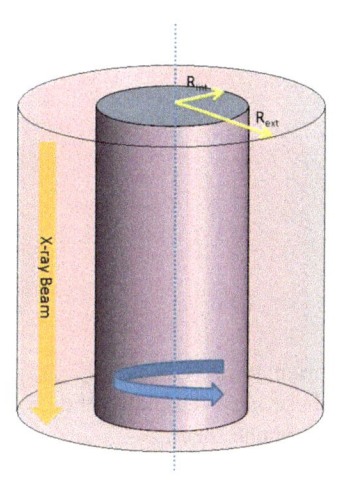

graphene/PCL composite. The composites were prepared by melt mixing using PCL ($M = 65,000$ Da) and 14 nm graphene nanoplatelets. The compounded mixture was formed in discs or cylinders to facilitate mounting in the x-ray shear cells shown in Figs. 3.22 and 3.23.

The Couette arrangement shown in Fig. 3.22 is to allow the three-dimensional nature of the shear flow to be explored. Details of the design are described elsewhere (Mitchell et al. 2015). As we can see in Fig. 3.24, the graphene nanoflakes have a lateral extent of more than 1 µm. Thus although we might anticipate the plane of the flake to lie parallel to the flow direction, it must lie normal to the velocity gradient, otherwise it will rotate and probably breakup.

The SAXS pattern (Fig. 3.25) shows the scattering typical of lamellar crystals which preferentially lie normal to the flow direction. The central streak we attribute to the graphene flakes which lie preferentially parallel to the flow direction. We know from previous work (See Chap. 5), that without the addition of the graphene nanoflakes, the sample would exhibit an isotropic distribution of lamellar crystal

Fig. 3.24 TEM micrographs of the graphene nanoflakes

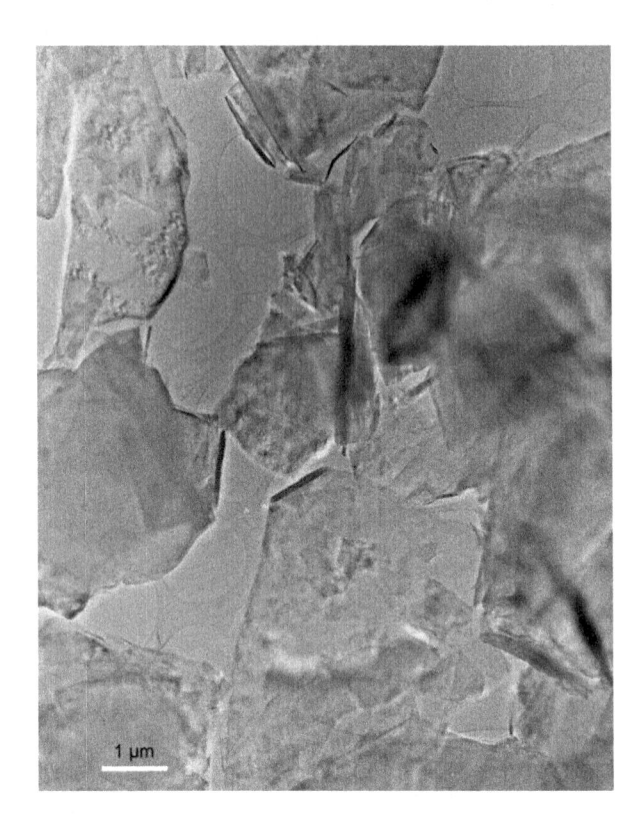

orientations. Clearly, the graphene nanoflakes have aligned in the shear flow, and these have templated the direction of the lamellar crystal growth, normal to the graphene sheet.

Lee et al. (2003) have studied the effect of graphene on the crystallization of poly(vinyl alcohol), where they found that samples containing 0.5 % wt graphene exhibited a higher level of crystallinity (up to 19 %).

3.5.4 Summary

This chapter has shown that the addition of nanoparticles to polymer matrices or the formation of nanoparticles within a polymer matrix, as in the case of high molecular weight chains can have a dramatic effect on the morphology of the polymer matrix. In some cases, the nanoparticles act as an efficient nucleating agent and as such can direct the crystallization of the polymer in the case of a crystallizable polymer. The impact of the nanoparticles, clearly depends on the nature of the nanoparticles in terms of shape and chemical composition, the particular polymer matrix and on the nature of the dispersion of the particles within the polymer. What is clear is that the ability to control the morphology of a polymer using the behaviour of the

3 Crystallization in Nanocomposites

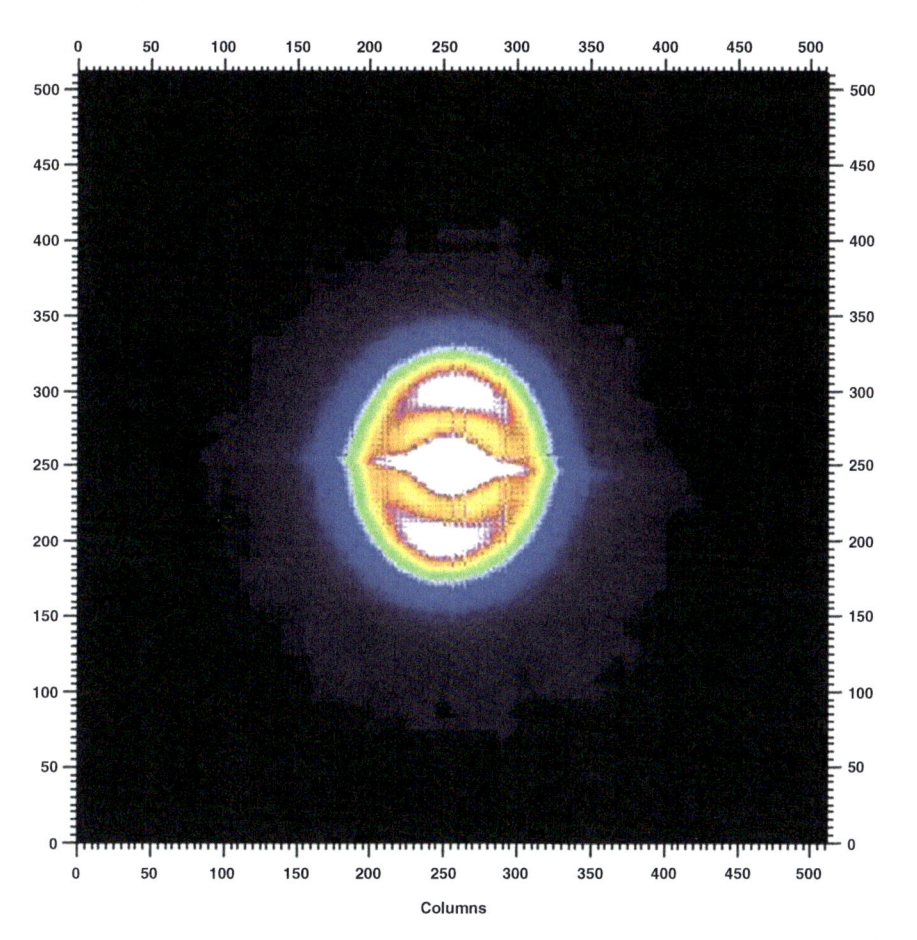

Fig. 3.25 SAXS pattern for a sample of PCL/graphene sheared for 1000 shear units at 1 s^{-1} using the arrangement shown in Fig. 3.22 at 100°C and cooled to room temperature at 10 C/min. The flow direction is vertical and the velocity gradient horizontal

nanoparticles provides a powerful approach to control the scales of structure in these novel polymer composites, and this is explored in the remaining chapters of this book.

References

Abo el Maaty MI, Bassett DC (2005) Evidence for isothermal lamellar thickening at and behind the growth front as polyethylene crystallizes from the melt. Polymer 46:8682–8688

Abo el Maaty MI, Bassett DC (2006) On the time for fold surfaces to order during the crystallization of polyethylene from the melt and its dependence on molecular parameters. Polymer 47:7469–7476

Amornsakchai T, Bassett DC, Olley RH, Hine PJ, Ward IM (2000) On morphologies developed during two-dimensional compaction of woven polypropylene tapes. J Appl Polym Sci 78:787–793

An Y, Holt JJ, Mitchell GR, Vaughan AS (2006) Influence of molecular composition on the development of microstructure from sheared polyethylene melts: molecular and lamellar templating. Polymer 47:5643–5656

Bafna A, Beaucage G, Mirabella F, Mehta S (2003) 3D hierarchical orientation in polymer–clay nanocomposite films. Polymer 44:1103–1115

Bhattacharyya AR, Sreekumar TV, Liu T, Kumar S, Ericson LM, Hauge RH, Smalley RE (2003) Crystallization and orientation studies in polypropylene/single wall carbon nanotube composite. Polymer 44:2373–2377

Bianco A, Cheng HM, Enoki T, Gogotsi Y, Hurt RH, Koratkar K, Kyotanic T, Monthioux M, Park CR, Tacson JMD, Zhang J (2013) All in the graphene family—a recommended nomenclature for two-dimensional carbon materials. Carbon 65:1–6

Bin Y, Kitanaka M, Zhu D, Matsuo M (2003) Development of highly oriented polyethylene filled with aligned carbon nanotubes by gelation/crystallization from solutions. Macromolecules 36:6213–6219

Broza G, Kwiatkowska M, Roslaniec Z, Schulte K (2005) Processing and assessment of poly (butylene terephthalate) nanocomposites reinforced with oxidized single wall carbon nanotubes. Polymer 46:5860–5867

Butler P (1999) Shear induced structures and transformations in complex fluids. Curr Opin Colloid Interface Sci 4:214–221

Chae HG, Sreekumar TV, Uchida T, Kumar S (2005) A comparison of reinforcement efficiency of various types of carbon nanotubes in polyacrylonitrile fiber. Polymer 46:10925–10935

Chen Y-H, Zhong G-J, Lei J, Li Z-M, Hsiao BS (2011) In situ synchrotron X-ray scattering study on isotactic polypropylene crystallization under the coexistence of shear flow and carbon nanotubes. Macromolecules 44:8080–8092

Cho JW, Kim JW, Jung YC, Goo NS (2005) Electroactive shape-memory polyurethane composites incorporating carbon nanotubes. Macromol Rapid Commun 26:412–416

Choi WJ, Kim SC (2004) Effects of talc orientation and non-isothermal crystallization rate on crystal orientation of polypropylene in injection-molded polypropylene/ethylene-propylene rubber/talc blends. Polymer 45:2393–2401

Du M, Guo B, Jia D (2010) Newly emerging applications of halloysite nanotubes: a review. Polym Int 59:574–582

Dykes LMC, Torkelson JM, Burghardt WR (2012) Shear-induced orientation in well-exfoliated polystyrene/clay nanocomposites. Macromolecules 45:1622–1630

Fong H, Liu WD, Wang C-S, Vaia RA (2002) Generation of electrospun fibers of nylon 6 and nylon 6-montmorillonite nanocomposite. Polymer 43:775–780

Fornes TD, Baur JW, Sabba Y, Thomas EL (2006) Morphology and properties of melt-spun polycarbonate fibers containing single-and multi-wall carbon nanotubes. Polymer 47:1704–1714

Garcia-Gutiérrez MC, Nogales A, Rueda DR, Domingo C, García Ramos JV, Broza G, Roslaniec Z, Schulte K, Davies RJ, Ezquerra TA (2006) Templating of crystallization and shear-induced self-assembly of single-wall carbon nanotubes in a polymer-nanocomposite. Polymer 47:341–345

García-Gutiérrez MC, Hernández JJ, Nogales A, Panine P, Rueda DR, Ezquerra TA (2008) Influence of shear on the templated crystallization of poly(butylene terephthalate)/single wall carbon nanotube nanocomposites. Macromolecules 41:844–851

Hernández JJ, García-Gutiérrez MC, Nogales A, Rueda DR, Ezquerra TA (2006) Small-angle X-ray scattering of single-wall carbon nanotubes dispersed in molten poly(ethylene terephthalate). Compos Sci Technol 66:2629–2632

Hernández JJ, García-Gutiérrez MC, Nogales A, Rueda DR, Kwiatkowska M, Szymczyk A, Roslaniec Z, Concheso A, Guinea I, Ezquerra TA (2009a) Influence of preparation procedure on the conductivity and transparency of SWCNT-polymer nanocomposites. Compos Sci Technol 69:1867–1872

Hernández JJ, García-Gutiérrez MC, Nogales A, Rueda DR, Ezquerra TA (2009b) Shear effect on crystallizing single wall carbon nanotube/poly(butylene terephthalate) nanocomposites. Macromolecules 42:4374–4376

Hine PJ, Ward IM, Olley RH, Bassett DC (1993) The hot compaction of high modulus melt-spun polyethylene fibres. J Mater Sci 28:316–324

Hine PJ, Ward IM, El Maaty MIA, Olley RH, Bassett DC (2000) The hot compaction of 2-dimensional woven melt spun high modulus polyethylene fibres. J Mater Sci 35:5091–5099

Hobbie EK, Wang H, Kim H, Lin-Gibson S, Grulke EA (2003) Orientation of carbon nanotubes in a sheared polymer melt. Phys Fluids 15:1196–1202

Hristozov D, Malsch I (2009) Hazards and risks of engineered nanoparticles for the environment and human health. Sustainability 1:1161–1194

Jeffery GB (1922) The motion of ellipsoidal particles immersed in a viscous fluid. Proc R Soc Lond Ser A 102:161

Jie L, Yujun H, Yimo H, Kai L, Jiaping W, Qunqing L, Shoushan F, Kaili J (2012) Direct identification of metallic and semiconducting single-walled carbon nanotubes in scanning electron microscopy. Nano Lett 12(8):4095–4101

Jimenez G, Ogata N, Kawai H, Ogihara T (1997) Structure and thermal/mechanical properties of poly(ε-caprolactone)-clay blend. J Appl Polym Sci 64:2211–2220

Jin F-L, Park S-J (2011) A review of the preparation and properties of carbon nanotubes-reinforced polymer composites. Carbon Lett 12:57–69

Jordan ND, Bassett DC, Olley RH, Hine PJ, Ward IM (2003) The hot compaction behaviour of woven oriented polypropylene fibres and tapes. II. Morphology of cloths before and after compaction. Polymer 44:1133–1143

Joussein E, Petit J, Churchman J, Theng B, Righi D, Delvaux B (2005) Halloysite clay minerals—a review. Clay Miner 40:383–426

Kalay G, Bevis MJ (1997) Processing and physical property relationships in injection-molded isotactic polypropylene. 2. Morphology and crystallinity. J Polym Sci B Polym Phys 35:265–291

Keller A, Kolnaar HWH (1997) Flow-induced orientation and structure formation. In: Meijer H (ed) Processing of polymers, vol 18. Wiley, Weinheim, pp 189–268

Kim G-M, Lee D-H, Hoffmann B, Kressler J, Stöppelmann G (2001) Influence of nanofillers on the deformation process in layered silicate/polyamide-12 nanocomposites. Polymer 42:1095–1100

Kojima Y, Usuki A, Kawasumi M, Okada A, Kurauchi T, Kamigaito O, Kaji K (1994) Fine structure of nylon–6–clay hybrid. J Polym Sci B Polym Phys 32:625–630

Koo CM, Kim JH, Wang KH, Chung IJ (2005) Melt-extensional properties and orientation behaviors of polypropylene-layered silicate nanocomposites. J Polym Sci B Polym Phys 43:158–167

Krishnamoorti R, Vaia RA (2001) Polymer nanocomposites: introduction. In: Vaia RA, Krishnamoorti R (eds) Polymer nanocomposites. ACS symposium series, vol 804, pp 1–5

Krishnamoorti R, Yurekli K (2001) Rheology of polymer layered silicate nanocomposites. Curr Opin Colloid Interface Sci 6:464–470

Kumaraswamy G, Issaian AM, Kornfield JA (1999) Shear-enhanced crystallization in isotactic polypropylene. 1. Correspondence between in situ rheo-optics and ex situ structure determination. Macromolecules 32:7537–7547

Lee S, Hong J-Y, Jang J (2003) The effect of graphene nanofiller on the crystallization behavior and mechanical properties of poly(vinyl alcohol). Polym Int 62:901–908

Lele A, Mackley M, Galgali G, Ramesh C (2002) In situ rheo-x-ray investigation of flow-induced orientation in layered silicate–syndiotactic polypropylene nanocomposite melt. J Rheol 46:1091–1110

Li J, He Y, Han Y, Liu K, Wang J, Li Q, Fan S, Jiang K (2012) Direct identification of metallic and semiconducting single-walled carbon nanotubes in scanning electron microscopy. Nano Lett 12:4095–4101

Li T, Xiang S, Ma P, Bai H, Dong W, Chen M (2015) Nanocomposite hydrogel consisting of Na-montmorillonite with enhanced mechanical properties. J Polym Sci B Polym Phys 53:1020–1026

Lim JY, Kim J, Kim S, Kwak S, Lee Y, Seo Y (2015) Nonisothermal crystallization behaviors of nanocomposites of poly(vinylidene fluoride) and multiwalled carbon nanotubes. Polymer 62:11–18

Linares A, Canalda JC, Cagiao ME, García-Gutiérrez MC, Nogales A, Martín-Gullón I, Vera J, Ezquerra TA (2008) Broad-band electrical conductivity of high density polyethylene nanocomposites with carbon nanoadditives: multiwall carbon nanotubes and carbon nanofibers. Macromolecules 41:7090–7097

Maiti P, Okamoto M (2003) Crystallization controlled by silicate surfaces in nylon 6-clay nanocomposites. Macromol Mater Eng 288:440–445

Medellin-Rodriguez FJ, Burger C, Hsiao BS, Chu B, Vaia R, Phillips S (2001) Time-resolved shear behavior of end-tethered nylon 6/clay nanocomposites followed by non-isothermal crystallization. Polymer 42:9015–9023

Medellin-Rodríguez FJ, Hsiao BS, Chu B, Fu BX (2003) Uniaxial deformation of nylon 6-clay nanocomposites by in-situ synchrotron x-ray measurements. J Macromol Sci Phys B42:201–214

Mingliang D, Baochun G, Demin J (2010) Newly emerging applications of halloysite nanotubes: a review. Polymer Int 59:574–582

Mitchell GR, Duraccio D (2011) Unpublished data

Mitchell GR, Davis FJ, Pezutto M (2015) Eur Polym J (in press)

Ning N-Y (2007) Crystallization behavior and mechanical properties of polypropylene/halloysite composites. Polymer 48(25):7374–7384

Nogales A, Thornley SA, Mitchell GR (2004a) Shear cell for in-situ WAXS, SAXS and SANS experiments on polymer melts under flow fields. J Macromol Sci Phys B43:1161–1170

Nogales A, Broza G, Roslaniec Z, Schulte K, Šics I, Hsiao BS, Sanz A, García-Gutiérrez MC, Rueda DR, Domingo C, Ezquerra TA (2004b) Low percolation threshold in nanocomposites based on oxidized single wall carbon nanotubes and poly(butylene terephthalate). Macromolecules 37:7669–7672

Nowacki R, Monasse B, Piorkowska E, Galeski A, Haudin JM (2004) Spherulite nucleation in isotactic polypropylene based nanocomposites with montmorillonite under shear. Polymer 45:4877–4892

Ogata N, Jimenez G, Kawai H, Ogihara T (1997) Structure and thermal/mechanical properties of poly(l-lactide)-clay blend. J Polym Sci B Polym Phys 35:389–396

Okamoto M, Nam PH, Maiti P, Kotaka T, Hasegawa N, Usuki A (2001) A house of cards structure in polypropylene/clay nanocomposites under elongational flow. Nano Lett 6:295–298

Olley RH, Bassett DC, Hine PJ, Ward IM (1993) Morphology of compacted polyethylene fibres. J Mater Sci 28:1107–1112

Paul DR, Robeson LM (2008) Polymer nanotechnology: nanocomposites. Polymer 49:3187–3204

Potts JR, Dreyer DR, Bielawski CW, Ruoff RS (2011) Graphene-based polymer nanocomposites. Polymer 52:5–25

Rashmi, Renukappa NM, Ranganathaiah C, Shivakumar KN (2011) Montmorillonite nanoclay filler effects on electrical conductivity, thermal and mechanical properties of epoxy-based nanocomposites. Polym Eng Sci 51: 1827–1836

Rasmussen AJ, Ebbesen M (2014) Characteristics, properties and ethical issues of carbon nanotubes in biomedical applications. Nanoethics 8:29–48

Ravindra K, Manasi G, Sheetal G, Bijoy KP (2012) Halloysite nanotubes and applications: a review. J Adv Sci Res 3:25–29

Razado-Colambo R, Avila J, Chen C, Nys JP, Wallart X, Asensio MC, Vignaud D (2015) Probing the electronic properties of graphene on C-face SiC down to single domains by nanoresolved photoelectron spectroscopies. Phys Rev B 92:035105

Rozanski A, Monasse B, Szkudlarek E, Pawlak A, Piorkowska E, Gałeski A, Haudin JM (2009) Shear-induced crystallization of isotactic polypropylene based nanocomposites with montmorillonite. Eur Polym J 45:88–101

Seki M, Thurman DW, Oberhauser JP, Kornfield JA (2002) Shear-induced orientation in the crystallization of an isotactic polypropylene nanocomposite. Macromolecules 35:2583

Siddiqui MA, Ahmed Z (2005) Mineralogy of the Swat kaolin deposits, Pakistan. Arab J Sci Eng 30:195–218

Somani RH, Hsiao BS, Nogales A, Srinivas S, Tsou AH, Šics I, Baltá-Calleja FJ, Ezquerra TA (2000) Structure development during shear flow-induced crystallization of i-PP: in-situ small-angle X-ray scattering study. Macromolecules 33:9385–9394

Somani RH, Hsiao BS, Nogales A, Fruitwala H, Srinivas S, Tsou AH (2001) Structure development during shear flow induced crystallization of i-PP: in situ wide-angle X-ray diffraction study. Macromolecules 34:5902–5909

Somwangthanaroj A, Lee EC, Solomon MJ (2003) Early stage quiescent and flow-induced crystallization of intercalated polypropylene nanocomposites by time-resolved light scattering. Macromolecules 36:2333–2342

Song BJ, Ahn JW, Cho KK, Roh JS, Lee DY, Yang YS, Lee JB, Hwang DY, Kim HS (2013) Electrical and mechanical properties as a processing condition in polyvinylchloride multi walled carbon nanotube composites. J Nanosci Nanotechnol 13:7723–7727

Sun T, Chen F, Dong C-X, Zhou Y, Wang D, Han CC (2009) Shear-induced orientation in the crystallization of an isotactic polypropylene nanocomposite. Polymer 50:2465–2471

Ting L, Shuangfei X, Piming M, Huiyu B, Weifu D, Mingqing C (2015) Nanocomposite hydrogel consisting of Na-montmorillonite with enhanced mechanical properties. J Polymer Sci B Polymer Phys 53(14):1020–1026

Uddin F (2008) Clays, nanoclays, and montmorillonite minerals. Metallur Mater Trans A 39A:2804–2814

Varlot K, Reynaud E, Kloppfer MH, Vigier G, Varlet J (2001) Clay-reinforced polyamide: preferential orientation of the montmorillonite sheets and the polyamide crystalline lamellae. J Polym Sci B Polym Phys 39:1360–1370

Wallace T, Center for Nanotechnology in Society, Arizona State University, Montmorillonite (MMT) Nanoclays in Ashpalt, [Online] Available from: https://nice.asu.edu/nano/montmorilonite-mmt-nanoclays-ashpalt [Accessed: 9th Sep 2015]

Wang K, Xiao Y, Na B, Tan H, Zhang Q, Fu Q (2005) Shear amplification and re-crystallization of isotactic polypropylene from an oriented melt in presence of oriented clay platelets. Polymer 46:9022–9032

Ward IM, Hine PJ (2004) The science & technology of hot compaction. Polymer 45:1413–1427

Wei W, Minulina R, Abdullayev E, Fakhrullin R, Mills D, Lvov Y (2014) Enhanced efficiency of antiseptics with sustained release from clay nanotubes. RSC Adv 4:488–494

White HM, Bassett DC (1998) On row structures, secondary nucleation and continuity in a-polypropylene. Polymer 39:3211–3218

Wick P, Louw-Gaume AE, Kucki M, Krug HF, Kostarelos K, Fadeel B, Dawson KA, Salvati A, Vázquez E, Ballerini L, Tretiach M, Benfenati F, Flahaut E, Gauthier L, Prato M, Bianco A (2014) Classification framework for graphene-based materials. Angew Chem Int Ed 53:2–7

Wilder JWG, Venema LC, Rinzler AG, Smalley RE, Dekker C (1998) Electronic structure of atomically resolved carbon nanotubes. Nature 391:59–62

Yalcin B, Valladares D, Cakmak M (2003) Amplification effect of platelet type nanoparticles on the orientation behavior of injection molded nylon 6 composites. Polymer 44:6913–6925

You Z, Mills-Beale J, Foley JM, Roy S, Odegard GM, Dai Q, Goh SW (2011) Nanoclay-modified asphalt materials: preparation and characterization. Construct Build Mater 25:1072–1078

Zhang Q, Wang Y, Fu Q (2003) Shear-induced change of exfoliation and orientation in polypropylene/montmorillonite nanocomposites. J Polym Sci B Polym Phys 41:1–10

Zhang WD, Shen L, Phang IY, Liu T (2004) Carbon nanotubes reinforced nylon-6 composite prepared by simple melt-compounding. Macromolecules 37:256–259

Zhao Y, Abdullayev E, Vasiliev A, Lvov YM (2013) Halloysite nanotubule clay for efficient water purification. J Coll Interface Sci 406:121–129

Chapter 4
Theoretical Aspects of Polymer Crystallization

Wenbing Hu and Liyun Zha

4.1 Introduction

A vast amount of natural and synthetic polymers are semicrystalline. From this standpoint alone, it is necessary to understand the process of polymer crystallization, which dominates the structure formation of semicrystalline polymers and thus allows us to control their morphologies and properties. Polymer crystallization is influenced by many factors, such as chemical structures, compositions, temperatures, and thermal history. The theoretical models serve as a powerful tool for us to comprehend these factors in polymer crystallization.

Thermodynamics and kinetics are the two most fundamental theoretical aspects of polymer crystallization. Thermodynamics addresses why, or under which circumstances, polymer crystallization will begin, or in the opposite direction polymer crystals will start to melt. Kinetics addresses how fast polymer crystallization will be initiated, be developed, and be further improved. Both aspects decide crystal morphologies that eventually influence the properties of polymer materials.

This chapter intends to make a brief survey on our current theoretical understanding about the basic thermodynamics and kinetics of polymer crystallization. On thermodynamics, we will introduce the melting point, metastable states, phase diagrams, mesophase formation, as well as those factors governing melting points. On this aspect, the mean-field lattice theory on the statistical thermodynamics of crystallizable polymer solutions will be focused. On kinetics, we will introduce the basic knowledge about crystal nucleation, crystal growth, and crystal annealing. On this aspect, the classical nucleation theory and the related kinetic equations will be focused.

W. Hu (✉) • L. Zha
School of Chemistry and Chemical Engineering, Nanjing University, Nanjing 210023, China
e-mail: wbhu@nju.edu.cn

Due to length limitation of this chapter, many theoretical approaches on the phenomenological aspects of polymer crystallization have to be skipped. The isothermal and non-isothermal kinetic analysis of overall crystallization appears as technically important in the data treatment of DSC measurements. Some theoretical considerations on the metastable aspects of crystal morphologies and their evolution under various circumstances appear as practically important and case sensitive (see Chap. 1). In this sense, a combination of this chapter with other contributions of this book will provide reader a broad cutting-edge knowledge about our basic understanding of polymer crystallization.

4.2 Thermodynamics of Polymer Crystallization

4.2.1 Basic Concepts

Polymers are either in the fully amorphous state or in the fully crystalline state, or in the states partial between them. When they are in the amorphous state of homogeneous solutions or melt, polymer chains are fully disordered, as described by a random-coil model. The random-coil model was first proposed by Kuhn (1934) as well as by Guth and Mark (1934) to predict the entropic elasticity of polymer chains, and then was used to describe the amorphous state of polymers by Flory (1953) and others.

Polymer crystallization and melting are typically first-order phase transitions between the amorphous phase and the crystalline phase. When these two phases are in thermodynamic equilibrium, two phase transitions are thermodynamically reversible under a certain temperature. This temperature is referred to the equilibrium melting point of polymer crystallization. The free energy changes of amorphous phase and crystalline phase under various temperatures are depicted in Fig. 4.1, illustrating the definition of the equilibrium melting point T_m^0.

In the bulk system of pure polymers, the free energy change of melting becomes zero when the system is under its equilibrium melting point, as given by

$$\Delta F_m = \Delta Q_m - T_m^0 \Delta S_m = 0 \tag{4.1}$$

The equilibrium melting point can accordingly be calculated by

$$T_m^0 = \Delta Q_m / \Delta S_m \tag{4.2}$$

In practice, to get over the nucleation barrier via thermal fluctuations, the initiation of primary crystal nucleation requires a crystallization temperature T_c something lower than T_m^0, and thus a supercooling for crystallization is defined as

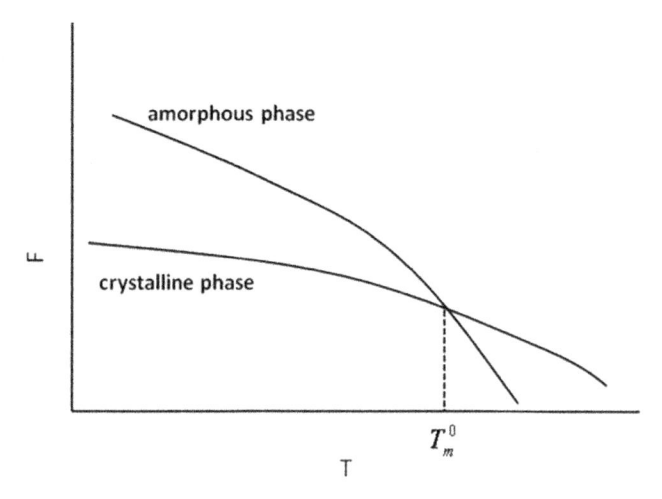

Fig. 4.1 Schematic free energy curves of amorphous phase and crystalline phase versus temperature. The temperature at which two curves intersect with each other is defined as the equilibrium melting point of polymers

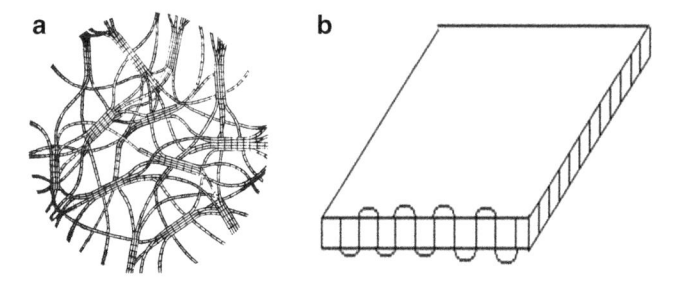

Fig. 4.2 Illustration pictures on the models of (**a**) fringed-micelle (Hermann et al. 1930) and (**b**) adjacent chain-folding (Keller 1957)

$$\Delta T = T_m^0 - T_c \tag{4.3}$$

In order to overcome the kinetic barrier under a certain supercooling, polymer crystallization commonly chooses a pathway favoring its kinetics, which will result in metastable semicrystalline states (Cheng 2008). It took a long time for people to figure out the structure features of the metastable semicrystalline polymers, with two dominant models, i.e., fringed-micelle and chain-folding models, as illustrated in Fig. 4.2. Here, the metastable state holds either a thermodynamic meaning as its local minimum in the free energy landscape, or a dynamic meaning with its negligible changes over the time window of our observations upon a very slow evolution towards more stable states.

In 1930, Hermann et al. set up the fringed-micelle model (Hermann et al. 1930; Hermann and Gerngross 1932) to describe the high elasticity of low-density

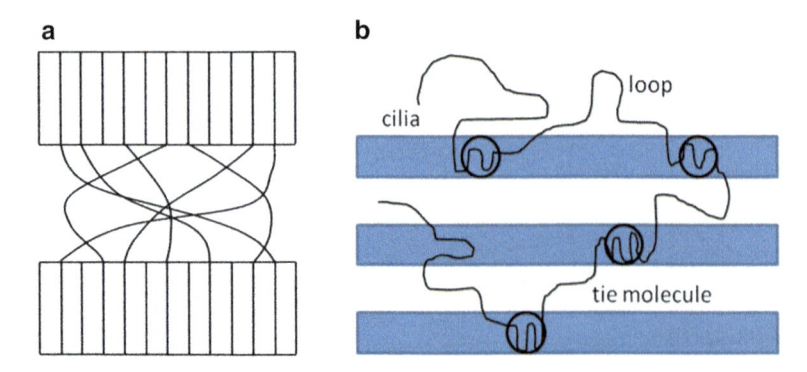

Fig. 4.3 Illustration pictures on the models of (**a**) Switchboard (Flory 1962) and (**b**) variable clusters (Hoffman 1983)

polyethylene products. But later on, this model could not explain the spherulite morphologies of polymer crystals often observed under optical microscopy. In 1957, Keller set up the adjacent chain-folding model (Keller 1957) on the basis of the facts that polymer stems in single lamellar crystal grown from dilute solutions were perpendicular to the lamellar surface, and the lamellar thickness was only in the scale of one tenth of chain lengths. This model was then confirmed by the small-angle neutron scattering (SANS) experiments of single crystal grown in dilute solutions (Spells et al. 1980) as well as by the integer folding of short chains in lamellar crystals grown in the melt (Arlie et al. 1966, 1967; Ungar et al. 1985; Organ and Keller 1987). For lamellar crystals grown in the long-chain melt, Flory proposed the "Switchboard" model (Flory 1962), which was then developed into the interzonal Switchboard model as discussed by Mandelkern (1964). Since both adjacent chain-folding and Switchboard models have their own interpretations on the same SANS experiment data, there was a hot debate in 68th Faraday Discussion Meeting. It is now well accepted that either model describes a certain aspect of structural features on lamellar crystals grown in the melt. The variable cluster model synthesizing both features of local chain-folding and global random-coil was discussed by Hoffman (1983). The loops, cilia, and tie molecules (Fig. 4.3) are restricted at the fold-end surfaces of lamellar crystals, which constitute the rigid amorphous phase near the crystalline region. The latter pretends to be the third phase besides the mobile amorphous phase and the crystalline phase, and appears as important in the thermal and mechanical properties of semicrystalline polymers (Wunderlich 2003).

As will be introduced in the following kinetic aspects of polymer crystallization, secondary intramolecular crystal nucleation favors chain-folding upon crystal growth, which dominates the lamellar feature of crystal morphologies. The metastable lamellar crystals intend to perform thickening into a more thermodynamically stable state upon annealing. Before they are able to reach the most stable state, their melting points appear as lower than the equilibrium melting point of infinitely large crystals. This is mainly because of a limited dimension in lamellar thickness.

4 Theoretical Aspects of Polymer Crystallization

The lowered melting point of lamellar crystals T_m is described by the well-known Gibbs–Thomson equation, as given by

$$T_m = T_m^0 - \frac{2\sigma_e T_m^0}{l \Delta h} \tag{4.4}$$

where σ_e is the free energy density of the fold-end surface, l is the lamellar thickness, and Δh is the heat of fusion.

A wide distribution of lamellar thickness in polymer crystals results in a broad range of melting temperature, shown as a wide melting peak in the heating curve of differential scanning calorimetry (DSC). We pragmatically define the peak temperature as the experimentally observed melting point of polymer crystals, which could be far below the equilibrium melting point due to the limited lamellar thickness.

Crystallization under various temperatures yields different corresponding lamellar thicknesses, exhibiting different variable melting points. This phenomenon allows us to derive the equilibrium melting point at the infinitely large crystal according to the Gibbs–Thomson equation, provided that there is no annealing effect upon heating the crystals for the measurement of their melting points. However, most of metastable crystals do perform annealing behaviors upon heating towards melting. Hoffman and Weeks (1965) supposed that lamellar crystals will thicken into a metastable state with the thickness several times larger than the minimum thickness, and derived the equilibrium melting point as the crossover point of the extrapolated T_m versus T_c curve at $T_m = T_c$. Although the fixed folds of thickness increase appear as a big assumption, this method has been widely applied to derive the equilibrium melting points of various semicrystalline polymers in the literature.

Besides crystallization, liquid–liquid (L–L) demixing is another basic phase transition in polymer solutions, and both of them are functions of polymer concentrations and temperatures. The schematic L–L binodal and liquid–solid (L–S) coexistence curves in polymer solutions are separately shown in Fig. 4.4. The

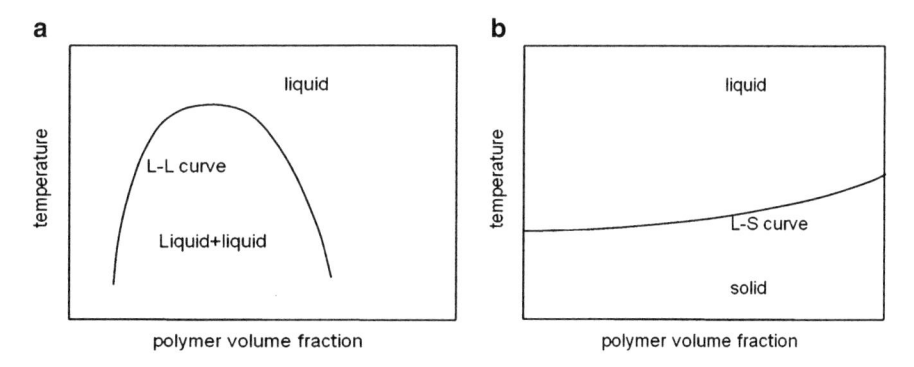

Fig. 4.4 Schematic pictures of (**a**) L–L binodal curve and (**b**) L–S coexistence curve in polymer solutions

Fig. 4.5 Schematic picture of typical phase diagrams combining both L–L binodal and L–S coexistence curves in polymer solutions

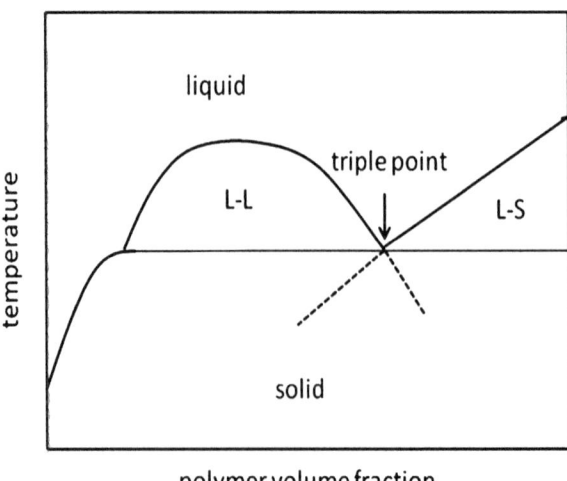

illustrated L–L binodal contains an upper critical solution temperature. Some solution binodals may contain a lower critical solution temperature, or even both.

When the L–S curve intersects with the L–L curve in the overlapping temperature windows, both curves will be terminated at the intersection point, which is referred to the monotectic triple point. The typical phase diagram in polymer solutions is shown in Fig. 4.5.

In between the amorphous phase and the crystalline phase, there sometimes occurs an important intermediate phase for polymers carrying anisotropic groups (called mesogen groups) either on the chain backbone or on the chain branches, which is referred to mesophase. The mesogen groups become orientational-ordered under suitable thermodynamic conditions. This ordered state could be the well-known liquid crystal (LC) state, as a typical mesomorphic state between the amorphous state and the crystalline-ordered state for LC polymers.

The transition temperature from LC mesophase to melt is often named as the clearing point or the isotropization point T_i. If the LC mesophase is thermodynamically stable, it will occur in both heating and cooling curves when detected by DSC measurements. This mesophase is also referred to the enantiotropic mesophase. If the mesophase is kinetically favored due to large supercooling required for the initiation of crystallization, it will occur only in the cooling curve but not in the heating curve. This mesophase is then referred to the monotropic mesophase. The free energy plots for the two kinds of mesophases are demonstrated in Fig. 4.6.

When L–S phase transition is accompanied by LC transition, the above phase diagram of polymer solution becomes more complicated as illustrated in Fig. 4.7.

According to the phase diagram schematically shown in Fig. 4.7, two types of liquid crystal ordering with different preparation methods can be categorized. One is the lyotropic liquid crystal approached by changing the concentration in solutions. The other is the thermotropic liquid crystal approached by changing the

4 Theoretical Aspects of Polymer Crystallization

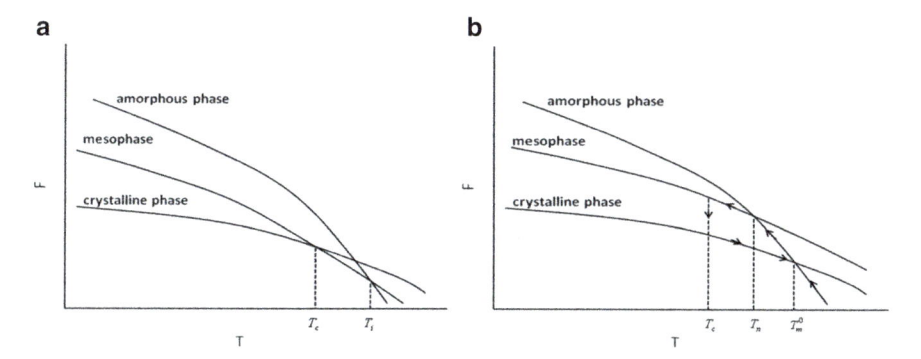

Fig. 4.6 Schematic pictures of (**a**) enantiotropic mesophase and (**b**) monotropic mesophase

Fig. 4.7 Schematic picture of typical phase diagram of polymer solutions combining L–L binodal (Iso), L–S coexistence (Cr), and L-liquid crystal-S coexistences (LC). The two *arrows* in the picture show the preparation methods of thermotropic liquid crystal and lyotropic liquid crystal, respectively (Keller 1992)

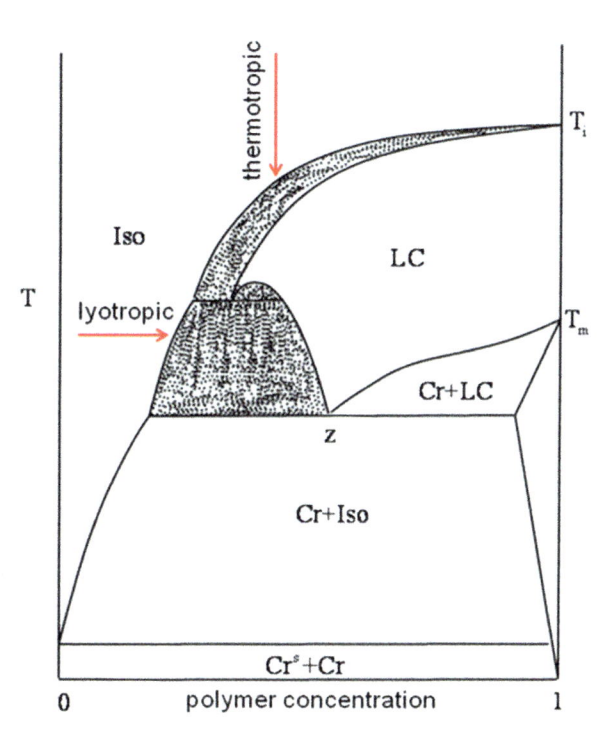

temperature in the concentrated solutions or in the bulk phase. Their conventional preparation methods are schematically shown in Fig. 4.7. According to Onsager's interpretation (Onsager 1949), the lyotropic liquid crystal results from an entropy-driven phase transition due to anisotropic excluded-volume interactions of mesogens. When the concentration of rod-like molecules becomes high enough, the space for anisotropic particles to move freely appears limited, resulting in an entropic loss. In this case, if parts of rodlike particles are aligned in parallel with each other in a domain of higher concentration, they release part of their space for the movement of other particles and thus increase the total entropy. This entropy-driven phase transition makes the ordered system stable. For the thermotropic liquid crystal, Maier and Saupe's theory (1959) considers that the thermotropic liquid crystal is a result of orientation-dependent dispersion interactions between mesogens. With the decrease of temperature, spontaneous orientational ordering lowers the attractive potential energy. Combination of the above two theories gave a better interpretation to the orientational ordering for the formation of liquid crystals as well as a better calculation of phase diagrams (Jähnig 1979; Ronca and Yoon 1982, 1984; ten Bosch et al. 1983a, b; Khokhlov and Semenov 1985; Gupta and Edwards 1993; Lekkerkerker and Vroege 1993).

4.2.2 Statistical Thermodynamics of Polymer Crystallization

Statistical thermodynamic theories provide a powerful tool to bridge between the microscopic chemical structures and the macroscopic properties. Lattice models have been widely used to describe the solution systems (Prigogine 1957). Chang (1939) and Meyer (1939) reported the earliest work related with the lattice model of polymer solution. The lattice model was then successfully established by Flory (1941, 1942) and Huggins (1942) to deal with the solutions of flexible polymers by using a mean-field approximation, and to derive the well-known Flory–Huggins equation.

In the lattice model of polymer solutions, polymer chain is simply represented by a number of consecutively occupied lattice sites, each site corresponding to one chain unit. The rest single sites are assigned to solvents. This simple lattice treatment of polymer solutions allows a very convenient way to calculate thermodynamic properties of flexible and semiflexible polymer solutions from the statistical thermodynamic approach. By the mean-field assumption, the entropy part and the enthalpy part of partition function can be separately calculated.

Following Flory's treatment of semiflexible polymer solutions (Flory 1956), the entropy part of partition function for polymer solutions is given by

$$Z_{\text{comb}} \approx \left(\frac{n}{n_1}\right)^{n_1} \left(\frac{n}{n_2}\right)^{n_2} \left(\frac{qz_c^{r-2}}{2e^{r-1}}\right)^{n_2} \tag{4.5}$$

where n_1 is the number of solvent molecules, n_2 is the number of polymer chains, each chain composed of r monomers, and $n = n_1 + n_2 r$ is the total lattice sites occupied by solvent and polymer chains. q is the coordination number of the lattice space. z_c is the conformational partition function defined as

$$z_c = 1 + (q-2)\exp\left(-\frac{E_c}{kT}\right) \tag{4.6}$$

4 Theoretical Aspects of Polymer Crystallization

where E_c is the energy penalty for the $q - 2$ noncollinear consecutive bonds along the polymer chain with a reference to the collinear connection, k is Boltzmann constant, and T is system temperature.

The partition function related with the mixing enthalpy of solvent and monomers is given by

$$z_m = \exp\left[-(q-2)\frac{n_1}{n} \times \frac{B}{kT}\right] \quad (4.7)$$

where B is the energy change for a pair of solvent and monomer before and after mixing as defined by

$$B = E_{12} - \frac{E_{11} + E_{12}}{2} \quad (4.8)$$

As polymer crystals are commonly featured with parallel packing of polymer chains, the driving force for crystallization can be modeled as the parallel packing of bonds in the lattice model (Hu 2000). Assuming the energy penalty for nonparallel packing of two bonds is E_p, deviating from the ground state with parallel-packed bonds, the partition function related to nonparallel packing of neighboring bonds around a chain bond is given by Hu and Frenkel (2005)

$$z_p = \exp\left\{-\frac{q-2}{2}\left[1 - \frac{2(r-1)n_2}{qn}\right]\frac{E_p}{kT}\right\} \quad (4.9)$$

Thus after integrating different parts, the total partition function of polymer solutions can be described as

$$Z = \left(\frac{n}{n_1}\right)^{n_1}\left(\frac{n}{n_2}\right)^{n_2}\left(\frac{q}{2}\right)^{n_2} e^{-(r-1)n_2} z_c^{(r-2)n_2} z_m^{rn_2} z_p^{(r-1)n_2} \quad (4.10)$$

The corresponding free energy density is calculated according to Boltzman's relationship, as given by

$$\frac{f(\phi)}{k_B T} = (1-\phi)\ln(1-\phi) + \frac{\phi}{r}\ln\phi$$
$$+ \phi\left(-\frac{\ln(qr/2)}{r} - (1-2/r)\ln z_c + (1-1/r) + (q-2)\frac{B}{k_B T} + (1-1/r)\frac{q-2}{2}\frac{E_p}{k_B T}\right)$$
$$- \phi^2\left((q-2)\frac{B}{k_B T} + (1-1/r)^2\frac{q-2}{q}\frac{E_p}{k_B T}\right)$$

$$(4.11)$$

Equation (4.11) can be used to calculate various thermodynamic properties of polymers, in particular, the equilibrium melting point. It can be used to calculate the mixing free energy and the binodal L–L curve, as well as the coexistence L–S curve in polymer solutions.

Following the protocols of molecular interactions set up in the lattice statistical thermodynamic theory introduced above, Monte Carlo simulations appear as a powerful tool in the study of both thermodynamics and kinetics of polymer crystallization (Hu and Frenkel 2005). Polymer motions in the simulations can be realized by a micro-relaxation model (Hu 1998). Its acceptance is judged by the Metropolis importance sampling algorithm (Metropolis et al. 1953), with the potential energy change in each step of micro-relaxation composed of three contributions, noncollinear connections of consecutive bonds along polymer chains (E_c) (Flory 1956), mixed pairs of polymer monomer and solvent (B) (Flory 1942; Huggins 1942), and nonparallel packing of neighboring bonds (E_p) (Hu 2000). The micro-relaxation model is highly efficient in relaxing local chain conformation and is a suitable approach for the study of polymer crystallization. Results from Monte Carlo simulations will be introduced along with theories of polymer crystallization in the following paragraphs.

4.2.3 Properties of Equilibrium Melting Points

4.2.3.1 Interaction Parameters

Although a precise measurement of the equilibrium melting point appears as a big challenge to the practical experiments, the statistical thermodynamic theory can predict equilibrium melting point on the basis of the mean-field assumption. Taking the fully ordered extended chains as the ground state, and considering bulk polymers with infinitely long polymer chains, $r \to \infty$, $n_1 = 0$, $n = rn_2$, one can get the free energy of the amorphous state from (4.11),

$$\frac{F}{nkT} = 1 - \ln z_c + \frac{(q-2)^2}{2q} \times \frac{E_p}{kT} \qquad (4.12)$$

At the equilibrium melting point, the free energy of the amorphous state is equal to that of the ground crystalline state (zero), so inserting the expression of z_c, one can derive

$$1 + (q-2)\exp\left(-\frac{E_c}{kT_m}\right) = \exp\left[1 + \frac{(q-2)^2}{2q} \times \frac{E_p}{kT_m}\right] \qquad (4.13)$$

Supposing that the first term "1" in the left-hand side of the equation is relatively small, one can omit it and derives the equilibrium melting point of bulk polymers, as given by

$$T_m \approx \frac{E_c + \frac{(q-2)^2}{2q}E_p}{k\ln(q-2) - k} \qquad (4.14)$$

From (4.14), we can clearly see that a larger E_c (a higher rigidity of polymer chains) favors a higher equilibrium melting point. For polyolefins, larger substitute groups benefit their higher melting points because internal rotation of polymer chains becomes more difficult. For instance, the melting point of polyethylene (–(–CH$_2$CH$_2$)$_n$–) is 146 °C, the melting point of polypropylene (–(–CH$_2$CH(CH$_3$))$_n$–) is 187 °C while for poly-3-methyl-1-butene (–(–CH$_2$CH(CH(CH$_3$)$_2$))$_n$–), its melting point is 304 °C. Apparently, more rigid groups on the backbones benefit their higher melting point as well. For instance, the melting point of polyethylene (–(–CH$_2$CH$_2$)$_n$–) is 146 °C, while with increasing benzene-ring density on the polymer chains the melting point of poly-p-xylene (–(–CH$_2$–ϕ–CH$_2$–)$_n$–) is 375 °C and that of polyphenylene (–(–ϕ–)$_n$–) is as high as 530 °C. Another example, the melting point of polyethylene adipate (–(CH$_2$CH$_2$–OCO–C$_6$H$_{12}$–OCO)$_n$–) is 52 °C, while the melting point of polyethylene terephthalate (–(CH$_2$CH$_2$–OCO–ϕ–OCO)$_n$–) is 265 °C, and that of polyethylene naphthalene-2,6-dicarbonoxylate (–(CH$_2$CH$_2$–OCO–ϕ–ϕ–OCO)$_n$–) reaches the highest 355 °C. Melting point of polymers with extremely high rigidity may be even higher than their thermal degradation temperature where their melt phase cannot be reached. On the other hand, a larger E_p favors also a higher equilibrium melting point. Longer side groups do not help the compact packing of backbone chains. Taking the melting points of (CH$_2$-CHR)$_n$ for example, if R group represents –CH$_3$, –CH$_2$CH$_3$, –CH$_2$CH$_2$CH$_3$, and –CH$_2$CH$_2$CH$_2$CH$_3$, respectively, the corresponding melting points are 187 °C, 138 °C, 130 °C, and −55 °C, respectively. The larger polarity of side groups also benefits the higher melting point of polymers. For instance, the melting point of polyethylene (–(–CH$_2$CH$_2$)$_n$–) is 146 °C, while the melting point of polyvinyl chloride (–(–CH$_2$CHCl)$_n$–) is 227 °C and that of polyacrylonitrile (–(–CH$_2$CHCN)$_n$–) is 265 °C. In the last case, the sequence regularity becomes a minor factor for the capability of crystallization.

The linear relationship between the melting points and E_p/E_c values in (4.14) is further confirmed by Monte Carlo simulations of polymers, and the results are shown in Fig. 4.8. The high consistency in the results between the lattice theory and Monte Carlo simulations validates the mean-field treatment in the lattice theory of polymer solutions.

4.2.3.2 Molecular Weights

In the equilibrium states of polymer crystals, the chain ends can be regarded as the crystalline defects in the infinitely large crystals formed by extended chains, which will apparently result in a depression of melting points with the increase of concentrations of chain ends, or in other words, with the decrease of chain lengths.

In the calculation of melting point of polymers with different molecular weights, Flory and Vrij (1963) divided the free energy change of melting into three parts. The first part is the free energy change for the melting of infinitely long polymers, the second part is the free energy change by introducing chain-end defects in the crystals, and the third part is the conformational entropy change when the infinitely

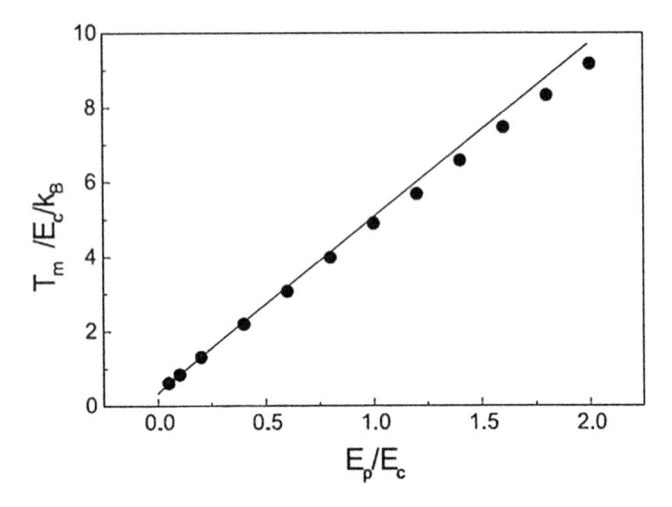

Fig. 4.8 Melting temperature ($T_m/E_c/k_B$) of bulk polymers with variable E_p/E_c values. *The Solid line* is calculated from (4.13) and the *circles* are simulation results of 32-mer polymer solutions with a concentration of 0.9375 in a 32^3 cubic lattice (Hu and Frenkel 2005)

long polymer chain is cut into the segments with limited chain length r. So the total free energy change of melting can be described as

$$\Delta F_m = r\Delta F_u + \Delta F_e - kT_m \ln r = 0 \quad (4.15)$$

ΔF_u is the free energy change of fusion, it can be calculated by

$$\begin{aligned}\Delta F_u &= \Delta h_u - T_m \Delta s_u \\ &= E_c + \frac{(q-2)^2}{2q} E_p - k_B T_m [\ln(q-2) - 1]\end{aligned} \quad (4.16)$$

ΔF_e is the extra free energy change due to the existence of chain ends, which can be calculated from the equilibrium condition $f=0$ in (4.11) by setting the chain length $r=2$ in the melt phase.

$$\Delta F_e = \frac{(q-2)(q-1)}{2q} E_p - k_B T_m (\ln q - 1) - 2\Delta f_u \quad (4.17)$$

The term $\ln(r)$ represents a change of conformational entropy upon cutting. Inserting (4.16) and (4.17) into (4.15), the equilibrium melting point predicted by the Flory–Vrij theory can be calculated.

One can also directly calculate the equilibrium melting point from (4.11). Under equilibrium melting temperature,

$$\mu^c = \mu^s \quad (4.18)$$

where μ^c and μ^s are the chemical potentials of polymers in crystals and in solutions, respectively. We assume that $\mu^c = 0$ by omitting any disorder defects in the crystals. When the chemical potentials in solid and liquid are equal, we can obtain the equation

$$(1-r)\frac{n_2 r}{n} + \ln\frac{qn}{2n_2} + (r-2)\ln\left[1 + (q-2)\exp\left(-\frac{E_c}{k_B T_m}\right)\right]$$
$$= \frac{(r-1)(q-2)}{2}\left[1 - \frac{2(r-1)n_2(n+n_1)}{qn^2}\right]\frac{E_p}{k_B T_m} + \frac{rn_1^2}{n^2}\frac{(q-2)B}{k_B T_m} \quad (4.19)$$

The equilibrium melting point of polymers can be calculated by taking into account all the parameters and solving (4.19).

Figure 4.9 compares the melting temperatures of bulk polymers with different chain lengths derived from the lattice theory, the Flory–Vrij equation and Monte Carlo simulations, respectively. The dimension of the temperature unit is reduced as $E_c/(k_B T_m)$. Although they hold various assumptions, their agreements are satisfying.

4.2.3.3 Comonomer Contents in Random Copolymers

Real polymers are not structurally uniform along the sequences of the chain, and more or less there exist various kinds of irregularities, such as different chemical compositions, different geometrical connections, or different stereo optical isomers

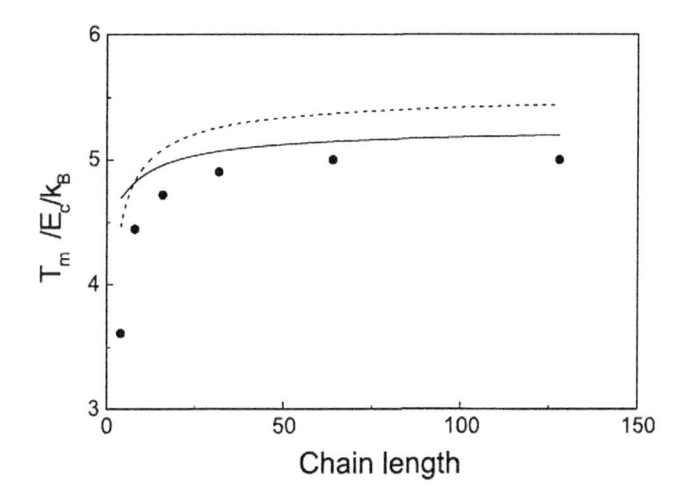

Fig. 4.9 Melting temperature ($T_m/E_c/k_B$) of polymers with different chain length. *The Solid line* is calculated from (4.19) and *dashed line* is results of Flory–Vrij equation. *The Circles* are results from Monte Carlo simulations of polymer solutions with a 0.9375 concentration in a 32^3 cubic lattice for short chains and a 64^3 cubic lattice for long chains (Hu and Frenkel 2005)

of monomers. Because of the spatial mismatch between irregular and regular sequences for the compact packing in crystalline order, the irregular sequences will result in a lowered melting point. If the regular sequences are referred to monomer A and the irregular sequences are referred to comonomer B, Flory (1955) considered this AB random copolymer as an ideal solution and its chemical potential deviation from the sequence-uniform homopolymer A is

$$\mu_A^{co} - \mu_A^0 = -RT_m \ln X_A \tag{4.20}$$

Thus, the chemical potential change of crystallization is

$$\mu_A^c - \mu_A^{co} = \Delta H_u - T_m \Delta S_u \tag{4.21}$$

The subscript "u" represents the unit of mole chain monomers.

For homopolymer composed of monomer A, the chemical potential change of crystallization is

$$\mu_A^c - \mu_A^0 = \Delta H_u^0 - T_m^0 \Delta S_u^0 = 0 \tag{4.22}$$

Considering $\Delta H_u^0 \approx \Delta H_u$ and $\Delta S_u^0 \approx \Delta S_u$, the melting point of copolymers can be derived as

$$\frac{1}{T_m} - \frac{1}{T_m^0} = \frac{R}{\Delta H_u} \ln X_A \tag{4.23}$$

The similar equation for stereo-optical-isomer copolymers with comonomers homogeneously distributed in the crystallites was also given by Coleman (1958).

Flory assumed all the comonomers staying only in the amorphous phase. By considering comonomers coexisting with monomers in the crystallites, Colson and Eby (1966), and later on Sanchez and Eby (1975) gave another expression for melting point of random copolymers, as given by

$$T_m = T_m^0 \left(1 - \frac{\Delta H_B}{\Delta H_u} X_B\right) \tag{4.24}$$

In (4.24), ΔH_B is the heat of fusion for each comonomer as a defect in the crystalline phase, and X_B is the mole fraction of comonomers.

4.2.4 Phase Diagrams of Polymer Solutions

From (4.11), the theoretical curves of L–S coexistence curve and L–L binodal curve can be separately calculated, provided by the absence of each counterpart.

4 Theoretical Aspects of Polymer Crystallization

At the equilibrium melting point, the chemical potential of polymers in the amorphous phase μ^s is equal to that in the crystalline phase μ^c, thus

$$\mu^s - \mu^0 = \mu^c - \mu^0 \tag{4.25}$$

where μ^0 is the chemical potential of the ground state where polymer chains are totally extended and packed in parallel with each other. Since in the crystalline phase, μ^c is almost the same as μ^0, we can thus derive that

$$\mu^s - \mu^0 = \frac{\partial \Delta F^s}{\partial n_2} = -kT \frac{\partial \ln Z}{\partial n_2} \approx 0 \tag{4.26}$$

It means

$$\frac{\partial \ln Z}{\partial n_2} = 0 \tag{4.27}$$

Thus, the equilibrium melting point can be calculated by solving the above equation. Changing the polymer concentrations, one can obtain the theoretical phase diagram of L–S coexistence.

The L–L binodal curve can be calculated by the chemical potential equivalence of components between the dense phase and the dilute phase after phase separation. That is,

$$\begin{cases} \Delta \mu_{1a} = \Delta \mu_{1b} \\ \Delta \mu_{2a} = \Delta \mu_{2b} \end{cases} \tag{4.28}$$

where the subscripts "a" and "b" are used to represent the dense and dilute phases, respectively. Before reaching the final result, the mixing free energy as defined as the free energy change of polymer solutions from that of bulk amorphous polymers before mixing should be calculated, as given by

$$\frac{\Delta F_{mix}}{kT} = \frac{F_{solution} - F_{bulk}}{kT} = -(\ln Z - \ln Z_{n_1 = 0})$$

$$= n_1 \ln \phi_1 + n_2 \ln \phi_2 + n_1 \phi_2 \left[(q-2)\frac{B}{kT} + \left(1 - \frac{2}{q}\right)\left(1 - \frac{1}{r}\right)^2 \frac{E_P}{kT} \right] \tag{4.29}$$

One can calculate the chemical potentials from (4.29) and insert them into (4.28), the equilibrium concentrations of dense and dilute phases under a certain temperature can be calculated. By changing temperature, the L–L binodal curve is thus obtained.

Monte Carlo simulations can calculate the phase diagrams of polymer solutions in a different way. In the lattice model of polymer solutions, each step of micro-relaxation in Monte Carlo simulations is determined by Metropolis sampling method with a potential energy barrier as described as

$$\frac{\Delta E}{kT} = \frac{cE_c + pE_p + bB}{kT} = \left(c + p\frac{E_p}{E_c} + b\frac{B}{E_c}\right) / \frac{kT}{E_c} \qquad (4.30)$$

where c, p, and b are the numbers of net changes in noncollinear connection, nonparallel packing and mixing pairs of polymer unit and solvent after and before each step, respectively. Figure 4.10 shows the parallel results of liquid–liquid demixing curves and liquid–solid coexistence curves in polymer solutions with different energy parameter sets, obtained from the lattice mean-field theory and Monte Carlo simulations. Theoretical calculations and Monte Carlo simulations agree well with each other, again validating the mean-field theory.

In Fig. 4.10, one can clearly see that the L–S curve is mainly determined by E_p/E_c, and the L–L curve is mainly controlled by B/E_c but is also slightly affected by E_p/E_c. By changing the values of E_p/E_c and B/E_c, the interplay between crystallization and liquid–liquid demixing can be studied in a combination of the lattice theory and parallel molecular simulations (Hu and Frenkel 2004; Ma et al. 2007, 2008), which has been introduced in the author's book (Hu 2013).

In the practical processing of polymers, many organic small molecules such as plasticizers, anti-UV agents and releasing agents are added as diluents, which also result in a depression of melting points. As deduced from Flory–Huggins equation, the chemical potential change in polymer solution is

$$\mu_2^L - \mu_2^0 = RT_m\left[\ln\phi_2 - (r-1)\phi_1 + r\chi\phi_1^2\right] \approx rRT_m\left(-\phi_1 + \chi\phi_1^2\right) \qquad (4.31)$$

The chemical potential change upon melting is

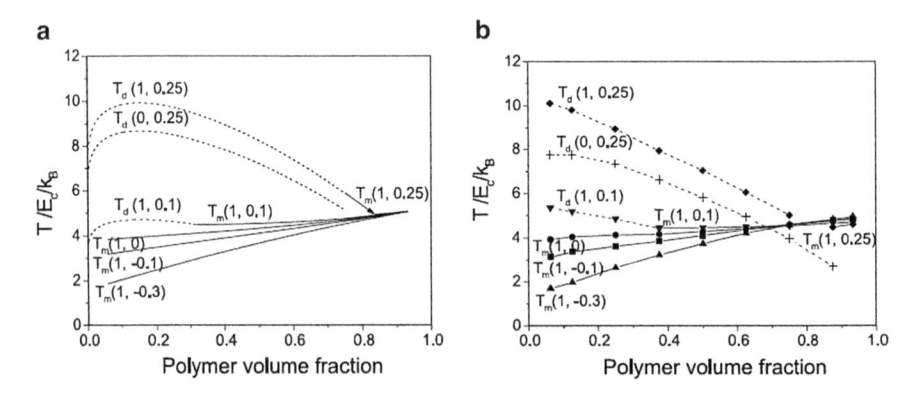

Fig. 4.10 Theoretical calculation (**a**) and Monte Carlo simulations (**b**) of liquid–liquid demixing curves (*dashed lines*) and liquid–solid curves (*solid lines*) of polymer solutions with different energy parameter combinations, denoted by T (E_p/E_c, B/E_c) (Hu et al. 2003a)

4 Theoretical Aspects of Polymer Crystallization

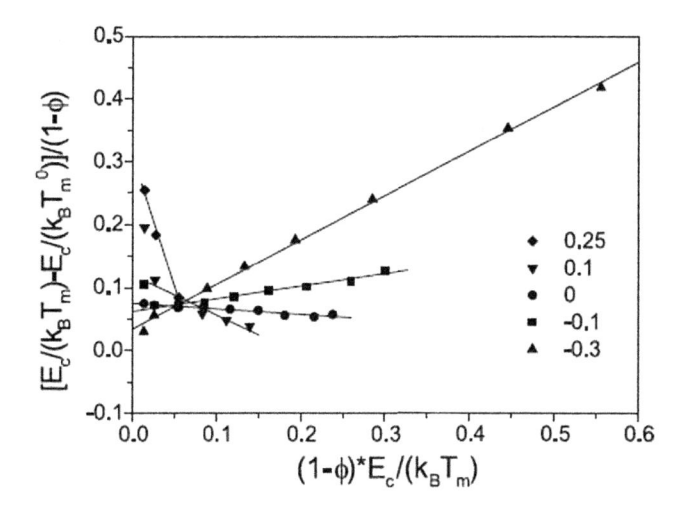

Fig. 4.11 Rescaled data from Fig. 4.10b according to (4.33). The bulk equilibrium melting temperature ($E_c/k_BT_m^0$) is chosen to be approximately 0.2. *Lines* are linear regressions of symbols at the same values of B/E_c as labeled (Hu et al. 2003a)

$$\mu_2^s - \mu_2^0 \approx r\Delta H_u\left(1 - \frac{T_m}{T_m^0}\right) \quad (4.32)$$

ΔH_u is the melting enthalpy of polymers per mole of monomers. When the amorphous phase and the crystalline phase become in equilibrium, their chemical potentials are equal. By combining (4.31) with (4.32) and making them equal to each other, the melting point of polymers in solution is derived as

$$\frac{1}{T_m} - \frac{1}{T_m^0} = \frac{R}{\Delta H_u}(\phi_1 - \chi\phi_1^2) \quad (4.33)$$

This equation was fitting well with experimental results (Prasad and Mandelkern 1989). It was also verified by the results of Monte Carlo simulations (Hu et al. 2003a). Figure 4.11 shows the linear relationship according to (4.33), with the data points adopted from Fig. 4.10b.

4.3 Kinetics of Polymer Crystallization

4.3.1 Crystal Nucleation

Polymer crystallization can be roughly divided into two sequential processes: crystal nucleation and crystal growth. When the sizes of ordered domains generated by thermal fluctuations become so large that the trend to increase the surface free energy can be overcome by the trend to decrease the body free energy, larger domains intend to be more stable (Kelton 1991). The critical sizes and the related free energy barriers depend on the crystallization temperature. Therefore, a certain supercooling is required for crystal nucleation. Considering nucleus as a sphere with a radius r, the free energy change of nucleation can be estimated as

$$\Delta G = -\Delta g \times \frac{4}{3}\pi r^3 + \sigma \times 4\pi r^2 \qquad (4.34)$$

Δg is the melting free energy of unit volume, and σ is the specific surface free energy. Schematic plot for the free energy change with the increasing radius of nucleus is shown in Fig. 4.12,

Δg in (4.34) is calculated as

$$\Delta g = \Delta h - T_c \Delta s \approx \Delta h - T_c \frac{\Delta h}{T_m} = \Delta h \frac{T_m - T_c}{T_m} \propto \Delta T \qquad (4.35)$$

Generally speaking, there are three different types of nucleation according to various dimensions, i.e., primary nucleation, secondary nucleation, and tertiary nucleation. Primary nucleation is a nucleus newly formed by thermal fluctuations, with six extra nucleus surfaces if the nucleus is considered to be cubic. Secondary nucleation is two-dimensional nucleation on the advancing surface of nucleus, with four extra surfaces produced. Secondary nucleation is easier than primary nucleation as its free energy barrier is lower. Tertiary nucleation is one-dimensional nucleation at the step edge of the spreading layer on the advancing surface of nucleus, with only two extra surfaces produced. Tertiary nucleation is so fast that it can rarely be observed. The schematic pictures of different types of nucleation are shown in Fig. 4.13.

Primary nucleation is the most observable phenomenon for the initiation of polymer crystallization, which can be categorized further into homogeneous nucleation and heterogeneous nucleation. In homogeneous nucleation, polymer nuclei can be treated as a cylindrical bunch of stems due to the anisotropic molecular structure, as depicted in Fig. 4.14.

Thus its free energy change during nucleation is

$$\Delta G = -\pi r^2 l \Delta g + 2\pi r l \sigma + 2\pi r^2 \sigma_e \qquad (4.36)$$

Fig. 4.12 Schematic curve of the free energy change as the radius of nucleus formed in the amorphous phase

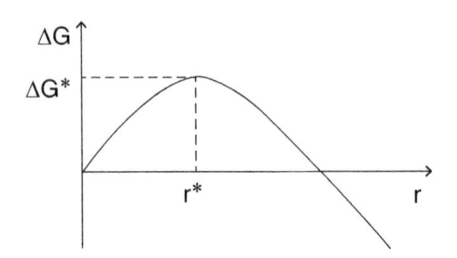

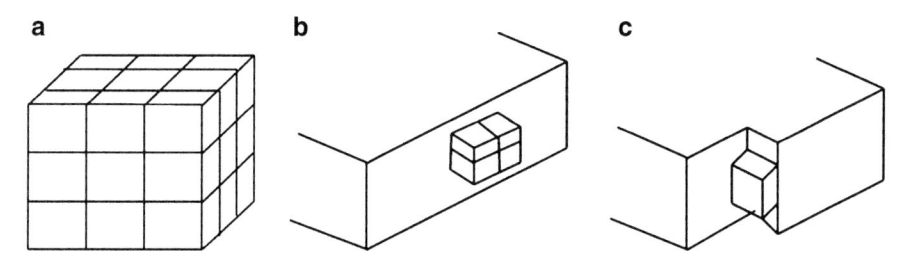

4 Theoretical Aspects of Polymer Crystallization

Fig. 4.13 Schematic pictures of (**a**) primary nucleation, (**b**) secondary nucleation, and (**c**) tertiary nucleation

Fig. 4.14 Schematic picture of nucleus treated as cylindrical bunch of stems

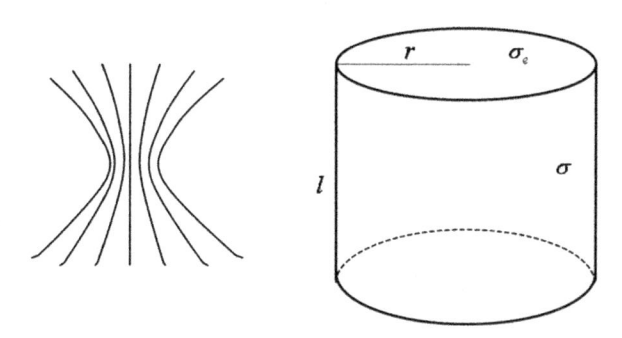

Here, r and l are the radius and the length of the cylinder, respectively; σ is the specific free energy on the lateral surface, and σ_e is the specific free energy on the end surface.

By taking the minimum of ΔG with respect to r and l, the critical free energy barrier for nucleation is derived as

$$\Delta G^* = \frac{8\pi\sigma^2\sigma_e}{\Delta g^2} \propto \Delta T^{-2} \tag{4.37}$$

The critical sizes are separately calculated as

$$r^* = \frac{2\sigma}{\Delta g} \propto \Delta T^{-1} \tag{4.38}$$

$$l^* = \frac{4\sigma_e}{\Delta g} \propto \Delta T^{-1} \tag{4.39}$$

So the length-to-radius ratio of the critical nucleus is

$$\frac{l^*}{r^*} = \frac{2\sigma_e}{\sigma} \tag{4.40}$$

Fig. 4.15 Schematic pictures of (**a**) intermolecular nucleation and (**b**) intramolecular nucleation

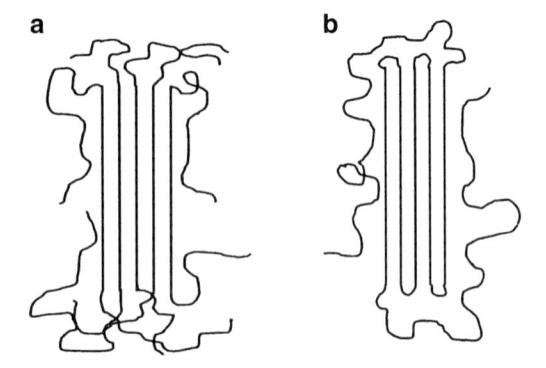

For polymers, homogeneous nucleation can be realized through two typical ways. One is the so-called *intramolecular nucleation* (Wunderlich 1976) featured with adjacent chain-folding, which can be called as chain-folding nucleation. The other is the so-called *intermolecular nucleation* composed of parallel stacking among neighboring chains, which can be called as fringed-micelle nucleation. Schematic illustration of these two models can be found in Fig. 4.15.

Let us first consider the intramolecular nucleation. The specific free energy on the lateral surfaces of PE crystals was estimated to be 11.8 erg/cm^2, and its specific free energy on the fold-end surface was about 90 erg/cm^2 (Hoffman and Miller 1997). By (4.40), the optimized aspect ratio of critical nucleus is 15.3. Now look at the intermolecular nucleation. The end surface has an extra free energy of about 245 erg/cm^2 as estimated by Zachmann (1967, 1969), due to the entropy loss of disordered chains. So the aspect ratio of critical nucleus in intermolecular nucleation is as high as 56.8, and it appears very difficult to produce such kind of fibril nucleus through thermal fluctuations for primary nucleation. Moreover, by (4.36), the higher end-surface free energy results in a higher nucleation barrier than the chain-folding model. So from both kinetics aspects of the critical nucleation barrier and of thermal fluctuations, the intramolecular nucleation is preferred in the process of primary nucleation.

Heterogeneous nucleation is primary crystal nucleation on the foreign surfaces of other materials such as catalysts, dusts, and container walls. Because less extra surface free energy is required, heterogeneous nucleation is much easier than homogeneous nucleation. Assuming that heterogeneous nucleus is a cubic nucleus, its free energy barrier is similar like homogeneous nucleation, again depending upon inverse square supercooling. When the foreign surface free energy is close to that of polymer crystals, the heterogeneous nucleation is more like a layer-by-layer growth on the foreign surface. We assume that a, b_0, and l are its width, depth, and length, respectively, as shown in Fig. 4.16, the free energy change of nucleation is given by

$$\Delta G = -ab_0 l \Delta g + 2b_0 l \sigma + al \Delta \sigma + 2ab_0 \sigma_e \quad (4.41)$$

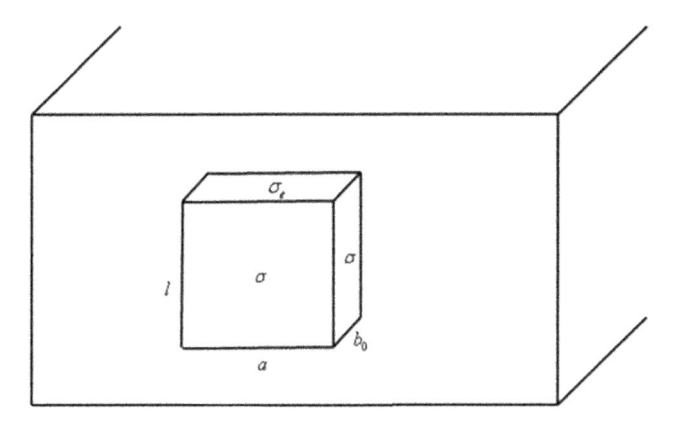

4 Theoretical Aspects of Polymer Crystallization

Fig. 4.16 Schematic picture of heterogeneous nucleation

In (4.41),

$$\Delta\sigma = \sigma + \sigma_{cs} - \sigma_{ms} \qquad (4.42)$$

σ_{cs} is the surface free energy between crystal nucleus and substrate, and σ_{ms} is the surface free energy between melt and substrate. If the free energy of substrate is almost the same as crystal nucleus, $\Delta\sigma \approx 0$, by minimizing ΔG of a and l separately, the critical Gibbs free energy is

$$\Delta G^* = \frac{4b_0\sigma\sigma_e}{\Delta g} \propto \Delta T^{-1} \qquad (4.43)$$

There is also another type of primary nucleation called self-nucleation investigated first by Blundell et al. (1966). The foreign surfaces for self-nucleation are provided by crystals of the same species which survived during thermal history. Since there is no extra surface free energy change during self-nucleation, it is also called athermal nucleation. This type of nucleation is an important source of memory effects for polymer crystallization.

The nucleation rate is dominated by two factors. One is the critical free energy barrier of nucleation. Its exponential dependence was first proposed by Volmer and Weber (1926). The other is the diffusion energy barrier for molecules crossing over the liquid–solid interfaces. Its exponential dependence was first proposed by Becker and Döring (1935). The quantitative expression of the prefactor in the kinetic equation of the nucleation rate is given by Turnbull and Fisher (1949) as

$$I = I_0 \exp\left(-\frac{\Delta E + \Delta G^*}{kT}\right) \qquad (4.44)$$

where ΔE is the activation barrier for short-range diffusion over the liquid–solid boundary, and I_0 is the prefactor. The critical free energy barrier is proportional to

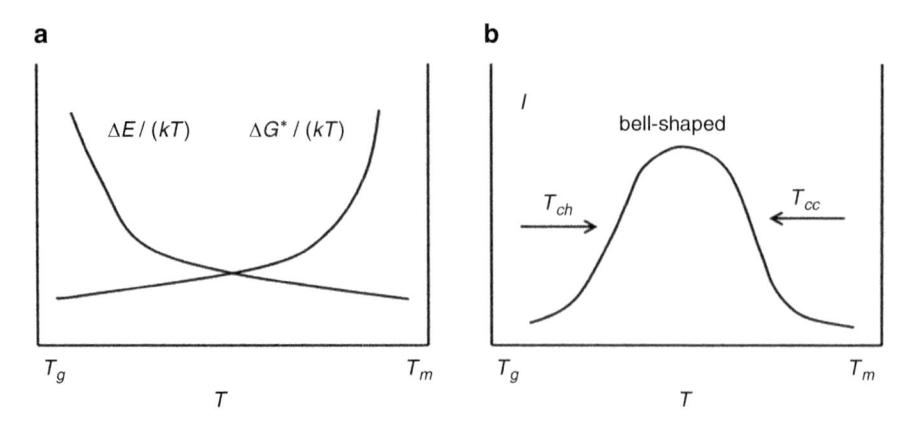

Fig. 4.17 Schematic illustration on the temperature dependence (**a**) of the critical free energy barrier and the activation barrier for diffusion, and (**b**) of the bell-shape curve of the nucleation rates

the inverse square supercooling of primary nucleation. So when temperature is high with a high free energy barrier, the nucleation rate is small. However, when temperature is low with a high activation barrier for polymer diffusion, the nucleation rate is again small. Thus, the temperature-dependence curve of the nucleation rate is somewhat like a bell shape between the glass transition temperature and the melting temperature, as illustrated in Fig. 4.17.

At high temperatures, the kinetic studies of polymer crystallization is mostly focused on the nucleation. In this region, heterogeneous nucleation takes place and the resulted mechanical properties of the semicrystalline polymers are usually hard and brittle. While in the region of low temperatures, as a result of high density of small crystallites, the semicrystalline polymers become soft and tough.

Intramolecular nucleation is preferred in polymer crystal nucleation, both primary and secondary. The secondary intramolecular nucleation explains why chain-folding is a kinetic preference, which results in lamellar shapes during crystal growth. The typical intramolecular nucleation was investigated by Monte Carlo simulations of crystal nucleation of a single-chain system (Hu et al. 2003b). Taking the extended single chain in a crystal composed of extended parallel polymer chains as the ground state, and assuming the number of melting bonds n, the free energy change of the chain is

$$\Delta F = \Delta f n + \sigma (N - n)^{2/3} \tag{4.45}$$

where $\Delta f n$ is the bulk free energy change and $\sigma(N-n)^{2/3}$ is the surface free energy change. The free energy change of one bond during melting is

$$\Delta f = \frac{q-2}{2} E_p - kT \ln(q-1) \tag{4.46}$$

4 Theoretical Aspects of Polymer Crystallization

The first term on the right-hand side means $q - 2$ parallel bonds around the bond in the ground state, and the denominator "2" is the symmetric factor. The number of total conformation of a chain with n melting bonds is $(q - 1)^n$, so the second term is the average conformational entropy change of each bond during melting.

When the system is in equilibrium, the free energy in the disordered state is equal to that in the ordered state, then

$$\Delta f_e = \sigma N^{-1/3} \tag{4.47}$$

And the equilibrium free energy barrier for primary nucleation is (Fig. 4.18)

$$\Delta F_e = \frac{4\sigma^3}{27\Delta f_e^2} \tag{4.48}$$

The free energy barriers for crystallization and melting of a single chain are separately calculated as follows,

$$\Delta F_c = \frac{4\sigma^3}{27\Delta f^2} \tag{4.49}$$

$$\Delta F_m = \frac{4\sigma^3}{27\Delta f^2} + \Delta f N - \sigma N^{2/3} \tag{4.50}$$

Equations (4.49) and (4.50) are confirmed by Monte Carlo simulations, as shown in Fig. 4.19. Single chains with different lengths will crystallize at the same temperature on cooling, but will melt at different temperatures on heating, with higher melting temperatures for higher chain lengths. In bulk polymers, the nucleation rates appear chain-length dependent, probably because the prefactor in the kinetic equation of nucleation could also be of chain-length dependence.

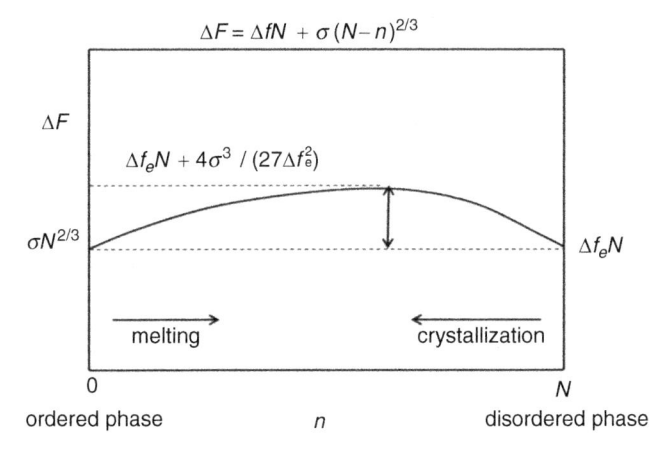

Fig. 4.18 Schematic free energy of single chain with number of melting bonds under thermal equilibrium state (Hu et al. 2003b)

Fig. 4.19 Free energy curves with various crystalline bonds at the fixed temperature $T = 2.174\,E_p/k_B$ in the single chain systems with different chain lengths as labeled (Hu et al. 2003b)

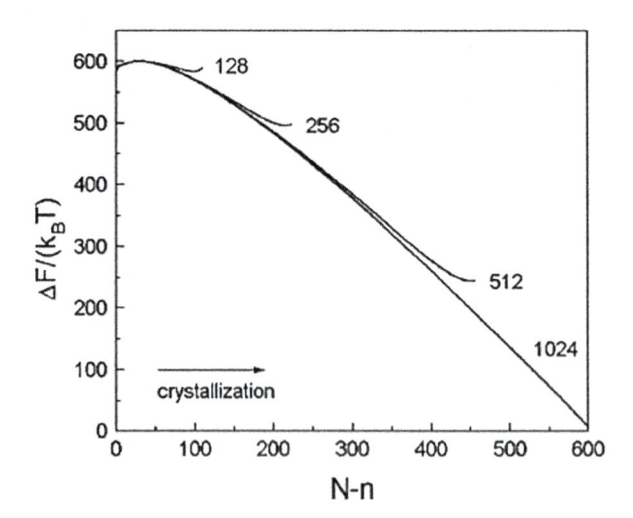

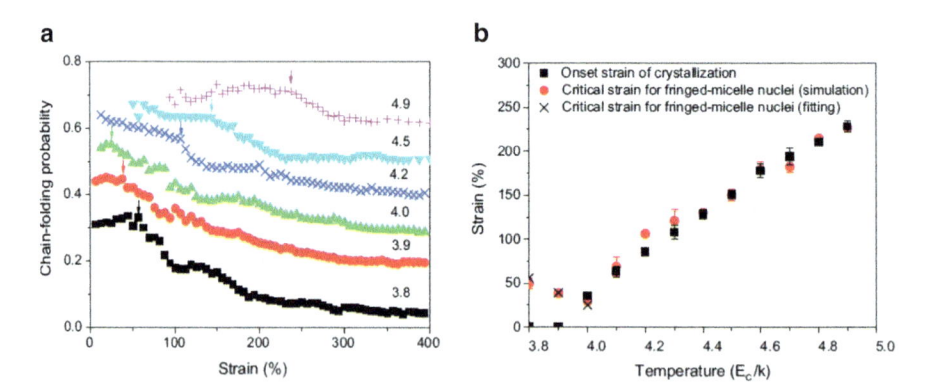

Fig. 4.20 (a) Strain-evolution curves of chain-folding probability of small crystalline clusters containing 50–200 parallel packed bonds under different temperatures. The *lines* are vertically shifted by 0, 0.15, 0.25, 0.35, 0.45, and 0.55, respectively. (b) Comparison between the onset strains of crystallization and the critical strain for fringed-micelle nuclei under different temperatures (Nie et al. 2013)

It should be noted that intermolecular nucleation could coexist with intramolecular nucleation. Intermolecular nucleation is often observed with short chains, rigid chains, polymerizing chains, or when chains are stretched. Recently, upon stretching network polymers, the transition from intramolecular nucleation to intermolecular nucleation was observed in Monte Carlo simulations (Nie et al. 2013). By analyzing the probability of adjacent chain-folding of those newly formed crystallites with a size between 50 to 200 parallel packed bonds at each step of stretching, an obvious reduction was observed in its evolution curve under each temperature as shown in Fig. 4.20a. The corresponding critical strain was considered to be the transition point under which intramolecular nucleation is the favorite and above which intermolecular nucleation becomes the dominant.

Compared the critical strain with the onset strain of crystallization during stretching under each temperature as shown in Fig. 4.20b, it is found that at low temperatures (the dimension-reduced $T \leq 4.0$) intramolecular nucleation dominates the initiation of polymer crystallization with a smaller strain than the critical value, and when temperature becomes higher, intermolecular nucleation begins to dominate the initiation of crystallization.

Recently, some researchers proposed a preordered structure in the polymer melt before nucleation. Imai and coworkers attributed the preordered structures to spinodal decomposition during orientational fluctuations at low temperatures for cold crystallization of PET (Imai et al. 1993, 1994). The main proof came from the observation of small-angle X-ray scattering (SAXS) signal before wide-angle X-ray diffraction (WAXD) during isothermal crystallization at low temperatures (Terrill et al. 1998). Sirota and Herhold (1999) and Kraack et al. (2000) observed a mesophase of chain cluster at the early stage of nucleation of long alkane chains and there is almost no supercooling for nucleation although the mesophase disappears as soon as chain-folding occurs in the crystallization of long-enough chains (Sirota 2000). The idea of nucleation initiated by spinodal decomposition was then prevailing in many theoretical models. Olmsted and coworkers (1998) thought spinodal decomposition is a result of coupling between orientational-order fluctuations and density fluctuations at low temperatures. The spinodal decomposition will enhance crystal nucleation at a certain supercooling. In molecular dynamics simulations, Gee and coworkers (2005) observed that the crystallization behaviors of PVDF under 600 K and PE under 450 K in a time scale of nanoseconds appear as spinodal decomposition. Milner calculated the free energy change of PE crystal nucleation through a rotated mesophase, and it is lower than the surface free energy of orthogonal crystalline phase. He thought this could be the free energy barrier for crystal nucleation (Milner 2011). However, the prior occurrence of SAXS signal could be attributed to a limited instrument sensitivity or improper experiment treatment. Howard Wang thought that the signal like that of spinodal decomposition could be expected as a result of improper over-reduction in the empty correction (Wang 2006). Wang et al. (2000) attributed the phenomenon to instrument sensitivity of WAXD for the small number of crystallites in the early stage of crystallization. Indeed, improving the sensitivity of WAXD by four magnitudes, the difference between SAXS and WAXD disappeared at low temperatures as observed by Heeley et al. (2003). The remained difference at high temperatures can be associated to heterogeneous nucleation rather than homogeneous nucleation.

4.3.2 Crystal Growth

4.3.2.1 Secondary Nucleation Models

After primary crystal nucleation, crystals begin to grow. The crystal growth may be diffusion-controlled or interface-controlled. If crystal grows under a large supercooling in dilute polymer solutions, the growth rate is mainly controlled by a long-distance diffusion from the far-away bulk solution to the crystal surface. This is referred to the diffusion-controlled mechanism. The linear crystal growth rate is

$$v = \frac{dr}{dt} \propto t^{-1/2} \tag{4.51}$$

The diffusion-controlled mechanism means that the crystal size is not linearly dependent on the growth time (Holland and Lindenmeyer 1962). In the commonly practical cases, crystal growth is controlled by the process at the advancing surface of the crystal. This is referred to the interface-controlled mechanism. When controlled by this mechanism, the linear crystal growth rate is independent of time. The interface-controlled mechanism can be further separated into three categories, i.e., secondary nucleation growth, screw dislocation growth, and surface roughing growth. The secondary nucleation is prevailing in the description of the kinetics of lamellar polymer crystal growth (Flory and McIntyre 1955; Burnett and Mcdevit 1957; Wunderlich and Cormier 1966; Wunderlich et al. 1967).

In experiments, for example, by small-angle laser scattering (SALS) on small crystallites, or by polarized light microscope (PLM) on large crystallites, it is found that the growth rate of lamellar crystals is independent of time. This behavior implies the surface-controlled mechanism for crystal growth. Thickening at the growth front is also observed in many experiments (Wunderlich and Mielillo 1968; Abo El Maaty and Bassett 2005; Mullin and Hobbs 2011). So the crystal growth process can be treated as two steps at the wedge-shaped growth front, secondary nucleation occurs first, followed with instant thickening until thickness becomes larger than the minimum thickness for further growth of lamellar crystals. Secondary nucleation dominates the temperature dependence of the growth rate, and thickening provides the driving force for crystal growth. As will be introduced below, Lauritzen-Hoffman model and its developments dominated the present understanding of growth kinetics of lamellar crystals. The intramolecular nucleation model recently provided a promising progress.

The linear crystal growth rate of lamella by lateral-surface advancing can be treated as the competition result between advancing rate and melting rate of the growth front. Taking intramolecular secondary nucleation as the rate-determining steps for both crystal growth and melting, they need to overcome the same nucleation energy barriers from mutually opposite directions. It is the difference of the two energy barriers ΔG that determines the linear growth rate, as derived according to (4.44) by

$$\begin{aligned} v &= v_{\text{growth}} - v_{\text{melting}} = v_{\text{growth}}\left(1 - \frac{v_{\text{melting}}}{v_{\text{growth}}}\right) \\ &= v_{\text{growth}}\left(1 - \exp\left(-\frac{\Delta G}{kT_c}\right)\right) \end{aligned} \tag{4.52}$$

Assuming ΔG is very small as reflected by the excess lamellar thickness above the minimum, and replacing the exponential term by the first two terms of Maclaurin expansion $(\exp(x) = 1 + x + x^2/2 + x^3/6 + \cdots + x^i/i! + \cdots)$, we can get the linear growth rate as

$$v \approx v_{\text{growth}} \frac{\Delta G}{kT_c} = v_{\text{growth}} (l - l_{\min}) \frac{b^2 \Delta g}{kT_c} \tag{4.53}$$

In (4.53), the term $l - l_{\min}$ determines the net free energy for crystal growth, b is the average distance between stems inside the crystal, and Δg is the free energy change for melting of unit volume. Thus, we can treat v_{growth} in (4.53) as the free energy barrier for crystal growth and the rest part as the driving force for crystal growth (Ren et al. 2010). Under low temperatures, $l > l_{\min}$ and the crystal will grow, while under high temperatures, $l < l_{\min}$ and the crystal will melt. So temperature variation can lead to a continuous switching between growth and melting at the lateral growth front of lamellar crystals (Ren et al. 2010).

Lauritzen–Hoffman (LH) theory is still the most widely used theoretical model in the explanation of the growth kinetics of lamellar crystals. Figure 4.21 schematically shows the basic assumptions of folded stems at the growth front in the LH theory, without considering the thickening at the lateral growth front.

The LH theory holds four basic assumptions as listed below.

1. The growth front of polymer crystal is smooth. Secondary nucleation begins with a first stem deposited at the growth front, and follows with lateral spreading until reaching the lateral edges of the front substrate.
2. The chain-folded length l is constant during crystal growth. The width, thickness and number of stems are separately a_0, b_0 and v.
3. The number of growth fronts holding v stems N_v is in a steady-state distribution.
4. Each stem should go through an activation state before entering the crystal lattice where the fraction in each stem successfully entering the crystal lattice is ϕ.

If the number of grown stems is v, the change of Gibbs free energy is

$$\Delta G = 2b_0 l \sigma + 2(v-1)a_0 b_0 \sigma_e - v a_0 b_0 l \Delta g \tag{4.54}$$

Fig. 4.21 Schematic picture of folded stems at the growth front considered in the LH theory

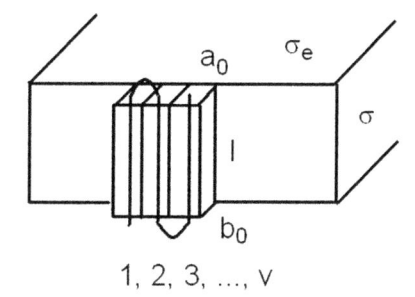

When $v = 1$, the growth rate for the first stem is

$$A_0 = \beta \exp\left(-\frac{2b_0 l \sigma - \phi a_0 b_0 l \Delta g}{kT}\right) \quad (4.55)$$

The melting rate of the first stem is

$$B_0 = \beta \exp\left(-\frac{(1-\phi) a_0 b_0 l \Delta g}{kT}\right) \quad (4.56)$$

If $v > 1$, the growth rate of the rest stems is

$$A = \beta \exp\left(-\frac{2a_0 b_0 \sigma_e - \phi a_0 b_0 l \Delta g}{kT}\right) \quad (4.57)$$

The melting rate of the rest stems is

$$B = B_0 \quad (4.58)$$

where β is a kinetic prefactor defined as

$$\beta = \frac{kT}{h} \exp\left(-\frac{\Delta E}{kT}\right) \quad (4.59)$$

By the assumption of a stable distribution of the growth fronts in the steady state, the growth flux of fronts is

$$S = N_0 \beta \exp\left[-(2b_0 \sigma - \phi a_0 b_0 \Delta g_f)\frac{l}{kT}\right]\left[1 - \exp\left(\frac{2a_0 b_0 \sigma_e - a_0 b_0 l \Delta g_f}{kT}\right)\right] \quad (4.60)$$

Different growth crystals contain different folded lengths. $S(l)$ is the crystal growth flux with folded length l. Thus, the average length of all the crystals is calculated as

$$\langle l \rangle = \frac{\int_{l_{\min}}^{\infty} lS(l)dl}{\int_{l_{\min}}^{\infty} S(l)dl} \quad (4.61)$$

The result of (4.61) is

$$\langle l \rangle = \frac{2\sigma_e}{\Delta g} + \frac{kT}{2b_0 \sigma} \times \frac{2 + (1 - 2\phi)a_0 \Delta g/(2\sigma)}{[1 - \phi a_0 \Delta g/(2\sigma)][1 + (1-\phi)a_0 \Delta g/(2\sigma)]} \quad (4.62)$$

where $2\sigma_e/\Delta g$ is the minimum length for steady growth. Equation (4.62) is used to predict the average thickness of lamellar crystals grown at different temperatures.

Besides an explanation of chain-folding lengths, the phenomenon of regime transitions can also be explained by the LH theory. If crystal growth is controlled by secondary nucleation, the temperature dependence of crystal growth rates can be dominated by

$$G = G_0 \exp\left(-\frac{U}{T - T_0}\right) \exp\left(-\frac{K_g}{T\Delta T}\right) \tag{4.63}$$

The first exponential term is attributed to short-range diffusion across the interface and the second term is attributed to secondary nucleation with the nucleation barrier proportional to the inverse supercooling. When temperature is high enough, crystal growth rate becomes mainly controlled by secondary nucleation. In this case, the curve of $lgG + U/(T - T_0)$ versus $-1/T/\Delta T$ can be divided into three linear regimes with the lowering of temperature, and the ratios of K_g among three regimes are

$$K_g(\mathrm{I}) : K_g(\mathrm{II}) : K_g(\mathrm{III}) = 2 : 1 : 2 \tag{4.64}$$

The above ratio is the so-called regime-transition phenomenon, which can be explained on the basis of secondary nucleation assumed in the LH theory. When crystallization takes place at a very high temperature, the rate of secondary nucleation i becomes the rate-determining step, and the followed surface spreading rate g is very large, so $i < g$. The secondary nucleation is supposed under the determination of the free energy barrier for depositing the first stem (Lauritzen and Hoffman 1960; Hoffman and Lauritzen 1961; Hoffman et al. 1976). Once the nucleus becomes stable, it will spread to the two lateral sides very fast with a layer thickness of b. The growth front with the width L is smooth until the next nucleus shows up. This temperature range is referred to Regime I and the crystal growth rate in this regime is

$$G_\mathrm{I} = ibL \tag{4.65}$$

Molecular simulations have reproduced regime-transition phenomena (Hu and Cai 2008). However, the growth front of Regime I is rather rough, favoring an alternative interpretation based on the intramolecular secondary nucleation model (Hu and Cai 2008).

When temperature is lower than Regime I, the rate of secondary nucleation becomes higher and $i \sim g$. Several nuclei will grow together and the growth front will no longer be smooth. This temperature range is referred to Regime II. The advancing rate is proportional to the square root of secondary nucleation rate i, as given by

$$G_\mathrm{II} = b(2gi)^{1/2} \tag{4.66}$$

If the temperature is further lower than Regime II, secondary nucleation rate becomes much larger than the spreading rate, $i > g$. In this case, several crystal layers will grow at the same time, each layer with several nuclei and the distance between two nuclei is L'. This temperature range is referred to Regime III and the crystal growth rate is

$$G_{III} = ibL' \tag{4.67}$$

In the LH theory, chain-folded length was assumed to be constant and thickening was omitted during crystal growth. Also the free energy barrier for secondary nucleation was assumed to be determined by the first stem, and no more stem was considered. These assumptions are not so reasonable and many other models (Armistead and Goldbeck-Wood 1992) have been proposed in order to reach a better understanding about the growth kinetics of lamellar crystals.

Wunderlich and Arakawa observed the layer structure of PE crystal under atmospheric pressures higher than 3 kbar and the crystallinity was almost 100 % (Prime and Wunderlich 1969). In the crystals, polymer chains were perpendicular to the lamella and the largest thickness was even larger than the molecular length, which indicated the existence of extended chains in the crystals (Olley and Bassett 1977). By further observation of the lamella growth front, Wunderlich found the wedge-shaped growth front and proposed a thickening-growth mechanism under high pressures (Wunderlich 1976). In his explanation, molecular nucleation or secondary nucleation firstly took place at the growth front, and later-on developed into extended-chain crystals by fast thickening. By observing the growth process of PE folded-chain crystal (FCC) and extended-chain crystal (ECC), Hikosaka developed the growth mechanism with chain-sliding diffusion for thickening, on the basis of the LH theory (Hikosaka 1987, 1990). In his equation of nucleation rate, the free energy barrier for short-range diffusion across the interface was also considered besides the free energy barrier of critical nucleus. The growth appears as two-dimensional, which holds both lateral and longitudinal growth of the chain stems. The two-dimensional nucleation growth mechanism can be used to explain the dependence of lamella thickness on supercooling near the triple point of high pressures (Hikosaka et al. 1995).

Wunderlich and Mehta put forward the concept of molecular nucleation (Wunderlich and Mehta 1974; Mehta and Wunderlich 1975; Wunderlich 1979; Cheng and Wunderlich 1986) in order to explain the molecular weight effect in crystallization of PE and some other polymers. At very high temperatures, the fraction of high molecular weights will crystallize first. It was proposed that each molecule entered the crystal with an additional nucleation barrier. Chain-folded secondary nucleus formed by molecules with enough lengths can only be stabilized over the critical size, while short-chain nucleus will be melted again, as illustrated in Fig. 4.22. The concept of molecular nucleation can be regarded as a patch on the LH model.

The free energy change of molecular nucleation at the growth front is

$$\Delta G = -abl\Delta g + 2bl\sigma + 2(n-1)ab\sigma_e + 2ab\sigma'_e \tag{4.68}$$

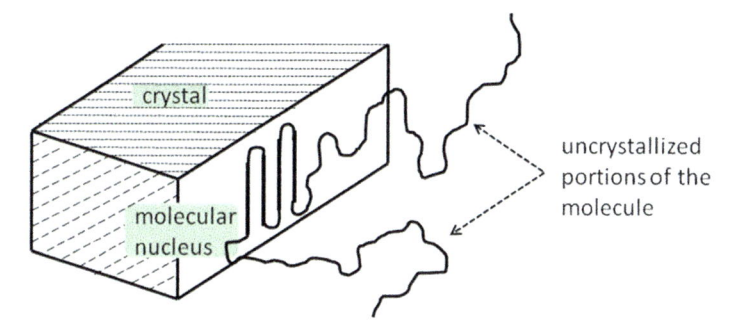

4 Theoretical Aspects of Polymer Crystallization

Fig. 4.22 Schematic picture of molecular nucleation (Wunderlich 2005)

σ_e is the surface free energy with chain cilia, thus the free energy change of critical nucleus is

$$\Delta G^* = \frac{4b\sigma\sigma_e}{\Delta g} + 2ab\sigma_e' \quad (4.69)$$

This critical free energy barrier is higher than normal secondary nucleation and thus the molecular nucleation can be the rate-determining step of crystal growth.

The intramolecular nucleation was then developed by supposing that all the secondary nucleation is mainly controlled by intramolecular nucleation (Hu et al. 2003b; Hu 2007) and the basic crystal lamella is resulted due to the preference of chain-folding in this unique style of secondary nucleation.

Assuming secondary nucleation of a single chain at the two-dimensional locally smooth growth front, the free energy change based on the classical nucleation theory is

$$\Delta F = \Delta f n + \sigma (N - n)^{1/2} \quad (4.70)$$

The free energy barrier for crystal nucleation is

$$\Delta F_c = \frac{\sigma^2}{4\Delta f} \quad (4.71)$$

And the equilibrium free energy barrier is

$$\Delta F_e = \frac{\sigma^2}{4\Delta f_e} \quad (4.72)$$

The free energy barrier for critical intramolecular nucleation under a certain temperature is independent of chain length, but not for the opposite direction, i.e., melting. Therefore, a critical molecular length exists for the equilibrium intramolecular melting at the growth front under each temperature. For a polymer sample with a polydispersity of molecular weights, only the fractions of chain lengths larger than the critical length can be stable and thus enter the crystal during secondary nucleation. This is the reason why molecular segregation occurs upon lamellar crystal growth. Molecular segregation can be observed only in polymer crystallization under very high temperatures.

Under a low temperature, the critical chain length for secondary nucleation may be much smaller than the chain length, and several events of intramolecular nucleation could happen along the same chains. If they occur in the same lamellar crystal, loops are formed; and if in different lamellar crystals, tie molecules are formed. The intramolecular nucleation model allows a statistical treatment on the semicrystalline texture.

The spreading right after the event of intramolecular nucleation may be stopped by the entanglements of long chains, collisions of nuclei at the same growth fronts, or the limited width at the growth front. Monte Carlo simulations (Hu et al. 2003c) have demonstrated that a single chain enters into the crystal growth front via several events of surface nucleation along the chain, as a result of limited growth-front sizes.

Besides molecular segregation (Hu 2005), many other phenomena unique for polymer crystal growth also favor the intramolecular nucleation model, such as co-crystallization of long and short chains (Cai et al. 2008; Jiang et al. 2015) as well as the interpretation of regime transitions (Hu and Cai 2008).

4.3.2.2 Other Non-nucleation Models

There are some models based on non-nucleation mechanisms for lamellar crystal growth. The (200) growth front of PE single crystals will become curvature when crystallization temperatures are very high, which could not be explained by the LH theory based on secondary nucleation growth on smooth surface. Also a pair of concave (110) surfaces were observed in the twin single crystal, which means the free energy barrier from side surfaces may not be the main problem for the advancing of the growth front (Sadler et al. 1986). Based on these observations, Sadler and Gilmer (1988) (SG) proposed the row model of continuous growth along the direction perpendicular to the growth front, as illustrated in Fig. 4.23.

In the SG model, crystal unit will be randomly added in or be removed from the growth front with a free energy change. On the one hand, the longer the growth stem, the larger the driving force for advancing the growth front. The driving force is proportional to the difference between the stem length and the critical length for a stable stem. On the other hand, the thicker the growth front, the longer time needed to reach the critical stem length because chain extending is pinned by the metastable chain-folding or loops. The possibility of stem lengths among all the conformations is only

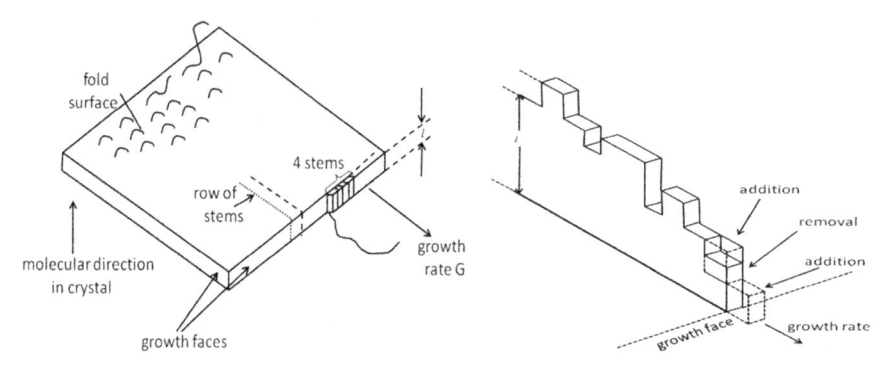

Fig. 4.23 Schematic pictures of the row model proposed by Sadler and Gilmer (1988)

$$P = e^{-kl} \tag{4.73}$$

Thus, the total growth rate is

$$G \propto e^{-kl}(l - l_{\min})\Delta g/(kT) \tag{4.74}$$

The stem lengths exhibit two opposite trends of contributions to the total growth rate, and the maximum of growth rate will be realized at a certain stem length, which explains the growth thickness of the lamellar crystals. The thickness can be calculated as

$$\langle l \rangle = l_{\min} + \frac{1}{k} \tag{4.75}$$

Like in the LH theory, the free energy barrier for crystal growth in the SG model is related directly with the lamellar thickness without considering further thickening after crystal growth. Also, the proposed thermal roughening may still be flat at the crystal edge, without the necessity of curvature at the growth surface.

In both LH and SG models, the growth front was supposed to directly reach the critical thickness in the growth process, and the thickening after growth was neglected, which appears not so reasonable. Keller and coworkers (Keller 1992; Keller et al. 1994) proposed a wedge-shaped growth-front model considering an obvious thickening in the crystal growth process of PE, as illustrated in Fig. 4.24.

In the wedge-shaped growth model, the melt may first grow into a mesophase (hexagonal phase) in the thinnest region of the growth front. The mesophase is stable because of large specific surface energy of small crystallites, which is referred as the finite-size effect. Then, the thin lamella thickens into the stable orthogonal phase, which decides the lamellar thickness. There is a triple point Q, as demonstrated in Fig. 4.25. If temperature is above the triple point temperature, the melt will grow into the orthogonal phase directly. If temperature is below the triple point, there will be mesophase.

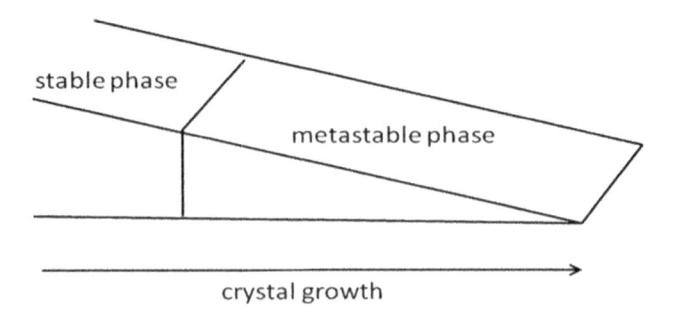

Fig. 4.24 Schematic picture of wedge-shaped growth front of PE lamella

Fig. 4.25 Schematic picture showing temperature against reciprocal thickness of polyethylene crystals

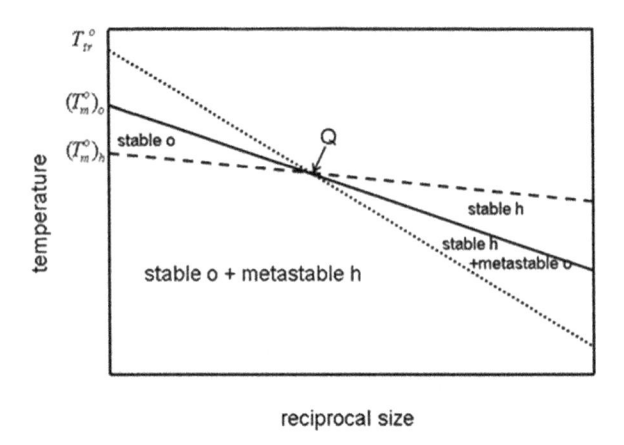

This model was later on expanded by Strobl (2000, 2005, 2006, 2009) to other polymers, in order to explain the experimental observations of his group, as illustrated in Fig. 4.26.

There is a linear relationship between the crystallization temperatures and the inverse lamella thicknesses, which is quite in accordance with Gibbs–Thomson equation. There is also a linear relationship between the melting temperatures and the inverse lamella thicknesses. Crossover of these two linear curves is considered to be the triple point of mesophase transition. Recently, the crossover was reproduced in the molecular simulations of lattice polymers, and the interpretation was updated to an uplimit of instant thickening at the lateral growth front of lamellar crystals (Jiang et al. 2016).

Allegra (Allegra 1977, 1980; Allegra and Meille 1999, 2005) proposed statistical thermodynamic theory for the mesophase of small crystallites or crystal cluster in the metastable disordered phase before crystallization. The cluster will first grow into a stable size and then joins into the crystal, as illustrated in Fig. 4.27. The thickness of lamella is decided by the cluster size. Zhang and Muthukumar (2007) performed simulations of clusters to form single crystals grown in dilute solutions, consistent with the experimental observations.

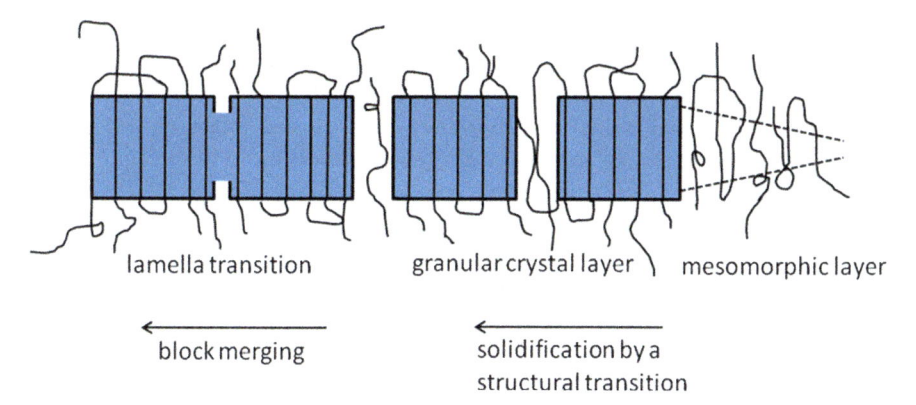

4 Theoretical Aspects of Polymer Crystallization

Fig. 4.26 Schematic picture of mesophase at the growth front in lamella (Strobl 2009)

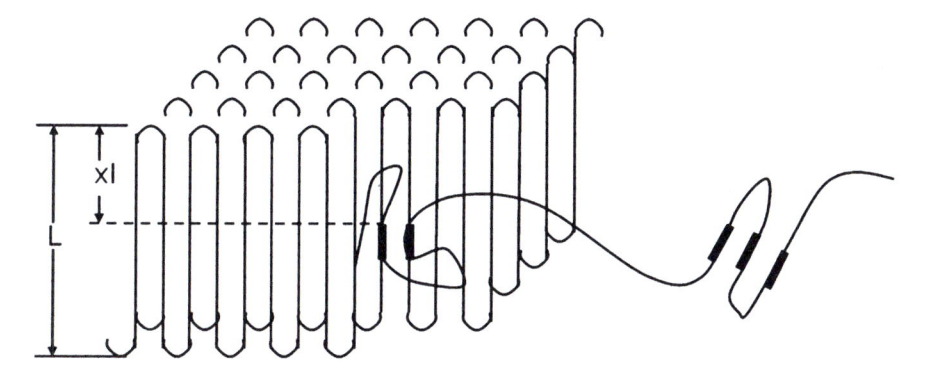

Fig. 4.27 Schematic picture of equilibrium-sized cluster growth model (Allegra and Meille 2005)

Muthukumar and coworkers (Welch and Muthukumar 2001; Muthukumar 2005) gave a thermodynamic explanation to the lamella thickness. He thought that the finite lamella thickness was a result of the largest thermodynamic stability of small crystallites. The crystalline chain will find the folded length related to the minimum free energy of the whole crystallite although there is a free energy barrier for the thickening of integer folding. The mechanism for crystal growth to select a limited lamellar thickness is recently addressed by the combination of secondary nucleation and instant thickening at the lateral growth front of lamellar crystals (Jiang et al. 2016).

4.3.3 Crystal Annealing

Annealing is a procedure to keep the temperature of a crystal body near its melting point so as to relax its inner stress and to remove defects. When annealing is used for polymer materials, it can make metastable polymer crystals more perfect and more stable. If the annealing temperature is low, it just makes the crystal more perfect by removing defects from crystal lattice. As a result, crystallinity is increased. If the annealing temperature is high, crystal thickening will happen and results in more stable crystals. If the annealing temperature is very high, even higher than the melting point of metastable crystals, crystals may melt and recrystallize into a more stable state unless the temperature is close to the equilibrium melting point of infinitely large crystals.

There are two mechanisms to explain lamellar thickening, one is *solid-chain sliding-diffusion* mechanism, and another is *melting-recrystallizaiton* mechanism. Peterlin (1963) proposed an activation energy barrier for sliding diffusion of folded chains in monolayer crystals, trying to explain why the folded length of polymer chains increases linearly with the logarithm of time. The sliding-diffusion mechanism was then developed into a more general theory by Sanchez and his collaborators (1973, 1974). Dreyfus and Keller (1970) proposed the fold-dislocation thickening model, in which the lamellar thickness can be doubly increased, as schematically demonstrated in Fig. 4.28. Another is the melting-recrystallization mechanism, which was firstly reviewed by Fischer (1969) and then introduced by Wunderlich (1976) in his famous book. This mechanism was confirmed by several experimental phenomena, such as decrease-then-increase of crystallinity in the annealing process (Matsuoka 1962).

The phenomenon that lamella thickness increases with the logarithm of time has been observed in many experiments (Fischer and Schmidt 1962; Wunderlich and Mielillo 1968). This continuous thickening of mobile high-molecular-weight polymer crystals was simulated by Monte Carlo simulations (Wang et al. 2012). Figure 4.29a demonstrates the wedge-shaped profile of the growth front resulted from the continuous thickening by chain-sliding diffusion at the growth front. Figure 4.29b provides three different growth-front profiles at different temperatures, which can be fitted into a logarithmic function of distances to the growth front. This function implies a logarithmic time dependence of crystal thickness. The prefactor in the function indicates an increasing thickening rate with increasing temperatures.

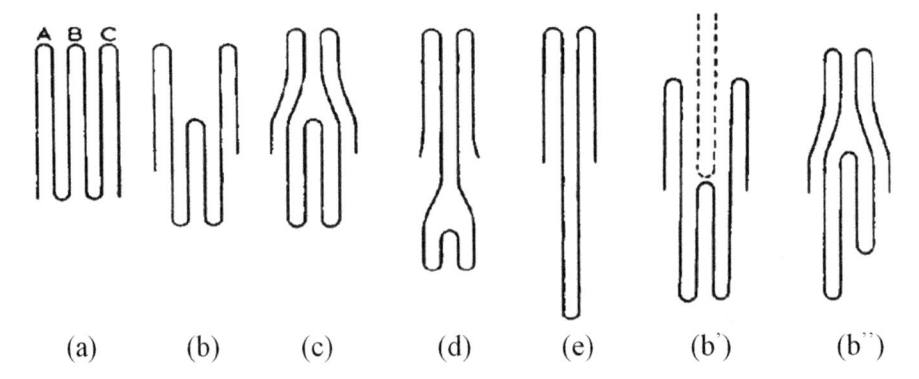

Fig. 4.28 Schematic picture of the fold-dislocation thickening model (Dreyfus and Keller 1970). (**a**) → (**e**) shows the thickening process of folded chain in a lamellar crystal. The space generated in this process should be filled by other stems (**b'**) or be discharged by merging of stems (**b''**)

4 Theoretical Aspects of Polymer Crystallization

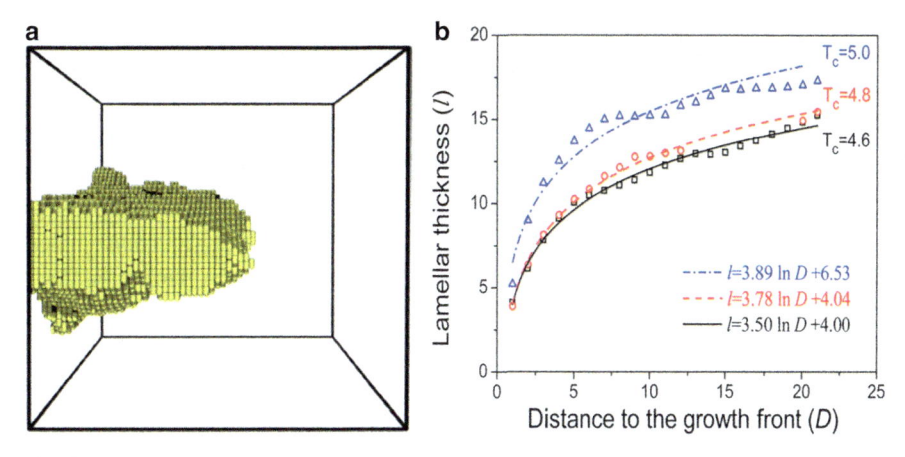

Fig. 4.29 (**a**) Snapshot of the wedge-shaped lamellar crystal of high-molecular-weight polymers grown for 35000MCc under $T = 4.6\,E_c/k_b$ in the 64^3 cubic lattice and polymer occupation density was 0.9375. Parameters were set as $E_p/E_c = 1$ for flexible chains and $E_f/E_c = 0.02$ to allow chain-sliding diffusion. The template was placed at the left end and only crystalline bonds were shown in *yellow color*. (**b**) Lamellar thickness as a function of distance to the growth front at three different temperatures $T = 4.6$, 4.8, and 5.0, respectively. The equations shown in the picture are the calculated logarithmic functions of distance D to the growth front (Wang et al. 2012)

This continuous thickening is obviously controlled by the chain-sliding diffusion in the crystals. The similar thickness profiles at the edges of lamellar polymer crystals have been observed in experiments (Reiter 2014).

The logarithmic-time dependence of crystal thickness can be deduced easily. By assuming a frictional barrier (ΔE_s) for chain-sliding diffusion proportional to the lamella thickness (l), thus the thickening rate of monolayer lamellar crystal under a certain temperature is

$$\frac{\mathrm{d}l}{\mathrm{d}t} \propto b e^{-al/k_b T} \quad (4.76)$$

Equation (4.76) can be solved by

$$l = c \ln t + d \quad (4.77)$$

In the above equations, a, b, c, and d are the coefficients.

4.4 Summary

We made a brief introduction about our current theoretical models of thermodynamics and kinetics of polymer crystallization. We first introduced basic thermodynamic concepts, including the melting point, the phase diagram, the metastable state, and the mesophase. The mean-field statistical thermodynamics based on a classic lattice model of polymer solutions, and its predictions on the melting points and phase diagrams were emphasized. Those molecular factors governing the melting points were also introduced. We then introduced crystal nucleation, crystal growth, and crystal annealing as the three basic stages of polymer crystallization. The classical nucleation theory as well as some recent ideas about primary nucleation were emphasized. On crystal growth, we introduced the secondary nucleation model, in particular, the well-known Lauritzen–Hoffman theory. The recently developed intramolecular nucleation model can be regarded as an updated version of secondary nucleation models. Some other models based on non-nucleation ideas were also introduced in a balanced way. The kinetics of crystal thickening, which is usually dominating crystal annealing, was also introduced under three typical circumstances.

Due to the length limitation, we have to skip many other theoretical aspects of polymer crystallization. We hope that our selected content above has already been strong enough to demonstrate the power of theoretical approaches, as a complemental to experimental approaches to gain a better understanding of the complicated polymer crystal morphologies.

References

Abo El Maaty MI, Bassett MD (2005) Evidence for isothermal lamellar thickening at and behind the growth front as polyethylene crystallizes from the melt. Polymer 46(20):8682–8688

Allegra G (1977) Chain folding and polymer crystallization: a statistical–mechanical approach. J Chem Phys 66(12):5453–5463

Allegra G (1980) Polymer crystallization: the bundle model. Ferroelectrics 30(1):195–211

Allegra G, Meille SV (1999) The bundle theory for polymer crystallisation. Phys Chem Chem Phys 1(22):5179–5188

Allegra G, Meille SV (2005) Pre-crystalline, high-entropy aggregates: a role in polymer crystallization? Adv Polym Sci 191:87–135

Arlie JP, Spegt PA, Skoulios AE (1966) Etude de la cristallisation des polymères I. Structure lamellaire de polyoxyéthylènes de faible masse moléculaire. Die Makromolekulare Chemie 99(1):160–174

Arlie JP, Spegt PA, Skoulios AE (1967) Etude de la cristallisation des polymères. II. Structure lamellaire et repliement des chaines du polyoxyéthylène. Die Makromolekulare Chemie 104(1):212–229

Armistead K, Goldbeck-Wood G (1992) Polymer crystallization theories. Adv Polym Sci 100:219–312

Becker R, Döring W (1935) Kinetische behandlung der keimbildung in übersättigten dämpfen. Ann Phys 416(8):719–752

Blundell D, Keller A, Kovacs A (1966) A new self-nucleation phenomenon and its application to the growing of polymer crystals from solution. J Polym Sci B Polym Lett 4(7):481–486

Burnett BB, Mcdevit W (1957) Kinetics of spherulite growth in high polymers. J Appl Phys 28(10):1101–1105

Cai T, Ma Y, Yin P, Hu W (2008) Understanding the growth rates of polymer co-crystallization in the binary mixtures of different chain lengths. J Phys Chem B 112(25):7370–7376

Chang TS (1939) Statistical theory of absorption of double molecules. Proc R Soc Ser A 169(939):512–531

Cheng SZ (2008) Phase transitions in polymers: the role of metastable states. Elsevier Science, Oxford

Cheng SZ, Wunderlich B (1986) Molecular segregation and nucleation of poly (ethylene oxide) crystallized from the melt. I. Calorimetric study. J Polym Sci B 24(3):577–594

Coleman BD (1958) On the properties of polymers with random stereo-sequences. J Polym Sci 31(122):155–164

Colson JP, Eby RK (1966) Melting temperatures of copolymers. J Appl Phys 37(9):3511–3514

Dreyfus P, Keller A (1970) A simple chain refolding scheme for the annealing behavior of polymer crystals. J Polym Sci B Polym Lett 8(4):253–258

Fischer E (1969) Zusammenhänge zwischen der Kolloidstruktur kristalliner Hochpolymerer und ihrem Schmelz-und Rekristallisationsverhalten. Kolloid-Zeitschrift und Zeitschrift für Polymere 231(1-2):458–503

Fischer E, Schmidt G (1962) Long periods in drawn polyethylene. Angew Chem Int Ed Engl 1(9):488–499

Flory PJ (1941) Thermodynamics of high polymer solutions. J Chem Phys 9(8):660

Flory PJ (1942) Thermodynamics of high polymer solutions. J Chem Phys 10(1):51–61

Flory PJ (1953) Principles of polymer chemistry. Cornell University Press, Ithaca

Flory PJ (1955) Theory of crystallization in copolymers. Trans Faraday Soc 51:848–857

Flory PJ (1956) Statistical thermodynamics of semi-flexible chain molecules. Proc R Soc Lond A Math Phys Sci 234(1196):60–73

Flory PJ (1962) On the morphology of the crystalline state in polymers. J Am Chem Soc 84(15):2857–2867

Flory P, McIntyre A (1955) Mechanism of crystallization in polymers. J Polym Sci 18(90):592–594

Flory P, Vrij A (1963) Melting points of linear-chain homologs. The normal paraffin hydrocarbons. J Am Chem Soc 85(22):3548–3553

Gee RH, Lacevic N, Fried LE (2005) Atomistic simulations of spinodal phase separation preceding polymer crystallization. Nat Mater 5(1):39–43

Gupta AM, Edwards SF (1993) Mean-field theory of phase transitions in liquid-crystalline polymers. J Chem Phys 98(2):1588–1596

Guth E, Mark H (1934) Zur innermolekularen statistik, insbesondere bei Ketten-molekiilen I. Monatshefte für Chemie/Chemical Monthly 65(1):93–121

Heeley E, Maidens A, Olmsted P, Bras W, Dolbnya I, Fairclough J, Terrill N, Ryan A (2003) Early stages of crystallization in isotactic polypropylene. Macromolecules 36(10):3656–3665

Hermann K, Gerngross O (1932) Die elastizitat des kautschuks. Kautschuk 8:181

Hermann K, Gerngross O, Abitz W (1930) Zur rontgenographischen strukturforschung des gelatinemicells. Zeitschrift für Physikalische Chemie B 10:371–394

Hikosaka M (1987) Unified theory of nucleation of folded-chain crystals and extended-chain crystals of linear-chain polymers. Polymer 28(8):1257–1264

Hikosaka M (1990) Unified theory of nucleation of folded-chain crystals (FCCs) and extended-chain crystals (ECCs) of linear-chain polymers: 2. Origin of FCC and ECC. Polymer 31(3):458–468

Hikosaka M, Okada H, Toda A, Rastogi S, Keller A (1995) Dependence of the lamellar thickness of an extended-chain single crystal of polyethylene on the degree of supercooling and the pressure. J Chem Soc Faraday Trans 91(16):2573–2579

Hoffman JD (1983) Regime III crystallization in melt-crystallized polymers: the variable cluster model of chain folding. Polymer 24(1):3–26

Hoffman JD, Lauritzen J (1961) Crystallization of bulk polymers with chain folding: theory of growth of lamellar spherulites. J Res Natl Bur Stand A65(4):297–336

Hoffman JD, Miller RL (1997) Kinetic of crystallization from the melt and chain folding in polyethylene fractions revisited: theory and experiment. Polymer 38(13):3151–3212

Hoffman JD, Weeks JJ (1965) X-Ray study of isothermal thickening of lamellae in bulk polyethylene at the crystallization temperature. J Chem Phys 42(12):4301–4302

Hoffman JD, Davis GT, Lauritzen JI Jr (1976) The rate of crystallization of linear polymers with chain folding. In: Hannay NB (ed) Treatise on solid state chemistry, vol 3. Plenum Press, New York

Holland V, Lindenmeyer P (1962) Morphology and crystal growth rate of polyethylene crystalline complexes. J Polym Sci 57(165):589–608

Hu W (1998) Structural transformation in the collapse transition of the single flexible homopolymer model. J Chem Phys 109(9):3686–3690

Hu W (2000) The melting point of chain polymers. J Chem Phys 113(9):3901–3908

Hu W (2005) Molecular segregation in polymer melt crystallization: simulation evidence and unified-scheme interpretation. Macromolecules 38(21):8712–8718

Hu W (2007) Intramolecular crystal nucleation. In: Reiter G, Strobl GR (eds) Lecture notes in physics: progress in understanding of polymer crystallization. Springer, Berlin, pp 47–63

Hu W (2013) Polymer physics: a molecular approach. Springer, Wien

Hu W, Cai T (2008) Regime transitions of polymer crystal growth rates: molecular simulations and interpretation beyond Lauritzen-Hoffman model. Macromolecules 41(6):2049–2061

Hu W, Frenkel D (2004) Effect of metastable liquid-liquid demixing on the morphology of nucleated polymer crystals. Macromolecules 37(12):4336–4338

Hu W, Frenkel D (2005) Polymer crystallization driven by anisotropic interactions. Adv Polym Sci 191:1–35

Hu W, Frenkel D, Mathot VB (2003a) Lattice-model study of the thermodynamic interplay of polymer crystallization and liquid–liquid demixing. J Chem Phys 118(22):10343–10348

Hu W, Frenkel D, Mathot VB (2003b) Intramolecular nucleation model for polymer crystallization. Macromolecules 36(21):8178–8183

Hu W, Frenkel D, Mathot VB (2003c) Sectorization of a lamellar polymer crystal studied by dynamic Monte Carlo simulations. Macromolecules 36(3):549–552

Huggins ML (1942) Thermodynamic properties of solutions of long-chain compounds. Ann N Y Acad Sci 43:1–32

Imai M, Kaji K, Kanaya T (1993) Orientation fluctuations of poly (ethylene terephthalate) during the induction period of crystallization. Phys Rev Lett 71(25):4162–4165

Imai M, Kaji K, Kanaya T (1994) Structural formation of poly (ethylene terephthalate) during the induction period of crystallization. 3. Evolution of density fluctuations to lamellar crystal. Macromolecules 27(24):7103–7108

Jähnig F (1979) Molecular theory of lipid membrane order. J Chem Phys 70(7):3279–3290

Jiang X, Li T, Hu W (2015) Understanding the growth rates of polymer co-crystallization in the binary mixtures of different chain lengths: Revisited. J Phys Chem B 119(30):9975–9981

Jiang X, Reiter G, Hu W (2016) How chain-folding crystal growth determines thermodynamic stability of polymer crystals. J Phys Chem B 120(3):566–571

Keller A (1957) A note on single crystals in polymers: evidence for a folded chain configuration. Philos Mag 2(21):1171–1175

Keller A (1992) Morphology of polymers. Pure Appl Chem 64(2):193–204

Keller A, Hikosaka M, Rastogi S, Toda A, Barham P, Goldbeck-Wood G, Windle A, Thomas E, Bassett D (1994) The size factor in phase transitions: its role in polymer crystal formation and wider implications [and discussion]. Philos Trans R Soc Lond Ser A Phys Eng Sci 348 (1686):3–17

Kelton K (1991) Crystal nucleation in liquids and glasses. Solid State Phys 45:75–177

Khokhlov A, Semenov A (1985) On the theory of liquid-crystalline ordering of polymer chains with limited flexibility. J Stat Phys 38(1–2):161–182

Kraack H, Sirota E, Deutsch M (2000) Measurements of homogeneous nucleation in normal-alkanes. J Chem Phys 112(15):6873–6885

Kuhn W (1934) Über die gestalt fadenförmiger moleküle in lösungen. Kolloid-Zeitschrift 68 (1):2–15

Lauritzen JI, Hoffman JD (1960) Theory of formation of polymer crystals with folded chains in dilute solution. J Res Natl Bur Stand A Phys Chem 64(1):73–102

Lekkerkerker H, Vroege G (1993) Lyotropic colloidal and macromolecular liquid crystals. Philos Trans R Soc Lond Ser A Phys Eng Sci 344(1672):419–440

Ma Y, Hu W, Wang H (2007) Polymer immiscibility enhanced by thermal fluctuations toward crystalline order. Phys Rev E 76(3):031801

Ma Y, Zha L, Hu W, Reiter G, Han CC (2008) Crystal nucleation enhanced at the diffuse interface of immiscible polymer blends. Phys Rev E 77(6):061801

Maier W, Saupe A (1959) A simple molecular statistical theory of the nematic crystalline-liquid phase. Zeitschrift für Naturforschung 14:882–889

Mandelkern L (1964) Crystallization of polymers. McGraw-Hill, New York

Matsuoka S (1962) The effect of pressure and temperature on the specific volume of polyethylene. J Polym Sci 57(165):569–588

Mehta A, Wunderlich B (1975) A study of molecular fractionation during the crystallization of polymers. Colloid Polym Sci 253(3):193–205

Metropolis N, Rosenbluth AW, Rosenbluth MN, Teller AH, Teller E (1953) Equation of state calculations by fast computing machines. J Chem Phys 21(6):1087–1092

Meyer K (1939) The compound entropies of systems with long-chain compounds and their statistical explanation. Zeitschrift für Physikalische Chemie B 44:383–391

Milner ST (2011) Polymer crystal-melt interfaces and nucleation in polyethylene. Soft Matter 7:2909–2917

Mullin N, Hobbs JK (2011) Direct imaging of polyethylene films at single-chain resolution with torsional tapping atomic force microscopy. Phys Rev Lett 107(19):197801

Muthukumar M (2005) Modeling polymer crystallization. Adv Polym Sci 191:241–274

Nie Y, Gao H, Ma Y, Hu Z, Reiter G, Hu W (2013) Competition of crystal nucleation to fabricate the oriented semi-crystalline polymers. Polymer 54(13):3402–3407

Olley R, Bassett D (1977) Molecular conformations in polyethylene after recrystallization or annealing at high pressures. J Polym Sci Polym Phys Ed 15(6):1011–1027

Olmsted PD, Poon WC, Mcleish T, Terrill N, Ryan A (1998) Spinodal-assisted crystallization in polymer melts. Phys Rev Lett 81(2):373–376

Onsager L (1949) The effects of shape on the interaction of colloidal particles. Ann N Y Acad Sci 51(4):627–659

Organ SJ, Keller A (1987) The onset of chain folding in ultralong n-alkanes: an electron microscopic study of solution-grown crystals. J Polym Sci B 25(12):2409–2430

Peterlin A (1963) Thickening of polymer single crystals during annealing. J Polym Sci B Polym Lett 1(6):279–284

Prasad A, Mandelkern L (1989) Equilibrium dissolution temperature of low molecular weight polyethylene fractions in dilute solution. Macromolecules 22(12):914–920

Prigogine I (1957) The molecular theory of solution. North-Holland, Amsterdam

Prime RB, Wunderlich B (1969) Extended-chain crystals. III. Size distribution of polyethylene crystals grown under elevated pressure. J Polym Sci A2 Polym Phys 7(12):2061–2072

Reiter G (2014) Some unique features of polymer crystallization. Chem Soc Rev 43:2055–2065

Ren Y, Ma A, Li J, Jiang X, Ma Y, Toda A, Hu W (2010) Melting of polymer single crystals studied by dynamic Monte Carlo simulations. Eur Phys J E 33(3):189–202

Ronca G, Yoon D (1982) Theory of nematic systems of semiflexible polymers. I. High molecular weight limit. J Chem Phys 76(6):3295–3299

Ronca G, Yoon D (1984) Theory of nematic systems of semiflexible polymers. II. Chains of finite length in the bulk. J Chem Phys 80(2):925–929

Sadler DM, Gilmer GH (1988) Selection of lamellar thickness in polymer crystal growth: a rate-theory model. Phys Rev B 38(8):5684–5693

Sadler DM, Barber M, Lark G, Hill MJ (1986) Twin morphology: 2. Measurements of the enhancement in growth due to re-entrant corners. Polymer 27(1):25–33

Sanchez I, Eby R (1975) Thermodynamics and crystallization of random copolymers. Macromolecules 8(5):638–641

Sanchez I, Colson J, Eby R (1973) Theory and observations of polymer crystal thickening. J Appl Phys 44(10):4332–4339

Sanchez I, Peterlin A, Eby R, Mccrakin F (1974) Theory of polymer crystal thickening during annealing. J Appl Phys 45(10):4216–4219

Sirota E (2000) Supercooling and transient phase induced nucleation in n-alkane solutions. J Chem Phys 112(1):492–500

Sirota E, Herhold A (1999) Transient phase-induced nucleation. Science 283(5401):529–532

Spells SJ, Sadler DM, Keller A (1980) Chain trajectory in solution grown polyethylene crystals: correlation between infra-red spectroscopy and small-angle neutron scattering. Polymer 21 (10):1121–1128

Strobl G (2000) From the melt via mesomorphic and granular crystalline layers to lamellar crystallites: a major route followed in polymer crystallization? Eur Phys J E 3(2):165–183

Strobl G (2005) A thermodynamic multiphase scheme treating polymer crystallization and melting. Eur Phys J E 18(3):295–309

Strobl G (2006) Crystallization and melting of bulk polymers: new observations, conclusions and a thermodynamic scheme. Prog Polym Sci 31(4):398–442

Strobl G (2009) Colloquium: laws controlling crystallization and melting in bulk polymers. Rev Mod Phys 81(3):1287–1300

ten Bosch A, Maissa AP, Sixou P (1983a) Molecular model for nematic polymers in liquid crystal solvents. J Chem Phys 79(7):3462–3466

ten Bosch A, Maissa AP, Sixou P (1983b) A Landau-de Gennes theory of nematic polymers. J Phys Lett 44(3):105–111

Terrill NJ, Fairclough PA, Towns-Andrews E, Komanschek BU, Young RJ, Ryan AJ (1998) Density fluctuations: the nucleation event in isotactic polypropylene crystallization. Polymer 39(11):2381–2385

Turnbull D, Fisher JC (1949) Rate of nucleation in condensed systems. J Chem Phys 17(1):71–73

Ungar G, Stejny J, Keller A, Bidd I, Whiting M (1985) The crystallization of ultralong normal paraffins: the onset of chain folding. Science 229(4711):386–389

Volmer M, Weber A (1926) Nucleus formation in supersaturated systems. Zeitschrift für Physikalische Chemie 119:277–301

Wang H (2006) SANS study of the early stages of crystallization in polyethylene solutions. Polymer 47(14):4897–4900

Wang ZG, Hsiao BS, Sirota EB, Agarwal P, Srinivas S (2000) Probing the early stages of melt crystallization in polypropylene by simultaneous small-and wide-angle X-ray scattering and laser light scattering. Macromolecules 33(3):978–989

Wang M, Gao H, Zha L, Chen EQ, Hu W (2012) Systematic kinetic analysis on monolayer lamellar crystal thickening via chain-sliding diffusion of polymers. Macromolecules 46 (1):164–171

Welch P, Muthukumar M (2001) Molecular mechanisms of polymer crystallization from solution. Phys Rev Lett 87(21):218302

Wunderlich B (1976) Macromolecular physics. 2. Crystal nucleation, growth, annealing. Academic, New York

Wunderlich B (1979) Molecular nucleation and segregation. Faraday Discuss Chem Soc 68:239–243

Wunderlich B (2003) Reversible crystallization and the rigid–amorphous phase in semicrystalline macromolecules. Prog Polym Sci 28(3):383–450

Wunderlich B (2005) Thermal analysis of polymeric materials. Springer, Berlin

Wunderlich B, Cormier C (1966) Seeding of supercooled polyethylene with extended chain crystals. J Phys Chem 70(6):1844–1849

Wunderlich B, Mehta A (1974) Macromolecular nucleation. J Polym Sci Polym Phys Ed 12 (2):255–263

Wunderlich B, Mielillo L (1968) Morphology and growth of extended chain crystals of polyethylene. Die Makromolekulare Chemie 118(1):250–264

Wunderlich B, Melillo L, Cormier C, Davidson T, Snyder G (1967) Surface melting and crystallization of polyethylene. J Macromol Sci B Phys 1(3):485–516

Zachmann H (1967) Der einfluss der konfigurationsentropie auf das kristallisations und schmelzverhalten von hochpolymeren stoffen. Kolloid-Zeitschrift und Zeitschrift für Polymere 216–217(1):180–191

Zachmann H (1969) Statistische thermodynamik des kristallisierens und schmelzens von hochpolymeren stoffen. Colloid Polym Sci 231(1):504–534

Zhang J, Muthukumar M (2007) Monte Carlo simulations of single crystals from polymer solutions. J Chem Phys 126(23):234904–234921

Chapter 5
Controlling Morphology Using Low Molar Mass Nucleators

Geoffrey R. Mitchell, Supatra Wangsoub, Aurora Nogales, Fred J. Davis, and Robert H. Olley

5.1 Introduction

Crystallisation is a hugely important process in physical sciences and is crucial to many areas of, for example, chemistry, physics, biochemistry, metallurgy and geology. The process is typically associated with solidification, for example, in the purification of solids from a heated saturated solution familiar to all chemistry undergraduates. Crystalline solids are also often the end result of cooling liquids, or in some cases gases, but in order to form require nucleation, in the absence of nucleation supercooling of liquids well below the melting point is possible (Cavagna 2009). The quality of crystals, as gauged by size and levels of order, is highly variable, and will depend on factors such as material purity and the rate of cooling; rapid cooling may result in poor crystallisation, or even the formation of amorphous materials with no long-range order. In geological systems rates of cooling may vary over many orders of magnitude, for example, obsidian is a largely amorphous material produced when lava is rapidly cooled (Tuffen et al. 2013),

---

G.R. Mitchell (✉)
Centre for Rapid and Sustainable Product Development, Institute Polytechnic of Leiria, Marinha Grande, Portugal
e-mail: geoffrey.mitchell@ipleiria.pt

S. Wangsoub
Department of Chemistry, Naresuan University, Phitsanulok, Thailand

A. Nogales
Instituto de Estructura de la Materia, CSIC, Madrid, Spain

F.J. Davis
Department of Chemistry, The University of Reading, Whiteknights, Reading RG6 6AD, UK

R.H. Olley
Electron Microscopy Laboratory, The University of Reading, Whiteknights, Reading RG6 6AF, UK

Fig. 5.1 Giant gypsum crystals in the Cueva de los Cristales in Chihuahua, Mexico, the scale is indicated by the person (photo A. Van Driessche)

while the gypsum crystals found in the Cueva de los Cristales in Chihuahua, Mexico can reach 10 m in length (Fig. 5.1) and are formed over hundreds of thousands of years. In this latter case, the formation of such large spectacular structures as shown in Fig. 5.1 can only be explained by a low nucleation rate (García-Ruiz et al. 2007; Van Driessche et al. 2011).

Crystallisation from solution is a core technology in major areas of the chemical industry and it is the critical initial step in many scientific programmes including structural studies of proteins and pharmaceuticals. The process of nucleation by which embryonic crystals form within a supersaturated solution remains largely unexplained. In classical nucleation theory, the volume excess free energy of the nuclei at a critical radius balances the surface excess free energy; nuclei larger than this critical radius grow, and smaller entities dissolve back into the solution. Recently, other pathways of transformation have been identified theoretically in which structure and density fluctuations are separated (Kashchiev et al. 2005). In terms of controlling morphology in polymers, nucleation of polymer crystals plays an important part in the production of several commodity plastics, giving the potential for control of the polymer microstructure and therefore the subsequent properties (Bassett 2006). Nucleating agents are thus of considerable commercial and scientific interest. One well-known example of a nucleating agent is 1,3:2,4-dibenzyline sorbitol (I) (DBS) the structure of which is shown in I; this is extensively used as a clarification agent with polypropylene. Its use leads to transparent materials with improved mechanical properties (Zweifel 2001). This chapter focuses on the use of nucleating agents to control the structure and morphology of polymers and focuses particularly on DBS and its derivatives (Nogales and Mitchell 2005 Nogales et al 2003b; Nogales et al. 2016). In particular the work describes the self-assembly of such materials to form a template which can direct the crystallisation of polymers such as poly(ε-caprolactone) (Siripitayananon 2004).

I

5.2 Organic Gelators

In general, gels are viscoelastic solid-like materials comprising an elastic cross-linked network and a solvent, which is the major component. The solid-like appearance of a gel is a result of the entrapment and adhesion of the liquid in the large surface area solid 3D matrix. DBS is an example of a low molar mass organic gelator (Terech and Weiss 1997) whose structure is shown in I; that is a gel derived from low molar mass compounds formed through self-aggregation of the small gelator molecules to produce entangled self-assembled fibrils (Smith 2009; Sangeetha and Maitra 2005). In the case of DBS typically these fibrils are of the order of about 10 nm in diameter and at sufficiently high DBS concentrations (generally less than 2 wt% depending on factors such as temperature and matrix polarity), the nanoscale fibrils form a three-dimensional network that promotes physical gelation in a wide variety of organic solvents and polymers.

The structures and properties of DBS-organogels formed in different solvents are well studied by Yamasaki and Tsutsumi (Yamasaki and Tsutsumi 1994, 1995; Yamasaki et al. 1995; Watase and Itagaki 1998). All of those studies show that DBS builds up helically twisted fibres in low viscosity solvents. This is accompanied by a tremendous change in the rheological properties of these materials, as self-assembly leads to gelation. This gelation is based on physical interaction between the molecules. Gelation depends on various factors, such as gelator concentration, temperature and solvent polarity. It has also been shown (Yamasaki and Tsutsumi 1995) that the DL-racemate is not able to self-aggregate; thus, chirality seems important in the self-organisation of DBS (though it might be expected that the L-enantiomorph would assemble in a similar form). Wangsoub et al. (2016a, b) have reported the properties and structure of gels formed from alkanes and dibenzylidene sorbitol, and the results are relevant in part to the use of DBS with polyethylene.

Gels of low molecular mass compounds are usually prepared by heating the gelator in an appropriate solvent and cooling the resulting isotropic supersaturated solution to room temperature. When the hot solution is cooled, the molecules start to condense and three situations are possible Fig. 5.2: (1) a highly ordered aggregation giving rise to crystals i.e. crystallisation, (2) a random aggregation resulting

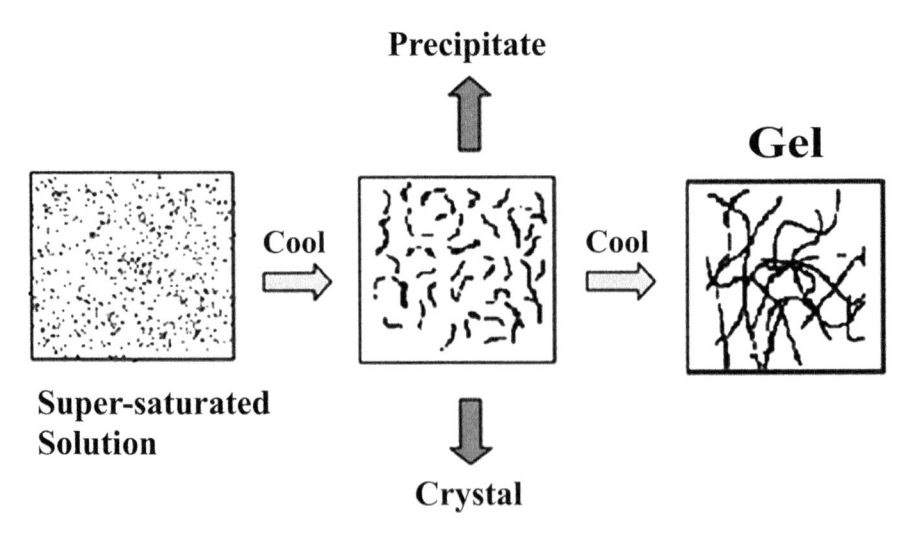

Fig. 5.2 Schematic representation of aggregation modes (redrawn after Sangeetha and Maitra 2005)

in an amorphous precipitate, and (3) and aggregation process intermediate between these two yielding a gel (1). The process of gelation involves self-association of the gelator molecules to form long, polymer-like fibrous aggregates, which get entangled during the aggregation process forming a matrix that traps the solvent mainly by surface tension. This process prevents the flow of solvent under gravity and the mass appears like a solid. At the microscopic level, the structures and morphologies of molecular gels have been investigated by conventional imaging techniques such as SEM, TEM and AFM. At the nanoscale, X-ray diffraction, small angle neutron scattering and X-ray scattering (SANS, SAXS) are required to elucidate the structures(Grubb et al 1975).

5.3 Synthesis of Sorbitol Derivatives

D-Sorbitol, or D-glucitol, is a sugar alcohol that is produced commercially by the reduction of glucose, but is also found naturally in many fruits. It finds many uses as an ingredient in the food industry and as such is designated E420 (Hanssen and Marsden 1984), though its use has been tainted with problems such as diarrhoea and flatulence. The dibenzylidene derivative is the main product of treatment with benzaldehyde, though mono and trisubstituted by-products may also be formed. Degradation and other studies confirmed that it is the 1,3:2,4 diacetal that is formed (Okesola et al. 2015). A range of other derivatives can be readily formed by the acid catalysed reaction of sorbitol with the appropriate substituted aldehyde (e.g. benzaldehyde, benzaldehyde d_6 or 4-chlorobenzaldehyde) in cyclohexane

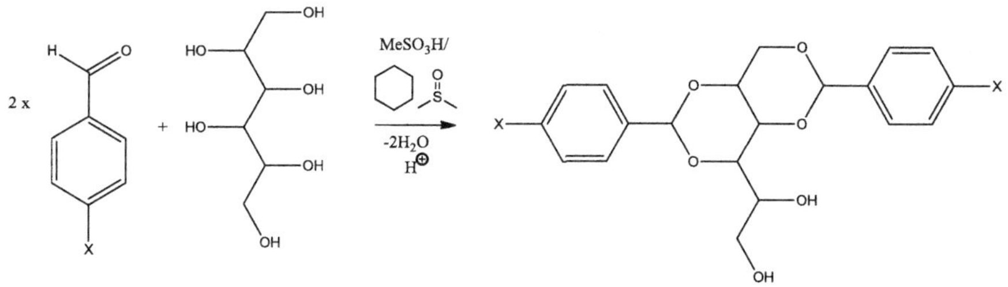

5 Controlling Morphology Using Low Molar Mass Nucleators

Scheme 5.1 Synthesis of Sorbitol derivatives

(Wangsoub et al. 2008). A typical procedure that for 1,3:2,4-di(4-chlorobenzylidene) sorbitol (Scheme 5.1) was as follows. An aqueous solution of D-sorbitol (0.1 mol, 70 % w/v) was placed into a round bottom flask equipped with Dean-Stark trap and condenser. To this was added 4-chlorobenzaldehyde (0.2 mol) and methanesulfonic acid (1 mL), cyclohexane (200 mL) and dimethylsulfoxide (6 mL). The mixture was heated to reflux with constant stirring. The mixture of cyclohexane and water was condensed and separated in the Dean-Stark arrangement. Once no further water was produced the reaction was stopped, and following neutralisation with triethylamine the white precipitate collected. Following purification by washing with dioxane and water 1,3:2,4-di(4-chlorobenzylidene) sorbitol was obtained as a white powder. The product was greater than 98 % pure by ^{1}H NMR, with a yield of 19.7 g, (46%) and a melting point of 187 °C.

5.4 DBS in Polymers

DBS and its derivatives can be easily dispersed in polymers using either melt mixing or mixing using a common solvent and drying. For the results shown here, the latter approach was used due to the small volume of material required. In this work we used the shear flow cell (Nogales et al. 2004) shown in Chap. 3 (Fig. 3.22) and for this, samples were prepared by co-solvent mixing, for example, with PCL we employed butanone, drying and then melt pressing into discs.

Figure 5.3 shows the phase behavior of the Cl-DBS/PCL system (Wangsoub et al. 2008). The line for the crystallisation temperature shows that the addition of the Cl-DBS increases the crystallisation temperature by ~10 °C due to the nucleating effect of the DBS derivative. This effect saturates at 1 % with Cl-DBS/PCL in contrast to ~3 % for DBS/PCL (Wangsoub et al. 2005). The upper curve shows the liquidus line separating the high temperature homogenous melt from the lower temperature 2-phase state of molten PCL and crystalline Cl-DBS fibrils. The liquidus line is shifted upwards by ~20 °C from the equivalent line for the DBS/PCL system (Wangsoub et al. 2005) indicating a reduced solubility for Cl-DBS in PCL.

Fig. 5.3 A part of the phase diagram for the PCL, Cl-DBS system with the liquidus line for Cl-DBS/PCL (*inverted filled triangles*) and the liquidus line for DBS/PCL (*inverted open triangles*), PCL melting point (*filled circles*) and the PCL crystallisation temperature on cooling from the single phase melt (*filled squares*). Redrawn from Wangsoub et al. (2008)

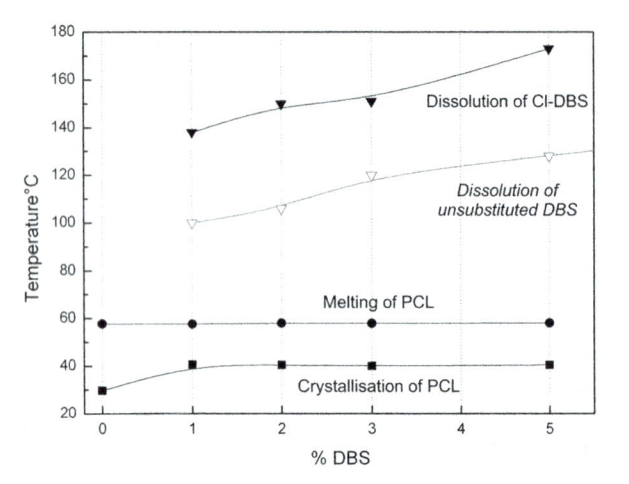

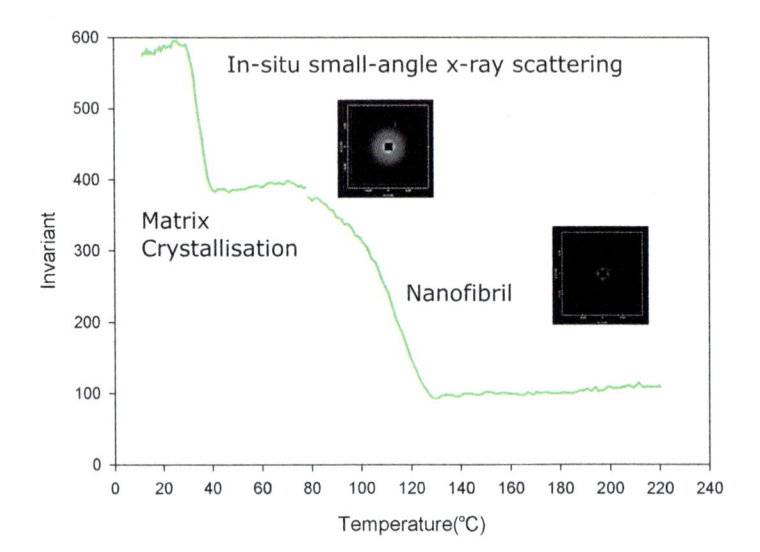

Fig. 5.4 A plot of the SAXS invariant Ω, (*5.1*) obtained from a sequence of time-resolving SAXS patterns recorded whilst cooling a 3 % Cl-DBS/PCL system from the higher temperature single phase region to room temperature at 10 °C/min (redrawn from Mitchell 2013). The *insets* show SAXS patterns taken at the temperatures corresponding to their position on the temperature scale

The phase behavior of the DBS or DBS derivative is dependent on both the chemical nature of the additive and the polarity of the polymer. For example, with DBS in PCL, the nucleating effect saturates at ~3 % whereas in polyethylene it saturates at a much lower concentration (Nogales et al. 2003a).

Figure 5.4 shows the results of a cooling experiment from the single homogenous phase of a composition of 3 % Cl-DBS in PCL in which we plot the invariant (5.1) calculated from the SAXS patterns recorded in a time-resolved manner. Each

SAXS pattern was recorded as a 2d dataset $I(Q, \alpha)$. The invariant Ω was calculated using (5.1) (Mitchell 2013):

$$\Omega = \int_0^{\pi/2} \int_{Q=0}^{Q_{\max}} \left|\underline{Q}\right|^2 I\left(\left|\underline{Q}\right|, \alpha\right) \sin\alpha \, dQ \, d\alpha \tag{5.1}$$

The invariant is related to the average of the electron density differences and for example in the early stages of crystallisation, it is proportional to the volume of crystals. In (5.1), $\underline{Q}$ is the scattering vector, whose magnitude is given by $4\pi\sin(\theta)/\lambda$, where 2θ is the scattering angle and λ is the wavelength of the incident X-rays. α is the angle between a specific direction and the scattering vector.

Initially the invariant has a low value and falls slightly with reduced temperature. At $T \sim 127\,°C$, the invariant increases sharply and this corresponds to formation of nanofibrils. Eventually at lower temperatures, the invariant plateaus and slightly falls. When the temperature reaches $40\,°C$, there is a second increase in the invariant which corresponds to the crystallisation of the PCL. It is clear that these two different ordering processes occur quite separately from each other. We can obtain more information about the state of the Cl-DBS in the 2-phase region by subjecting the material to a shear flow which leads to alignment of the Cl-DBS fibrils.

Figure 5.7a shows the SAXS pattern recorded in the 2-phase region. The horizontal streak corresponds to highly anisotropically shaped objects aligned parallel to the shear flow direction. Analysis of this scattering shows that it arises from objects with a radius of ~14 nm and a length greater than 100 nm (Wangsoub and Mitchell 2009). It is clear that a high level of anisotropy of these objects is achieved at modest shear rates. The level of anisotropy that can be achieved will also depend on the polymer and its molecular weight distribution.

Wangsoub et al. have reported the effects of the molecular weight distribution in the Polyethylene/DBS system (Wangsoub et al. 2016a, b). Samples were subjected to the same shear/temperature profile of 1000 shear units at 10 s^{-1} at temperatures considerably higher than the melting point of each material. After cooling to room temperature at 10 °C/min the samples were examined using wide-angle X-ray scattering and the azimuthal profiles for the 110 reflection are shown in Fig. 5.5. The lowest two molecular weight matrix polymers, PE190 and GA, show only a modest level of anisotropy as judged from the small variation in intensity with azimuthal angle, whereas the highest two molecular weight matrix polymers, GX and PE26, show a marked level of anisotropy. We attribute this to the stress which develops during shearing in the melt and its effect of the orientation of the DBS fibrils. In the curve for 1%DBS/PE26, the high molecular weight matrix polymer, there is evidence for a range of lamellar tilt angles, in the form of a four point pattern which usually occurs when the matrix polymer is under stress in the typical shish kebab structures (Pople et al. 1999; Olley et al. 2014).

We have tested the effect of the DBS on the long period of the polyethylene GX and as Fig. 5.6 shows for samples crystallised at different temperatures the long

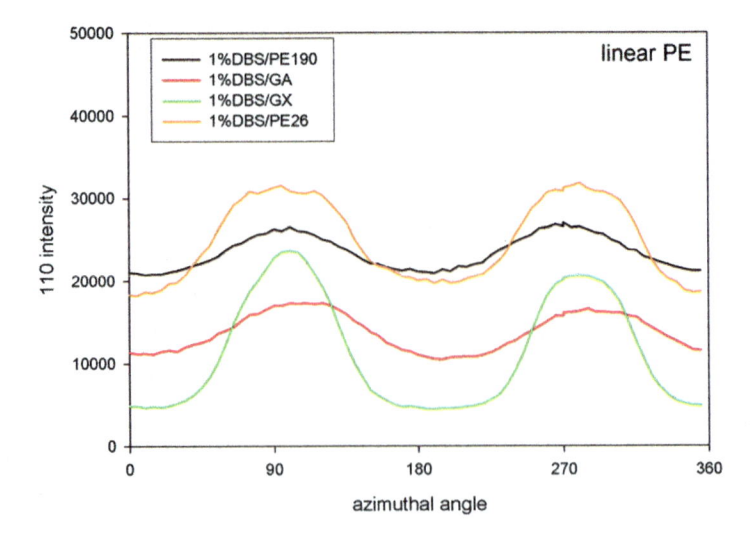

Fig. 5.5 WAXS Azimuthal profiles for the 110 reflection in Linear Polyethylene samples with different molecular weight distributions as listed in Table 5.1 containing 1 % DBS crystallised from a sheared melt as described in the text

Fig. 5.6 The long periods for a series of polyethylenes crystallised at differing temperatures for polyethylenes GX with and without DBS as indicated in the key

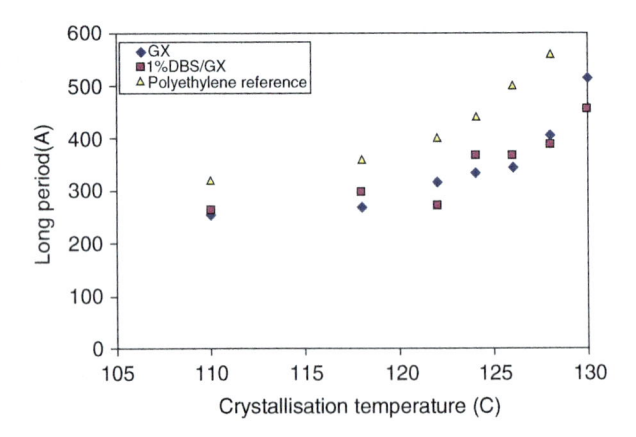

Table 5.1 Molecular weight characteristics of polymers used in the experiments reported in Fig. 5.5

Polymer	Mw	Mn	Mv	Type
Polymer code	Mw	Mn	Mv	Type
PE190	18,700	5410	10,000	Linear
GA	70,950	6250	45,000	Linear
GX	163,100	17,650	105,000	Linear
PE26	424,150	97,650	350,000	Linear

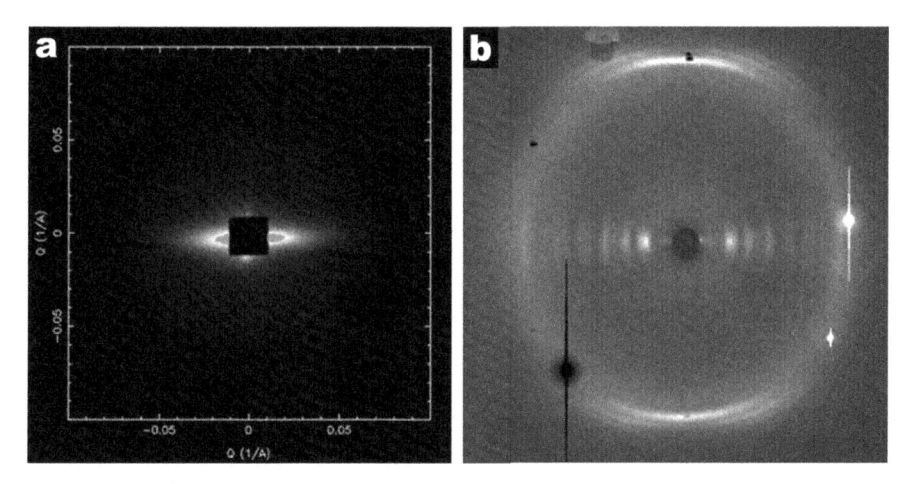

Fig. 5.7 (a) SAXS and (b) WAXS patterns recorded for a sample of Cl-DBS/PCL during shear flow 10 s^{-1} in the 2-phase region of the phase diagram

period for GX alone or GX/DBS is essentially the same. In other words the role of the DBS is to direct the crystallisation of the polymer, the lamellar structure remaining unaffected.

Figure 5.7b shows the WAXS pattern for the equivalent system to that shown in Fig. 5.7a. The figure is a difference between the pattern recorded in the 2-phase region and that recorded in the homogenous high temperature phase. As the scattering from the molten polymer and the crystallised DBS will be additive, we can subtract the polymer scattering to give a clearer view of the pattern of the additive.

It is clear that the Cl-DBS exhibits an ordered crystalline phase. It is interesting that the DBS fibrils are crystalline. This gives them different physical characteristics to other fibrillar gel systems such as self-assembling polypeptide gels and the worm-like micellar systems. The crystal structure leads to very rapid crystal growth in one direction, whereas in the lateral direction, the crystal growth is slower and more fragile. As discussed elsewhere this may be a consequence of the chiral nature of the structures. Optical microscopy studies have shown that the DBS nucleates crystallisation of the DBS from the single phase melt of polymer and DBS. Shear flow may lead to breakage of fibrils whereas the self-assembled micellar systems are virtual polymers and can pass through each other by breaking and reforming. There is good experimental evidence that shear flow leads to some breakage of fibrils. Of course reheating to the single phase region and subsequent crystallisation would lead to repair but loss of any anisotropy which had developed.

Prolonged shearing of the system shown in Fig. 5.7 leads to some unexpected changes. The sequence of SAXS patterns in Fig. 5.8 shows the effect increasing shear strain. We attribute the appearance of additional peaks in the pattern to a narrowing of the fibril diameter distribution. We have analysed the results using a Monte Carlo methodology to fit to Gaussian Distribution of Cylinder radii

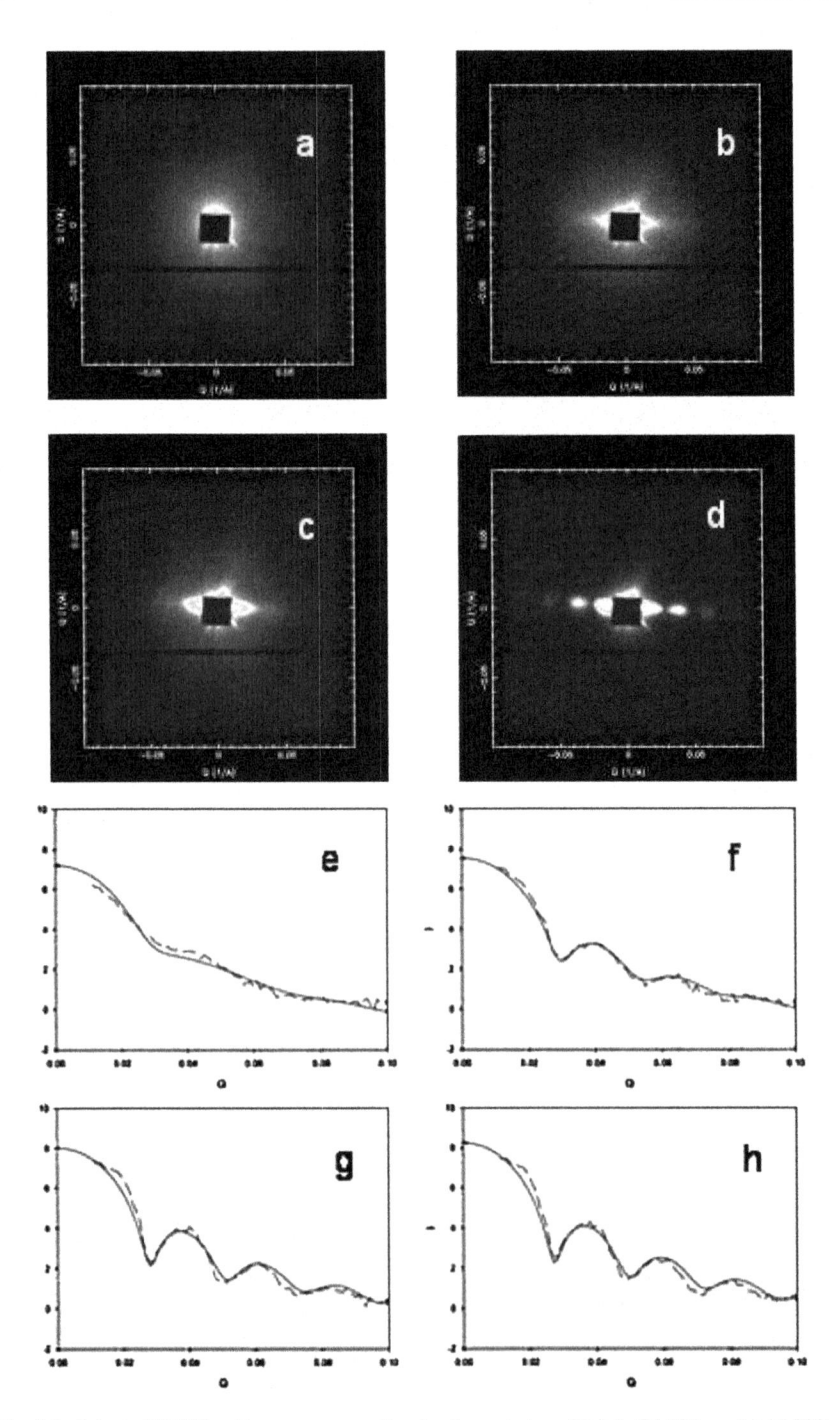

Fig. 5.8 Selected SAXS patterns corresponding to shear strains of (**a**) 0, (**b**) 100 su, (**c**) 10,000 su and (**d**) 28,000; (**e–h**) Equatorial sections (*broken lines*) corresponding to shear strains of (**e**) 1000 su (**f**) 4000 su, (**g**) 13,000 su and (**h**) 28,000 su taken from patterns from the same series as shown in (**a–d**) but for the shear strains indicated. The *full lines* in the sections correspond to the best-fit

(Wangsoub and Mitchell 2009). The extracted results are shown in Fig. 5.9 with plots of the mean radius which is largely unchanged and the Gaussian radius distribution b. The width reduces sharply and reaches a plateau value between 5000 and 10,000 shear units (shear rate × time of shearing).

Of other sugar alcohols, debenzylidene xylitol is widely used as a gelling agent; moreover, the three enantiomeric carbons in xylitol are arranged like a sequence of three carbons in xylitol. Dibenzylidene xylitol has been specified in the literature as a nucleating agent for polymers, but was not found to have any significant directing effect in polypropylene or polycaprolactone.

In contrast to the generation of row nuclei in polymers with a high molecular weight tail in the distribution, cessation of shear flow does not immediately lead to the relaxation of the oriented row nuclei. The aligned DBS fibrils remain aligned for quite some period (hours) which will doubtlessly in some cases lead to difficulty with experiments whilst attempting to remove prior history. Of course it may also prompt some novel methods of developing the anisotropy.

5.5 DBS Directing Crystallisation

We have shown in Sect. 5.4 how by using shear flow in conjunction with a polymer containing a DBS derivative we can obtain aligned nanoscale fibrils which are arranged parallel to the shear flow. This can be achieved while the matrix polymer is in the molten state. If we now crystallise the sample, we can exploit these extended objects as row nuclei to direct the crystalline morphology of the matrix polymer. Figure 5.10 shows a SAXS pattern recorded at Diamond on beamline I22 using the shear flow cell with Cl-DBS/PCL sample, after a cycle of shearing and cooling. The horizontal streak around the beam stop corresponds to the highly extended fibrils about 14–15 nm radius and greater than 100 nm long. The features above and below the beam stop are typical of small-angle scattering from semi-crystalline polymers and relate to the lamellar crystal morphology. Clearly the level of preferred orientation of both the nanofibrils and the crystal lamellae is very high.

Figure 5.11 shows the same type of information but revealed using electron microscopy coupled with differential etching. The extended object in the centre of the image is the impression of the DBS fibrils left in the polymer matrix after it was removed by etching. We can see that the main object is made up of smaller fibrils fused together. The highly aligned lamellae, seen here edge on, are easily visible and their geometric arrangement to the DBS nanofibrils is clear. In the following two images, the directing influence of the DBS nanofibrils can be easily observed. In Fig. 5.9b to the left beyond the end of the impression left by the nanofibril we can

←

Fig. 5.8 (continued) model for a set of polydisperse cylinders. The fitting parameters are plotted in Fig. 5.9. The horizontal black bar in the lower half of each pattern arises from a defect in the detector system during the measurements

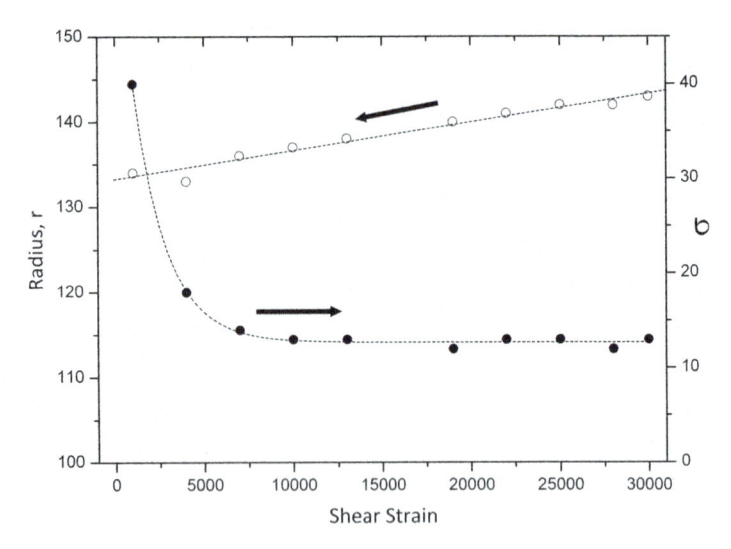

Fig. 5.9 Plots of the mean diameter and the standard deviation of the Gaussian radius distribution as a function of increasing shear strain obtained from SAXS data using a Monte Carlo approach (redrawn from Wangsoub and Mitchell (2009))

Fig. 5.10 Shows the SAXS pattern recorded for the 3 % Cl-DBS/PCL system after shearing at 80 °C at a shear rate of 10 s^{-1} This pattern was obtained at the Diamond Light Source beam line I22 (redrawn from Mitchell 2013)

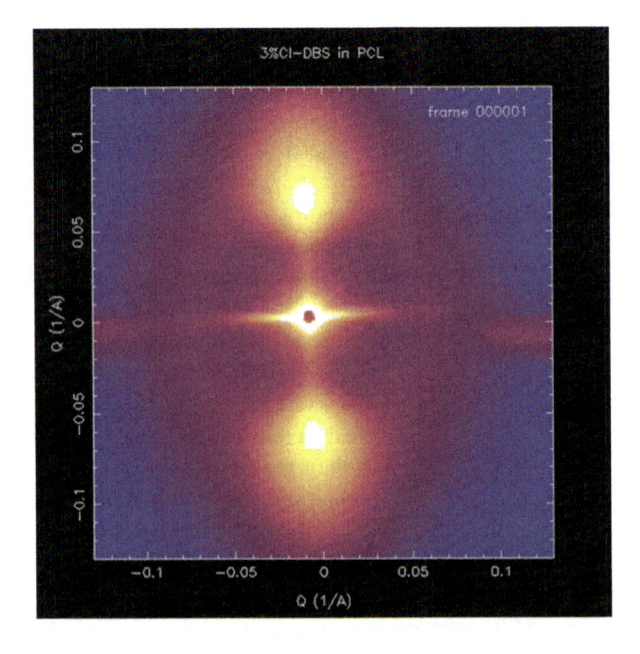

see a more random arrangement of lamellae. In fact we can now see some face on. Furthermore between the two nanofibrils which are not quite parallel with each other we can the two sets of directed lamellae impacting in the centre portion (Fig. 5.12).

5 Controlling Morphology Using Low Molar Mass Nucleators 157

Fig. 5.11 Transmission electron micrograph of a replica of a differentially etched interior surface of polyethylene sample containing DBS

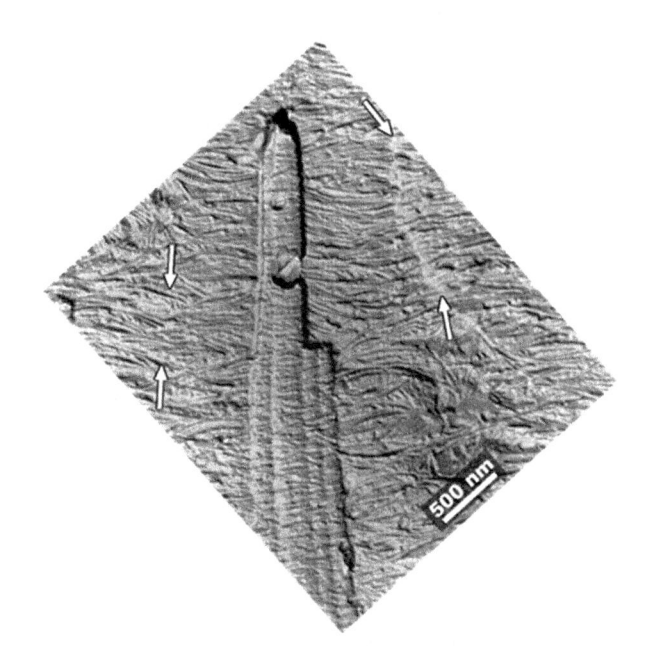

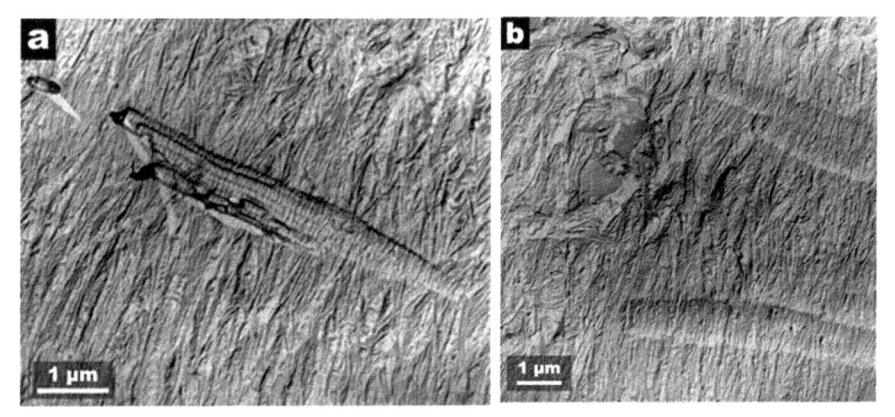

Fig. 5.12 (**a**) Detail of a single cluster of DBS fibres (**b**) Polyethylene development from DBS fibres, *right*, compared with general development, *left*

Figure 5.13 shows the interior surface of DBS/PE system (Wangsoub et al 2016b) prepared by cutting a sheared disc in half. We are now looking in a direction parallel to the aligned nanofibrils. The holes correspond to where the nanofibrils were, and subsequent etching has broadened the hole from its original size. We can see the excellent dispersion of the DBS nanofibrils.

Fig. 5.13 A scanning electron micrograph of the surface of section cut through a sheared disc which was crystallised from a sheared melt of PE/DBS. The surface is perpendicular to the alignment of the DBS fibrils

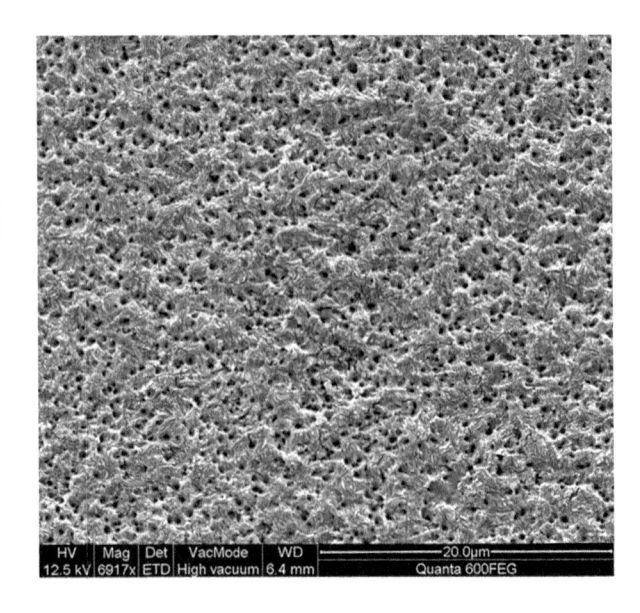

5.6 Model of Directed Crystallisation

In the case of highly aligned fibrils, consideration of the templating process reduces to a 2-d matter. The number density of fibrils is given by Wangsoub et al. (2008)

$$N = (f-f_c)/\pi r_f^2$$

where f is the fraction of the additive and f_c is the upper limit of solubility of the additive in the polymer matrix. We consider all fibrils to have the same radius r_f.

If we associate a templated zone of radius r_t with each fibril, the fraction of templated material, f_t, (Fig. 5.14) is given by:

$$f_t = (f - f_c)\frac{\pi r_t^2}{\pi r_f^2}$$

Clearly, this is a very simple model without any distribution of fibril dimensions and the templated zone size does not account for a random spatial distribution of fibrils. However, we do have an estimate of the fibril radius of ~7.5 nm from the SAXS measurements and we have estimated for polyethylene that the templated zone has a radius ~100 nm. This suggests that the value of $(f-f_c)$ needs to be greater than 0.014 to achieve full templating. Clearly the solubility of the sorbitol derivative in the polymer matrix is critical in determining the actual amount of additive required to achieve this. We previously estimated that the solubility limit of DBS in PCL was ~0.01 and we attribute in large part the enhanced templating of Cl-DBS to a reduction in the solubility limit in PCL. This is shown in the DSC and

Fig. 5.14 Schematic of model to examine the templating process with nanoscale fibrils of DBS in a polymer. This is section taken normal to the fibril axis

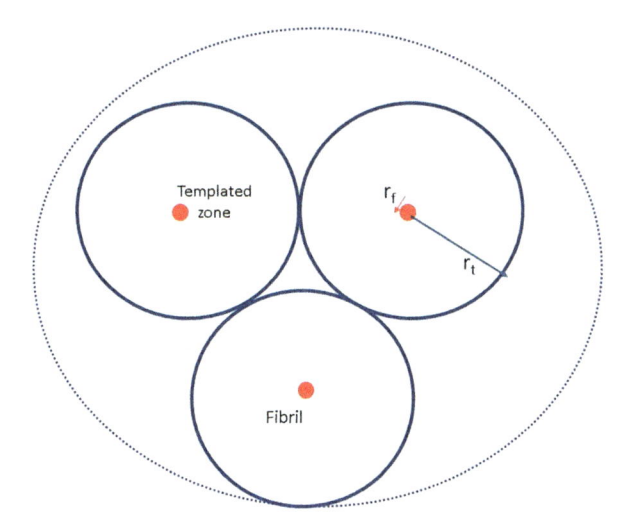

liquidus line. In the examples of the morphology shown in Figs. 5.8, 5.9 and 5.10, the directing is more or less perfect and the hence the orientation parameter $\langle P_2 \rangle_L$ describing the preferred orientation of the lamellae with respect to the flow axis is given by the fraction of material templated:

$$\langle P_2 \rangle_L = -0.5(f - f_c)\frac{\pi r_t^2}{\pi r_f^2}$$

The factor -0.5 (Lovell and Mitchell 1981) takes account of the fact that the lamellae are arranged normal to the flow axis. The efficacy of the templating or directing process depends on the number density of the DBS fibrils. For a given amount of DBS, the number density can be optimised by making the nanofibrils as thin as possible. The processes described in this chapter are effective because of the lateral nanoscale of the fibrils.

5.7 Summary

We have shown that the inclusion of small amounts of DBS or selected derivatives leads to the formation of a well-dispersed system of nanofibrils. These are particularly effective in directing the crystallisation of polyethylene, polypropylene and poly(ε-caprolactone) due to the high number density of the fibrils in the sample. It may well be possible to extend to other polymer systems. Such work is underway. The approach offers distinct advantages over the other approaches described in other chapters in this volume. It avoids the use of viscous high molecular fractions to produce row nuclei. By using a nanoparticle it may be possible to achieve other orientations of the nanoparticles than parallel to the flow axis. The latter takes place because of the high level of anisotropy of the nanofibrils. The high aspect ratio of the nanofibrils means that any deformation of the polymer melt is likely to lead to a high level of orientation of the nanofibrils leading to a high level of preferred orientation of the lamellar crystals.

Acknowledgements This work was supported by the Fundação para a Ciência e a Tecnologia (Portugal) through project PTDC/CTM-POL/7133/2014 and through the Project reference UID/Multi/04044/2013.

We thank Naresuan University for supporting SW during her Ph.D. programme and the Faculty of Science at Naresuan University for funding subsequent short visits when much of the work described here was performed.

This chapter contains various data recorded at international synchrotron and neutron scattering facilities. We are indebted to those facilities for access and to the beamline scientists for their involvement in these experiments; Dr. Sigrid Bernstoff (Elettra), Dr. Francois Fauth (ESRF), Dr. Sergio Funari (Hasylab), Dr. Jen Hiller and Dr. Nick Terrill (Diamond), Dr. Steve King, Dr. Sarah Rogers, Dr. Ann Terry and Dr. Richard Heenan (ISIS).

References

Bassett DC (2006) Linear nucleation of polymers. Polymer 47:5221–5227
Cavagna A (2009) Supercooled liquids for pedestrians. Phys Rep 476(4–6):51–124
García-Ruiz JM, Villasuso R, Ayora C, Canals A, Otálora F (2007) Formation of natural gypsum megacrystals in Naica, Mexico. Geology 35(4):327–330
Grubb DT, Dlugosz J, Keller A (1975) Direct observation of lamellar morphology in polyethylene. J Mater Sci 10:1826–1828
Hanssen M, Marsden J (1984) E for additives: the complete E number guide. Thorsons, Wellingborough
Kashchiev D, Vekilov PG, Kolomeisky AB (2005) Kinetics of two-step nucleation of crystals. J Chem Phys 122:244706
Lovell R, Mitchell GR (1981) Molecular orientation distribution derived from an arbitrary reflection. Acta Cryst A37:135–137
Mitchell GR (2013) Characterisation of safe nanostructured polymers. In: Silvestre C, Cimmino S (eds) Ecosustainable polymer nanomaterials for food packaging. Taylor and Francis, Boca Raton, Print ISBN: 978-90-04-20737-0, eBook ISBN: 978-90-04-20738-7
Nogales A, Mitchell GR (2005) Development of highly oriented polymer crystals from row assemblies. Polymer 46:5615–5620
Nogales A, Olley RH, Mitchell GR (2003a) Directed crystallisation of synthetic polymers by low-molar-mass self-assembled templates. Macromol Rapid Commun 24:496–502
Nogales A, Mitchell GR, Vaughan AS (2003b) Anisotropic crystallization in polypropylene induced by deformation of a nucleating agent network. Macromolecules 36:4898–4906
Nogales A, Thornley SA, Mitchell GR (2004) Shear cell for in-situ WAXS, SAXS and SANS experiments on polymer melts under flow fields. J Macromol Sci Phys B43:1161–1170
Nogales A, Olley RH, Mitchell GR (2016) On morphology of row structures in polyethylene generated by shear alignment of dibenzylidene sorbitol. J Polym Res (Submitted)
Okesola BO, Vieira VMP, Cornwell DJ, Whitelaw NK, Smith DK (2015) 1,3:2,4-Dibenzylidene-D-sorbitol (DBS) and its derivatives—efficient, versatile and industrially relevant low-molecular-weight gelators with over 100 years of history and a bright future. Soft Matter 11:4768–4787

Olley RH, Mitchell GR, Moghaddam Y (2014) On row-structures in sheared polypropylene and a propylene-ethylene copolymer. Eur Polym J 53:37–49

Pople JA, Mitchell GR, Sutton SJ, Vaughan AS, Chai C (1999) The development of organised structures in polyethylene crystallised from a sheared melt, analyzed by WAXS and TEM. Polymer 40:2769–2777

Sangeetha NM, Maitra U (2005) Supramolecular gels: functions and uses. Chem Soc Rev 34:821–836

Siripitayananon J, Wangsoub S, Olley RH, Mitchell GR (2004) The use of a low-molar-mass self-assembled template to direct the crystallisation of poly (epsilon-caprolactone). Macromol Rapid Commun 25:1365–1370

Smith DK (2009) Lost in translation? Chirality effects in the self-assembly of nanostructured gel-phase materials. Chem Soc Rev 38:684–694

Terech P, Weiss RG (1997) Low molecular mass gelators of organic liquids and the properties of their gels. Chem Rev 97(8):3133–3160

Tuffen H, James M, Castro J, Schipper CI (2013) Exceptional mobility of an advancing rhyolitic obsidian flow at Cordon Caulle volcano in Chile: observations from Cordón Caulle, Chile, 2011-2013. Nat Commun 4:2709. doi:10.1038/ncomms3709

Van Driessche AES, García-Ruíz JM, Tsukamoto K, Patiño-Lopez LD, Satoh H (2011) Ultraslow growth rates of giant gypsum crystals. Proc Natl Acad Sci U S A 108:15721–15726

Wangsoub S, Mitchell GR (2009) Shear controlled crystal size definition in a low molar mass compound using a polymeric solvent. Soft Matter 5:525

Wangsoub S, Olley RH, Mitchell GR (2005) Directed crystallisation of poly(ε-caprolactone) using a low-molar-mass self-assembled template. Macromol Chem Phys 206:1826–1839

Wangsoub S, Davis FJ, Mitchell GR, Olley RH (2008) Enhanced templating in the crystallisation of poly(ε-caprolactone) using 1,3:2,4-di(4-chlorobenzylidene) sorbitol. Macromol Rapid Commun 2008(29):1861–1865

Wangsoub S, Davis FJ, Harris PJF, Mitchell GR, Olley RH (2016a) Structure and morphology of high liquid content gels formed from alkanes and dibenzylidene sorbitol. Phys Chem Chem Phys (Submitted)

Wangsoub S, Olley RH, Mitchell GR (2016b) Templating the crystallisation of polyethylene using dibenzylidene sorbitol. Macromol Chem Phys (Submitted)

Watase M, Itagaki H (1998) Thermal and rheological properties of physical gels formed from benzylidene-D-sorbitol derivatives. Bull Chem Soc Jpn 71(6):1457–1466

Yamasaki S, Tsutsumi H (1994) Microscopic studies of 1,3: 2,4-di-O-benzylidene-D-sorbitol in ethylene glycol. Bull Chem Soc Jpn 67:906–911

Yamasaki S, Tsutsumi H (1995) The dependence of the polarity of solvents on 1,3: 2,4-di-*o*-benzylidene-D-sorbitol gel. Bull Chem Soc Jpn 68:123–127

Yamasaki S, Ohashi Y, Tsutsumi H, Tsujii K (1995) The aggregated higher-structure of 1,3: 2,4-di-*o*-benzylidene-D-sorbitol in organic gels. Bull Chem Soc Jpn 68:146–151

Zweifel H (2001) Plastics additives handbook. Hanser, Munich, Chapter 18

Chapter 6
Crystallization in Nanoparticles

Aurora Nogales and Daniel E. Martínez-Tong

6.1 Introduction

Under particular circumstances, some polymers may crystallize. However, a polymer melt consists of an assembly of polymer chains that are coiled and mutually interpenetrated, and therefore it is impossible to attain an ideal crystalline state of extended straight chains with the endgroups assembled in planar interfaces, purely due to kinetic reasons. The required complete disentangling would need a very long time, as it is associated with an extremely high entropic activation barrier. Instead, a polymer melt cooled below the equilibrium melting point produces a system which is only in part crystalline. Semicrystalline polymers consist of a complex puzzle of crystalline lamellae, crystal-amorphous interphases, stacks of crystalline lamellae, liquid pockets, rigid amorphous phases, and fringed micellar crystals. The arrangement of these structural elements in a given polymer depends on different factors, including chemical structure, chain flexibility, thermal history, and orientation. The structure of semicrystalline polymers exhibits characteristic features depending on the length scales. In the 10^{-8} m scale, in highly crystalline polymers, such as polyethylene (PE) (Wunderlich 1973) and in intermediate crystalline polymers, such as polyethylene terephthalate (PET) (Santa Cruz et al. 1991) or poly(etherketone)s (Bassett et al. 1988), there is an alternation between crystalline regions (lamellar crystals) and amorphous regions (interlamellar amorphous regions). With the exception of highly crystalline polymers, this alternation does not extend to the whole volume of the sample (Santa Cruz et al. 1991). The lamellae are packed into stacks which are separated by broad amorphous regions. The stacks can assemble

---

A. Nogales (✉) • D.E. Martínez-Tong
Instituto de Estructura de la Materia (IEM-CSIC), C/Serrano 121,
Madrid 28006, Madrid, Spain
e-mail: aurora.nogales@csic.es

© Springer International Publishing Switzerland 2016
G.R. Mitchell, A. Tojeira (eds.), *Controlling the Morphology of Polymers*,
DOI 10.1007/978-3-319-39322-3_6

themselves into superstructures, generally with spherical symmetry (spherulites) which can reach microns or even several millimeters.

Nowadays, current interest in the properties of polymers confined into nanometer scale is very intense, both from the fundamental and practical perspectives (Soles and Ding 2008). Polymers are extensively used in processes to fabricate nano-objects like wires of nanometer-scale diameters (Martín and Mijangos 2009; Chen et al. 2012; Martín et al. 2012) nanoimprinting (Soles and Ding 2008), and nanoscale polymeric particles (Landfester 2001). Confined polymers are central to a broad range of advanced materials and emerging nanotechnologies (Coakley and Mcgehee 2004; Chen et al. 2012), with applications including biomaterials (Yu et al. 2011; Bonanno and Segal 2011), micro- and optoelectronics (Yuan et al. 2011; Tong et al. 2011; Di Benedetto et al. 2008), energy capture/storage (Li et al. 2012; Cai et al. 2012; Li and Malardier-Jugroot 2011; Guan et al. 2011; Mohapatra et al. 2009), among others. Besides cutting-edge fabrication strategies, control over the changes in properties induced by nanoscale confinement is a central issue to be taken into account. The rapid development of nano-science and nano-technology in the twentieth and twenty-first centuries raises a lot of questions about the structure and surface properties of nano-materials (Svorcik et al. 2013). The prefix *nano*- is used in a broad way to describe systems where one or several physical dimensions have been reduced to length scales between 1 and 100 nm, and also to the application of concepts and understanding of properties (physical, chemical, biological, mechanical, ...) that derive as a result of the reduced length scale (Gates et al. 2005; Martín Gago et al. 2008). The possible change in properties, when a material is nanostructured, is understood as **confinement effects** (Alcoutlabi and McKenna 2005).

As for smaller molecules (Grigoriadis et al. 2011), crystallization becomes even more sluggish when the chains are confined at the nanoscale level (Vanroy et al. 2013; Liu and Chen 2010). Confinement is able to induce different morphologies (Asada et al. 2012; Maillard and Prud'homme 2008), reduce crystallization rates by several orders of magnitude (Despotopoulou et al. 1996; Massa et al. 2003), and in some cases, inhibition of crystallization within the experimental time scale (Capitán et al. 2004). Therefore, in order to widen the applicability of semicrystalline polymers at the nanoscale, the effect of confinement on the crystallinity and morphology should be considered.

Polymers confined in nanometer-sized geometries have been intensively investigated in the last decades, aiming both to achieve a deeper understanding of finite-size effects in soft matter and to improve the performance of nano-devices and hybrid materials (Napolitano et al. 2013). When designing and preparing nanostructures, one or several physical dimensions can be affected. For example, if a nanostructured polymer system is prepared in a way that two dimensions have macroscopic dimensions, while the remaining is in the nanometer scale, the resulting material is said to be one dimensionally confined (1D-confinement). This is the case of *polymer thin films*. If two dimensions have nanometer scales, one can talk about two dimensional confinement (2D-confinement), which is the case of *nanowires* and nanotubes. Finally, when all three dimensions are in the nanometer length scale, such as in ***nanoparticles***, the polymer is said to be three-dimensionally confined (3D-confinement). Moreover, confined polymer structures can be further processed to obtain *à la carte* nano-features, such as nanostructured surfaces on polymer thin films (Rebollar et al. 2011), alignment of polymer nanowires by percolation into nanoporous membranes (Garcia-Gutierrez et al. 2013; Martin et al. 2014), or even physical responsive polymer nanoparticles (Deng et al. 2013).

Among the most studied confining geometries are thin films, probably due to their presence in a large number of technological applications, and also because it provides an easy control of the level of confinement and tunability of interfacial interactions between polymers and substrates (Rotella et al. 2011). In this geometry, the sample thickness becomes a crucial parameter controlling crystallization (Vanroy et al. 2013), while surface effects mostly lead to a competition between adsorption to the solid surface and nucleation of polymer crystals (Reiter and Sommer 1998; Vanroy et al. 2013; Napolitano and Wübbenhorst 2007; Asada et al. 2012; Bertoldo et al. 2010). Also, polymer thin films represent highly metastable forms of matter, presenting unexpected properties (Reiter and DE Gennes 2001) which have been widely discussed by the scientific community in the past years (Forrest et al. 1997; Tsui and Zhang 2001; Schönhals et al. 2002; McKenna 2003; Wübbenhorst and Lupascu 2005; Napolitano et al. 2013; Napolitano and Cangialosi 2013; Boucher et al. 2012). It is also possible to attain information of the role of interfacial interactions by intercalating a polymer thin film in between two adsorbing layers (Napolitano et al. 2013) although the introduction of burying interfaces limits the possible experimental protocols; for example, metallization of both polymer faces does not permit an optical access to the polymer surface.

On the other hand, using confined polymers into droplets and nanoparticles might serve as starting point in order to prepare *nanocrystals* and also to better understand the mechanisms involved during nucleation and crystallization in confined geometries (Massa and Dalnoki-Veress 2004; Li 2009) without including any preferential confined dimension. In this chapter, we present the crystallization behavior of polymers in confined nanoparticles.

6.2 Generation of Polymer Nanoparticles

The formation of structured polymeric nanoparticles is of great importance for many applications. In general, polymer nanoparticles can be prepared from several heterophase methods. One of the most known methods, developed extensively by Landfester et al. during the last decade, is the so-called miniemulsion method (Landfester 2001, 2009; Kietzke et al. 2007).

An emulsion consists of well-dispersed droplets of a substance into a continuous phase. This is formed by mixing two immiscible phases that are subjected to high shear, resulting in small, homogeneous, and narrowly distributed nanodroplets. The miniemulsion can be stabilized by means of surfactants. The protocol is sketched in Fig. 6.1.

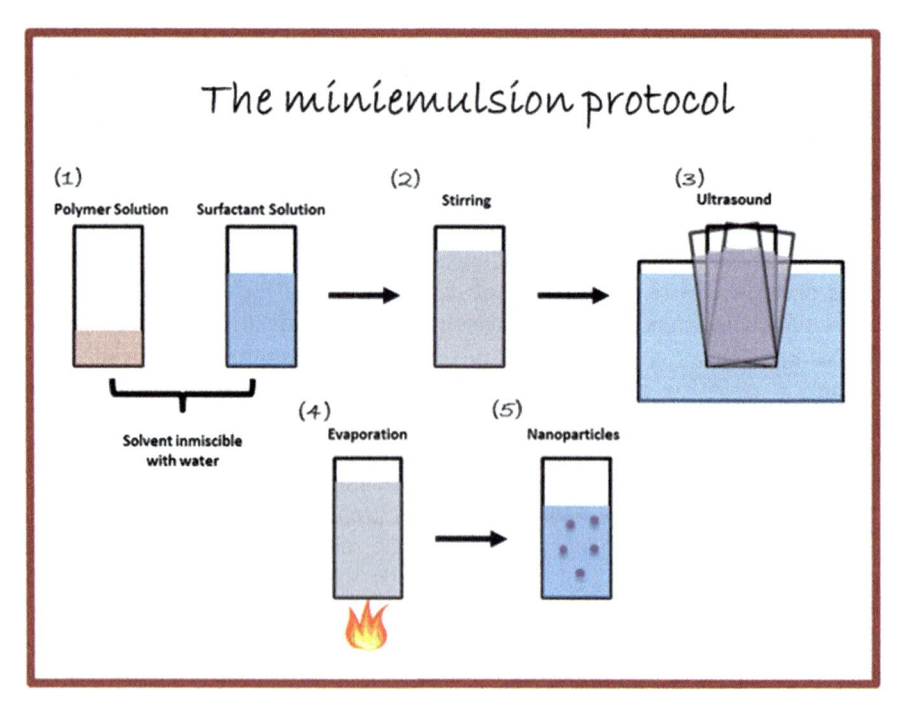

Fig. 6.1 Sketch of the miniemulsion protocol to generate polymer nanoparticles

By using this method, polymer nanoparticles of different polymers have been reported (Martínez-Tong et al. 2013, 2014; Landfester 2001). For amorphous polymers, the shape of the nanoparticles generated by this technique is spherical (Fig. 6.2). Variations of physical properties like the glass transition temperature of the polymer confined into the shape of this nanoparticles have been reported in the literature (Martínez-Tong et al. 2013, 2014).

Another approach for obtaining polymer nanoparticles is the so-called reprecipitation method, previously proposed by Shimizu et al. (2007). This method relays on crashing out hydrophobic polymer chains in solution by displacing a solvent with a non-solvent, generally water. Both polymer solvent and non-solvent must be miscible. As in the miniemulsion technique, reprecipitation can be used by starting from the bulk polymer of known molecular weight as the precursor material; only the nanoparticle diameter needs to be adjusted during the process by controlling solution concentration. The method is schematically described in Fig. 6.3.

Figure 6.4 shows an AFM image of polyethylmethacrylate (PEMA) nanoparticles prepared by reprecipitation.

6 Crystallization in Nanoparticles

Fig. 6.2 Atomic Force Microscopy (AFM) topography image of polystyrene (PS) nanoparticles prepared by the miniemulsion method after spin casting the miniemulsion on a silicon wafer. The starting material is the bulk polymer

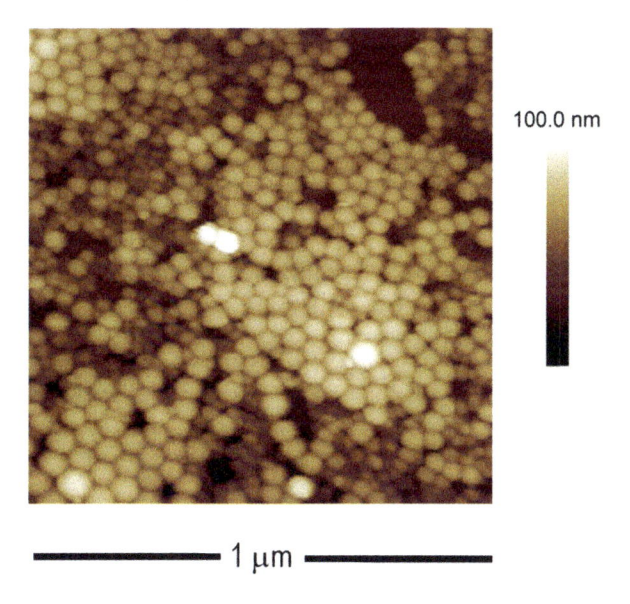

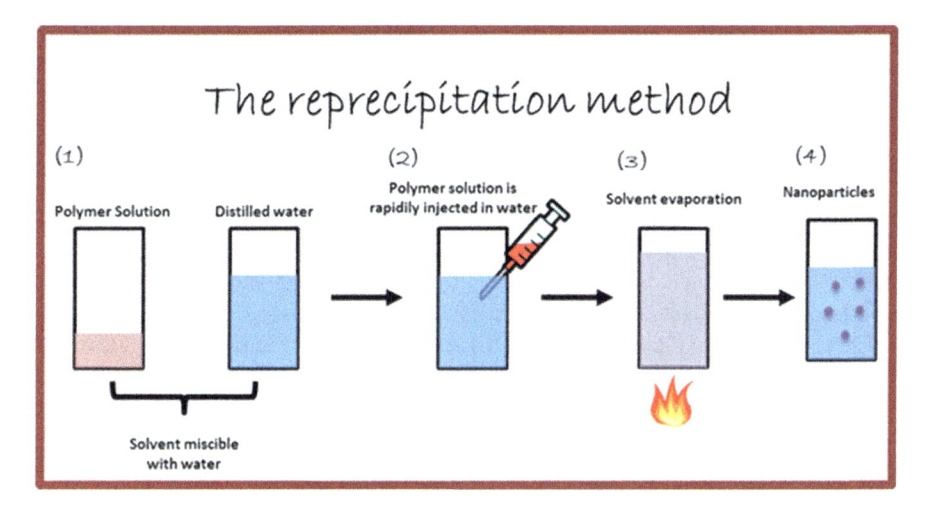

Fig. 6.3 Sketch of the reprecipitation protocol to generate polymer nanoparticles

6.3 Modification of the Crystalline Morphology by Confinement into Nanoparticles

By the two previously mentioned methods it is possible to nanostructure, within the shape of particles, polymers that are either inherently semicrystalline or that are able to crystallize. One example of this type of polymers is poly(lactic acid), from now on abbreviated as PLA. PLA is an aliphatic thermoplastic polymer, commonly

Fig. 6.4 AFM topography image of PEMA nanoparticles prepared by the reprecipitation method after spin casting on a silicon wafer

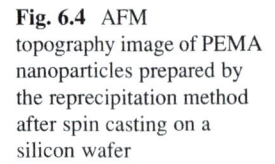

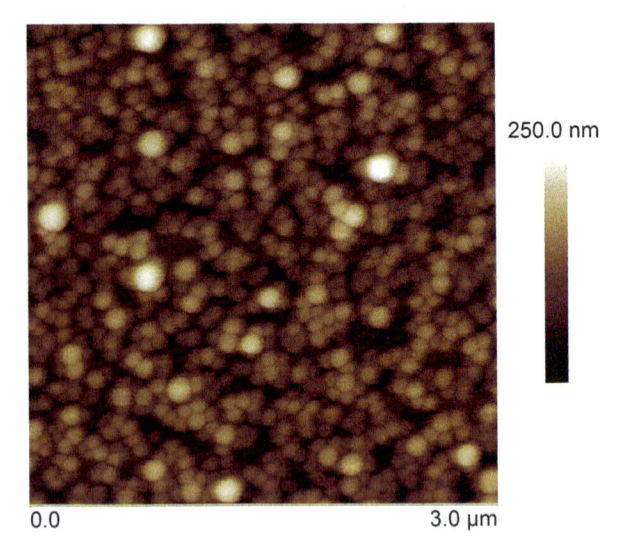

Fig. 6.5 (*Left*) Sketch of a PLLA repeating unit. (*Right*) Sketch of a D,L-dimer

made from α-hydroxy acids which are considered biodegradable, biocompatible, and compostable (Garlotta 2001; Anderson and Shive 2012). Although the term acid is generally included in its name, PLA is a polyester instead of a *polyacid* (Garlotta 2001). Generally, PLA grades are copolymers of poly(L-lactic acid) (PLLA) and poly(D,L-lactic acid) (PDLLA), which are produced from L-lactides and D,L-lactides, respectively (Martin and Avérous 2001). The sketch of these molecules is shown in Fig. 6.5.

The letters D and L are related to Dextrorotation (D) and Levorotation (L) and refer to the polymer property of rotating plane polarized light. If the light rotates clockwise as it approaches an observer, this is known as dextrorotation (light with a rotation to the right). If the light rotates counterclockwise as it approaches the observer, then the light exhibits levorotation (rotation to the left). The ratio of L- to D,L- enantiomers is known to affect the properties of PLA, such as the melting temperature and the degree of crystallinity (Martin and Avérous 2001).

PDLLA (NatureWorks, PLA 2002D, D-content 4.25 %, $\rho = 1.24$ g/cm^3) nanoparticles can be prepared either by the miniemulsion method or the reprecipitation method. In the present case, for the miniemulsion method PDLLA was dissolved in chloroform (CHCl$_3$), at a 0.2 wt% concentration. The polymer solution was added to a 1 wt% aqueous surfactant solution of sodium dodecyl sulfate (SDS). Pre-emulsification was obtained by stirring at room temperature for

6 Crystallization in Nanoparticles

60 min and, afterwards, the stirred mixture was ultrasonicated for 15 min in an ultrasound bath. This allowed obtaining a miniemulsion. Evaporation of the organic solvent, under stirring the miniemulsion at 66 °C for 180 min, yields a dispersion of polymeric nanospheres in a non-solvent medium. To eliminate the excess of SDS, suspensions were dialyzed against distilled water, using a dialysis membrane. The size of the obtained nanospheres is governed principally by the concentration of the polymeric solution and by the concentration of the surfactant, as previously reported (Martínez-Tong et al. 2013).

PDLLA nanoparticles were also prepared by the reprecipitation method. Specifically, 15 mg of PDLLA were dissolved in 5 mL of tetrahydrofuran (THF). The solution was left under stirring at room temperature for 30 min. Afterwards, in order to remove any macroscopic residues, the solution was filtered. Finally, the polymer solution was rapidly injected into a beaker filled with distilled water. This emulsion was left under stirring at room temperature for 90 min and then for 120 min at 66 °C to allow complete removal of the solvent. Figure 6.6 shows AFM topography images of the prepared PDLLA nanoparticles.

Figure 6.6a, b show AFM topography images of the PDLLA nanoparticles prepared by the miniemulsion and reprecipitation method, respectively. In both cases, the nanoparticles consist of polymer nanospheres, without signs of

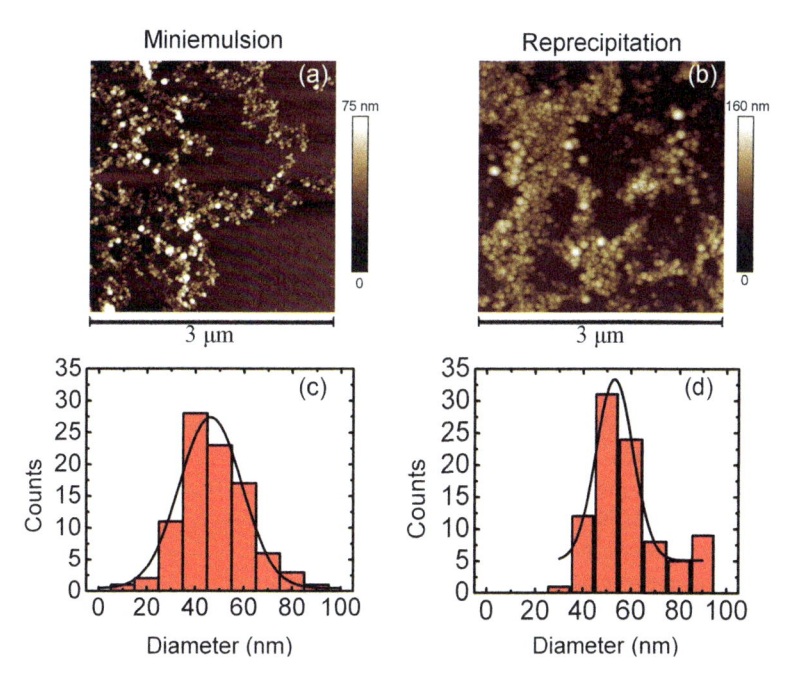

Fig. 6.6 AFM topography images and size distribution of PDLLA nanoparticles prepared by the miniemulsion method (**a, c**) and reprecipitation method (**b, d**). Size distribution was quantified on the AFM images by measuring the nanoparticles diameters. *Continuous black line* represents a Gaussian fit to the distribution

coalescence and/or ripening. From the AFM images the size distributions of the polymer nanoparticles could be quantified and the results are shown in Fig. 6.6c, d. For both preparation methods, a similar mean value of the PDLLA nanoparticles diameter (d) was found, namely $d_{\text{miniemulsion}} = (46 \pm 1)$ nm, and $d_{\text{reprecipitation}} = (53 \pm 3)$ nm. To take into consideration the width of the size distributions, it is possible defining a *quality factor* of the resulting nanospheres by $Q = W/\langle d \rangle$, where W is the width of the distribution (nm) and $\langle d \rangle$ is the mean diameter (nm). The closer the factor Q is to zero, the better is the resulting preparation in terms of size monodispersity. In this work, we found small Q values for both preparation methods, $Q_{\text{miniemulsion}} = 0.57 \pm 0.06$ and $Q_{\text{reprecipitation}} = 0.28 \pm 0.07$, indicating that size monodispersity was within acceptable values.

6.3.1 Crystallization in Nanoparticles

Figure 6.7 shows the X-ray diffraction patterns of the nanoparticles, annealed at different temperatures, together with the diffraction from the bulk treated at the same temperatures for comparison. In general, as the samples were annealed, the development of diffraction peaks is evidenced. This indicates that all samples were able to crystallize. Specifically, bulk and miniemulsion nanoparticles patterns are quite similar between each other for the whole temperature range. At 30 °C, these two samples show an amorphous halo, indicating their amorphous nature. As temperature increased up to 70 °C, the development of a diffraction peak, around $2\theta = (17.0 \pm 0.3)°$, is seen. This peak has been reported to be associated with the diffraction from the (200) and/or (110) planes of the α form of PDLLA (Mano et al. 2004). At higher temperatures, the diffraction patterns show the development of several other peaks. This behavior indicates that as the samples were annealed crystallization continues. At 110 °C, diffraction peaks show their highest intensity. Comparing with the literature (Mano et al. 2004; Zhang et al. 2008; Wang et al. 2014; Wei et al. 2014; Wasanasuk et al. 2011), we have indexed the peaks as shown at the 110 °C temperature in Fig. 6.7. Results show that the PDLLA has crystallized in the ordered α-phase (Wei et al. 2014; Xiong et al. 2013; Wasanasuk et al. 2011), which is expected when the polymer is crystallized at temperatures below 100 °C. Indexing results are summarized in Table 6.1.

The diffraction patterns of PDLLA reprecipitation nanoparticles are shown in the right column of Fig. 6.7. In this case, patterns are noisier in comparison to the others. This might be related to the amount of material enclosed in the aluminum sheets, which in this case was lesser. As stated previously, as temperature increases crystallization evolves; however, comparing with bulk and miniemulsion nanoparticles there are important differences. At 30 °C, besides the amorphous halo, an increase of the diffraction signal is observed at small values of 2θ (2°–5°). This result can be related to the possible existence of some ordering in the PDLLA chains within the nanoparticles, which lengthscale is above of WAXS range. It has been reported that poly(lactic acid) polymers can show a mesophase (Zhang

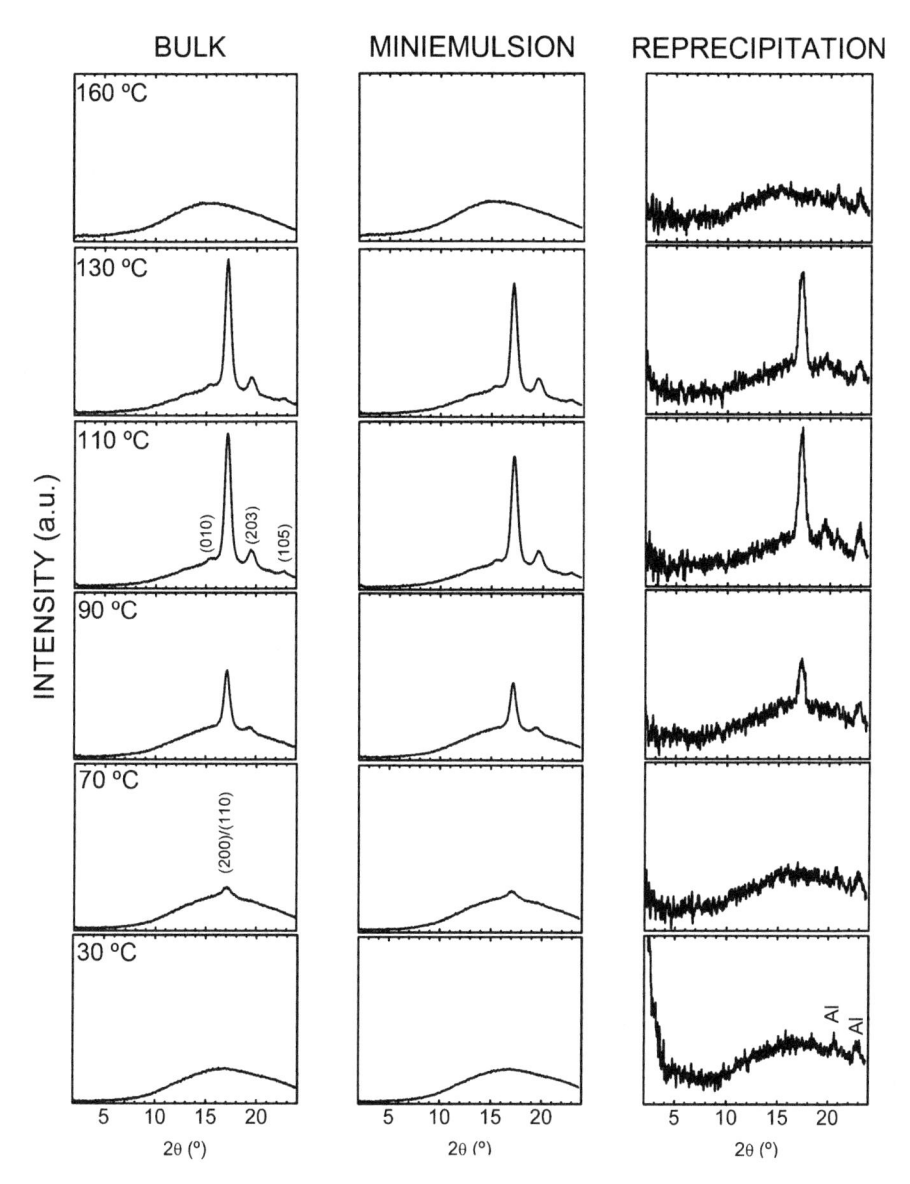

Fig. 6.7 Wide Angle X-ray Scattering (WAXS) diffraction patterns for bulk PDLLA (*left*) and PDLLA miniemulsion (*center*) and reprecipitation (*right*) nanoparticles

Table 6.1 Crystalline peaks and 2θ position, as observed at 130 °C (Fig. 6.7)

Index	Bulk	Miniemulsion	Reprecipitation
(010)	$(15.4 \pm 0.3)°$	$(15.3 \pm 0.3)°$	–
(200)/(110)	$(17.0 \pm 0.3)°$	$(17.1 \pm 0.3)°$	$(17.2 \pm 0.3)°$
(203)	(19.5 ± 0.5)	(19.5 ± 0.5)	(19.5 ± 0.5)
(105)	$(22.7 \pm 0.3)°$	$(22.7 \pm 0.3)°$	$(22.8 \pm 0.3)°$

et al. 2010; Wasanasuk and Tashiro 2011), with distinct chain packing and chain conformation. Mesophases in semicrystalline polymers can be obtained by the melt-quenched methods (Strobl 2006); for example, isotactic polypropylene can be solidified into an intermediate state between crystal and amorphous states when a thin specimen of molten state is rapidly quenched (Qiu et al. 2007). Also, Stoclet and collaborators have shown that it is possible to induce a mesophase in PDLLA when an external strain was applied (Stoclet et al. 2010a, b). Collecting all these facts, we argue that the preparation procedure used in the reprecipitation method might have led to the formation of a mesophase in the resulting PDLLA nanoparticles. We recall that in this method, the polymer solution is rapidly injected into a non-solvent. In the solution, the polymer chains have maximum mobility, which can be somehow compared to the behavior in the melt state. As the solvent is quickly removed from the nanoparticles, while the polymer is being transferred to a non-solvent medium (chains have no mobility), it is possible to assume that the polymer chains are being subjected to a quenching-like procedure that ultimately leads to the formation of the mesophase. It can also be argued that during the fast precipitation, the polymer chains retain residual stresses and thus a sort of strain-induced mesophase takes place. These strains could arise from the physical procedure involved in going from separated chains to nanospheres.

The preparation method argument just explained can be also justified by comparing the results between reprecipitation and miniemulsion PDLLA nanoparticles. In the miniemulsion method, nanodroplets of solvent/polymer are formed simultaneously during ultrasonication. After the formation, solvent is slowly evaporated and thus nanoparticles do not suffer a melt-quench like procedure. Also, the solvent/polymer nanodroplets serve as precursors to the resulting polymer nanoparticles and thus, during solvent evaporation process, polymer chains have enough mobility to lose possible strains. Since the solvent is evaporated at a temperature in which no Bragg peaks are observed, and since there is not enough mobility to crystallize, the resulting nanoparticles are amorphous, without any mesophases.

The WAXS signature of the *so-called mesophase* in the PDLLA reprecipitation nanoparticles vanishes with temperature, in the range 30–60 °C. At 70 °C (Fig. 6.7) the increase a low 2θ is not observed anymore. Figure 6.8 shows the Differential Scanning Calorimetry (DSC) trace of the reprecipitation nanoparticles. Glass transition (T_g), crystallization and melting temperatures are highlighted throughout the curve.

The glass transition temperature of the nanoparticles is observed around 60 °C indicating no change in comparison to the bulk (Bitinis et al. 2011). Comparing WAXS and DSC results, we observe that the mesophase signature in the diffraction pattern disappears at temperatures above T_g. Also, in the DSC curve, just below T_g an endothermic peak can be seen, highlighted with the symbol *. DSC endothermic peaks in semi-crystalline polymers are related to first-order transitions, such as the melting of the crystalline structure. Gathering results, it is possible to speculate that this peak is related to the disordering of the polymer chains in the mesophase, that happens just when the polymer start getting mobility due to the incipient glass

Fig. 6.8 DSC trace (*left*) and heat flow derivative (*right*) of the reprecipitation PDLLA nanoparticles

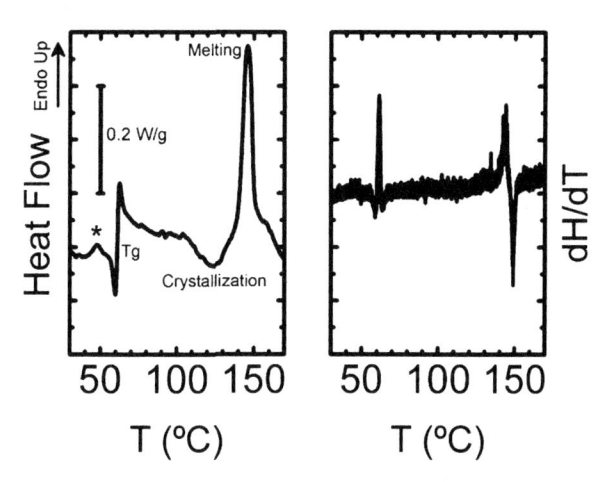

transition. At temperatures above T_g, the calorimetric curve shows a crystallization peak. The onset of crystallization can be observed around 90 °C and its maximum at 125 °C. Finally, the polymer nanoparticles melt at 145 °C. DSC results of crystallization and melting of the nanoparticles agree with the WAXS patterns.

6.3.2 From Polymer Nanoparticles to Polymer Nanocrystals

WAXS results in the previous section showed that polymer nanoparticles of PDLLA were able to crystallize when heated above certain temperatures. The diffraction patterns also show that nanoparticles prepared via the miniemulsion protocol showed a crystallization behavior quite similar to the bulk polymer. Based on these results, we have evaluated the morphological change of PDLLA nanoparticles prepared by the miniemulsion method, when subjected to thermal treatments.

Figure 6.9 shows $(3 \times 3)\,\mu m$ AFM topography images of the deposited PDLLA nanoparticles annealed at 75 °C for different times (t). This temperature is above T_g of the PDLLA and according to the WAXS patterns crystallization must take place. Annealing was performed on a hot stage, by heating the Si wafers on where the particles were originally deposited. Each annealing treatment was carried out on independent wafers.

Image at $t = 0$ s corresponds to the deposited nanoparticles without annealing, showing their spherical geometry. After annealing for 5 min, the nanoparticles lose their spherical shape, turning into *islands*, which show a bigger diameter and lower height in comparison to the original nanoparticles. This is an indication of agglomeration and coalescence of neighbor particles, as previously reported for nanospheres of amorphous polymers, where the height:width ratio increased from 1:1 to 1:5 (Martínez-Tong et al. 2013, 2014). The morphology of the nanoparticles

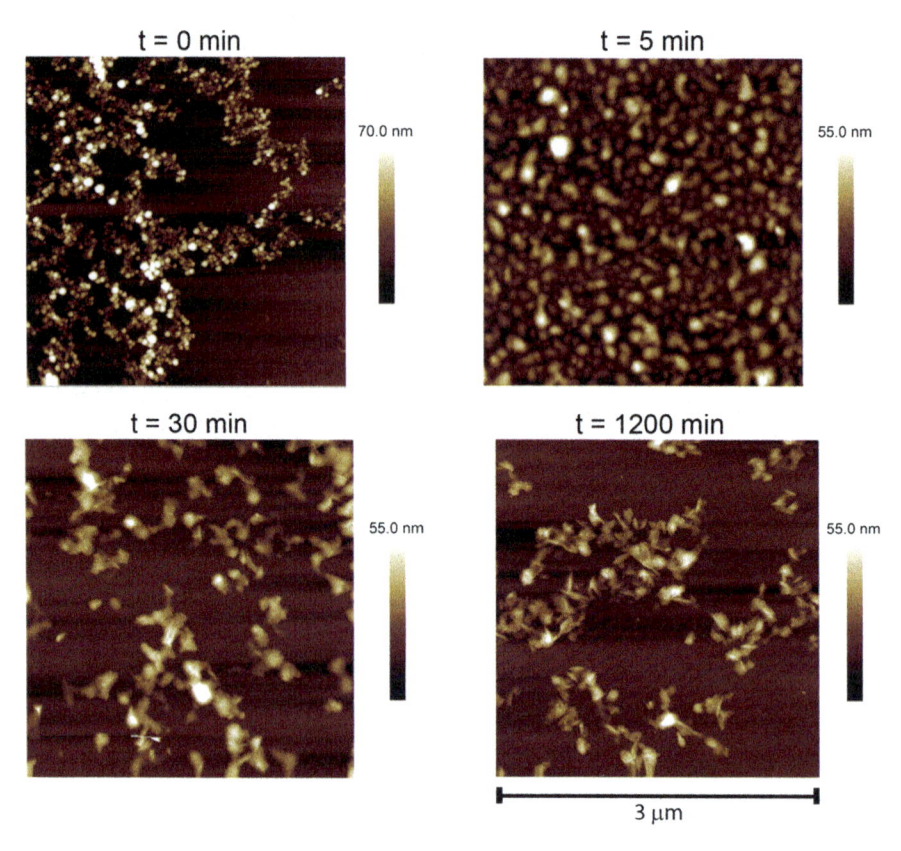

Fig. 6.9 AFM images of PDLLA nanoparticles prepared by the miniemulsion protocol, annealed at 75 °C for different times (*t*)

suffers further changes after 30 min at 75 °C. As time increases, the original smooth edges of the islands observed at 5 min start reshaping into well-defined straight contours. Also, in some cases it is possible to see the appearance of needle-like features inside the nanoparticles. This fact is enhanced at the longest annealing time (1200 min).

Figure 6.10 shows a $(1 \times 1)\,\mu m$ AFM topography image of the PDLLA nanoparticles annealed at 75 °C for 1200 min. In this case, a smaller scan was used (in comparison to images in Fig. 6.9), which allowed getting more insight into the details of the nanocrystals. As stated before, the original spherical nanoparticles present now well-defined edges, as expected for crystalline structures. These crystalline structures seem to be formed by the agglomeration of several nanoparticles, since at the chosen annealing temperature PDLLA is expected to have mobility. Two different structures are observed in the AFM image of Fig. 6.10. First, needle-like crystals (red arrow) resemble the edge-on crystalline structures in

Fig. 6.10 AFM topography image of PDLLA nanoparticles annealed at 75 °C for 1200 min

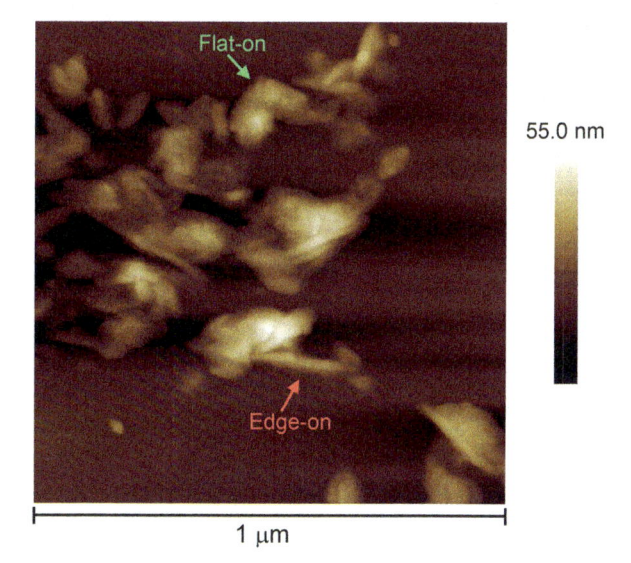

polymer thin films. On the other hand, the green arrow highlights the structures resembling flat-on crystals, compared to thin polymer films (Wang et al. 2008; Xu et al. 2005).

The needle-like crystals prepared from the PDLLA nanoparticles show widths between 15 and 25 nm, heights between 15 and 40 nm and lengths between 100 and 300 nm. On the other hand, flat-on like crystals show sides between 50 and 150 nm and heights in the 15–60 nm range. Based on the geometrical sizes of the crystals obtained from the PDLLA nanoparticles, it is possible to consider them as **nanocrystals**. In principle, these nanocrystals could be recovered from the silicon waffers by washing it in distilled water. Afterwards, they could be used as additives in composites and/or polymer blends. Also, the heights of these nanostructures are comparable to the ones obtained for ultra-thin films of PLLA (Maillard and Prud'homme 2008). This allows thinking in possible comparisons in the crystallization between the two confined geometries.

6.4 Summary

We have shown that by various physicochemical methods it is possible to nanostructure bulk polymers in the shape of nanoparticles, that in the case of amorphous polymers, have spherical shape. If the original bulk polymer is able to crystallize, the shape of the nanoparticle is modified by the process of crystallization. However, it is possible to limit the size of the crystal due to the confinement imposed by the nanoparticle shape.

6.5 Methods

Atomic Force Microscopy (AFM). Size and shape of the polymer nanoparticles were characterized by atomic force microscopy (AFM). Samples were drop casted on (100) Si wafers. A Multimode 8 AFM with a Nanoscope V controller (Bruker) was used under tapping mode with NCHV probes (Bruker). Square images with 512×512 pixels resolution were taken. Analysis of size and shape of nanoparticles was performed with the Nanoscope Analysis 1.50 software (Bruker).

Wide Angle X-ray Scattering (WAXS). 2D-WAXS investigations were carried out in transmission geometry using a Bruker AXS Nanostar X-ray scattering instrument. The instrument uses CuKα radiation (1.54 Å) produced in a sealed tube. The sample chamber was under vacuum and controlled temperature. For WAXS experiments, we used lyophilized nanoparticles packed in aluminum sheets. Bulk sample was measured as received also packed in aluminum sheets. The scattered X-rays were detected on a two-dimensional multiwire area detector (Bruker Hi-Star) and converted to one-dimensional scattering by radial averaging and represented as a function of the momentum transfer vector q ($=4\pi/\lambda \sin \theta$) in which θ is half the scattering angle and λ the wavelength of the incident X-ray beam. The sample-to-detector distance was 10 cm. Patterns were collected during 5 min at each temperature.

Differential Scanning Calorimetry (DSC). Calorimetric measurements were carried out by means of a Perkin-Elmer DSC8500 instrument equipped with an Intracooler 2 sub-ambient device and calibrated with purity indium standards. In order to measure the transitions of the nanoparticles, the external block temperature was set at -100 °C. A lyophilized solid powder of nanoparticles was used as sample with weight $c.a.$ 2 mg. The powder was enclosed in aluminum pans and heated from -20 to 200 °C at a rate of 20 °C/min. Bulk sample was measured as received, enclosed in aluminum pans.

Bibliography

Alcoutlabi M, McKenna GB (2005) Effects of confinement on material behaviour at the nanometre size scale. J Phys Condens Matter 17:R461

Anderson JM, Shive MS (2012) Biodegradation and biocompatibility of PLA and PLGA microspheres. Adv Drug Deliv Rev 64(Suppl):72–82

Asada M, Jiang N, Sendogdular L, Gin P, Wang Y, Endoh MK, Koga T, Fukuto M, Schultz D, Lee M, Li X, Wang J, Kikuchi M, Takahara A (2012) Heterogeneous lamellar structures near the polymer/substrate interface. Macromolecules 45:7098–7106

Bassett DC, Olley RH, Al Raheil IAM (1988) On crystallization phenomena in PEEK. Polymer 29:1745–1754

Bertoldo M, Labardi M, Rotella C, Capaccioli S (2010) Enhanced crystallization kinetics in poly(ethylene terephthalate) thin films evidenced by infrared spectroscopy. Polymer 51:3660–3668

Bitinis N, Verdejo R, Cassagnau P, Lopez-Manchado MA, Lopez Manchado MA (2011) Structure and properties of polylactide/natural rubber blends. Mater Chem Phys 129:823–831

Bonanno LM, Segal E (2011) Nanostructured porous silicon-polymer-based hybrids: from biosensing to drug delivery. Nanomedicine 6:1755–1770

Boucher VM, Cangialosi D, Yin H, Schonhals A, Alegria A, Colmenero J (2012) Tg depression and invariant segmental dynamics in polystyrene thin films. Soft Matter 8:5119–5122

Cai YB, Gao CT, Xu XL, Fu Z, Fei XZ, Zhao Y, Chen Q, Iu XZ, Wei QF, He GF, Fong H (2012) Electrospun ultrafine composite fibers consisting of lauric acid and polyamide 6 as form-stable phase change materials for storage and retrieval of solar thermal energy. Solar Energy Mater Solar Cell 103:53–61

Capitán MJ, Rueda DR, Ezquerra TA (2004) Inhibition of the crystallization in nanofilms of poly (3-hydroxybutyrate). Macromolecules 37:5653–5659

Chen D, Zhao W, Russell TP (2012) P3HT nanopillars for organic photovoltaic devices nanoimprinted by AAO templates. ACS Nano 6:1479–1485

Coakley KM, Mcgehee MD (2004) Conjugated polymer photovoltaic cells. Chem Mater 16:4533–4542

Deng R, Liang F, Li W, Yang Z, Zhu J (2013) Reversible transformation of nanostructured polymer particles. Macromolecules 46:7012–7017

Despotopoulou MM, Miller RD, Rabolt JF, Frank CW (1996) Polymer chain organization and orientation in ultrathin films: a spectroscopic investigation. J Polym Sci B 34:2335–2349

Di Benedetto F, Camposeo A, Pagliara S, Mele E, Persano L, Stabile R, Cingolani R, Pisignano D (2008) Patterning of light-emitting conjugated polymer nanofibres. Nat Nanotechnol 3:614–619

Forrest JA, Dalnoki-Veress K, Dutcher JR (1997) Interface and chain confinement effects on the glass transition temperature of thin polymer films. Phys Rev E 56:5705

Garcia-Gutierrez M-C, Linares A, Martin-Fabiani I, Hernandez JJ, Soccio M, Rueda DR, Ezquerra TA, Reynolds M (2013) Understanding crystallization features of P(VDF-TrFE) copolymers under confinement to optimize ferroelectricity in nanostructures. Nanoscale 5:6006–6012

Garlotta D (2001) A literature review of poly(lactic acid). J Polym Environ 9:63–84

Gates BD, Xu Q, Stewart M, Ryan D, Willson CG, Whitesides GM (2005) New approaches to nanofabrication: molding, printing, and other techniques. Chem Rev 105:1171–1196

Grigoriadis C, Duran H, Steinhart M, Kappl M, Butt H-J, Floudas G (2011) Suppression of phase transitions in a confined rodlike liquid crystal. ACS Nano 5:9208–9215

Guan FX, Wang J, Yang LY, Tseng JK, Han K, Wang Q, Zhu L (2011) Confinement-induced high-field antiferroelectric-like behavior in a poly(vinylidene fluoride-co-trifluoroethylene-co-chlorotrifluoroethylene)-graft-polystyrene graft copolymer. Macromolecules 44:2190–2199

Kietzke T, Neher D, Kumke M, Ghazy O, Ziener U, Landfester K (2007) Phase separation of binary blends in polymer nanoparticles. Small 3:1041–1048

Landfester K (2001) The generation of nanoparticles in miniemulsions. Adv Mater 13:765–768

Landfester K (2009) Miniemulsion polymerization and the structure of polymer and hybrid nanoparticles. Angew Chem Int Ed 48:4488–4507

Li CY (2009) Polymer single crystal meets nanoparticles. J Polym Sci B 47:2436–2440

Li X, Malardier-Jugroot C (2011) Synthesis of polypyrrole under confinement in aqueous environment. Mol Simul 37:694–700

Li JJ, Tan SB, Ding SJ, Li HY, Yang LJ, Zhang ZC (2012) High-field antiferroelectric behaviour and minimized energy loss in poly(vinylidene-co-trifluoroethylene)-graft-poly(ethyl methacrylate) for energy storage application. J Mater Chem 22:23468–23476

Liu Y-X, Chen E-Q (2010) Polymer crystallization of ultrathin films on solid substrates. Coord Chem Rev 254:1011–1037

Maillard D, Prud'homme RE (2008) Crystallization of ultrathin films of polylactides: from chain chirality to lamella curvature and twisting. Macromolecules 41:1705–1712

Mano JF, Wang Y, Viana JC, Denchev Z, Oliveira MJ (2004) Cold crystallization of PLLA studied by simultaneous SAXS and WAXS. Macromol Mater Eng 289:910–915

Martin O, Avérous L (2001) Poly(lactic acid): plasticization and properties of biodegradable multiphase systems. Polymer 42:6209–6219

Martin J, Campoy-Quiles M, Nogales A, Garriga M, Alonso MI, Goni AR, Martin-Gonzalez M (2014) Poly(3-hexylthiophene) nanowires in porous alumina: internal structure under confinement. Soft Matter 10:3335–3346

Martín J, Mijangos C (2009) Tailored polymer-based nanofibers and nanotubes by means of different infiltration methods into alumina nanopores. Langmuir 25:1181–1187

Martín J, Maiz J, Sacristan J, Mijangos C (2012) Tailored polymer-based nanorods and nanotubes by "template synthesis": from preparation to applications. Polymer 53:1149–1166

Martín Gago JA, Briones Llorente C, Casero Junquera E, Serena Domingo PA (2008) Nanotecnologia y Nanociencia: Entre la ciencia ficción del presente y la tecnología del futuro. Fundación Española para la Ciencia y la Tecnología, Madrid

Martínez-Tong DE, Soccio M, Sanz A, García C, Ezquerra TA, Nogales A (2013) Chain arrangement and glass transition temperature variations in polymer nanoparticles under 3D-confinement. Macromolecules 46:4698–4705

Martínez-Tong DE, Cui J, Soccio M, García C, Ezquerra TA, Nogales A (2014) Does the glass transition of polymers change upon 3D confinement? Macromol Chem Phys 215:1620–1624

Massa MV, Dalnoki-Veress K (2004) Homogeneous crystallization of poly(ethylene oxide) confined to droplets: the dependence of the crystal nucleation rate on length scale and temperature. Phys Rev Lett 92:255509

Massa MV, Dalnoki-Veress K, Forrest JA (2003) Crystallization kinetics and crystal morphology in thin poly(ethylene oxide) films. Eur Phys J E 11:191–198

McKenna GB (2003) Status of our understanding of dynamics in confinement: perspectives from confit 2003. Eur Phys J E 12:191

Mohapatra SR, Thakur AK, Choudhary RNP (2009) Effect of nanoscopic confinement on improvement in ion conduction and stability properties of an intercalated polymer nanocomposite electrolyte for energy storage applications. J Power Sources 191:601–613

Napolitano S, Cangialosi D (2013) Interfacial free volume and vitrification: reduction in Tg in proximity of an adsorbing interface explained by the free volume holes diffusion model. Macromolecules 46:8051–8053

Napolitano S, Wübbenhorst M (2007) Deviation from bulk behaviour in the cold crystallization kinetics of ultrathin films of poly(3-hydroxybutyrate). J Phys Condens Matter 19:205121

Napolitano S, Capponi S, Vanroy B (2013) Glassy dynamics of soft matter under 1D confinement: how irreversible adsorption affects molecular packing, mobility gradients and orientational polarization in thin films. Eur Phys J E 36:1–37

Qiu J, Wang Z, Yang L, Zhao J, Niu Y, Hsiao BS (2007) Deformation-induced highly oriented and stable mesomorphic phase in quenched isotactic polypropylene. Polymer 48:6934–6947

Rebollar E, Pérez S, Hernández JJ, Martín-Fabiani I, Rueda DR, Ezquerra TA, Castillejo M (2011) Assessment and formation mechanism of laser-induced periodic surface structures on polymer spin-coated films in real and reciprocal space. Langmuir 27:5596–5606

Reiter G, de Gennes PG (2001) Spin-cast, thin, glassy polymer films: highly metastable forms of matter. Eur Phys J E 6:25

Reiter G, Sommer J-U (1998) Crystallization of adsorbed polymer monolayers. Phys Rev Lett 80:3771

Rotella C, Wübbenhorst M, Napolitano S (2011) Probing interfacial mobility profiles via the impact of nanoscopic confinement on the strength of the dynamic glass transition. Soft Matter 7:5260

Santa Cruz C, Stribeck N, Zachmann HG, Balta Calleja FJ (1991) Novel aspects in the structure of poly(ethylene terephthalate) as revealed by means of small angle x-ray scattering. Macromolecules 24:5980–5990

Schönhals A, Goering H, Schick C (2002) Segmental and chain dynamics of polymers: from the bulk to the confined state. J Non Cryst Solids 305:140

Shimizu H, Yamada M, Wada R, Okabe M (2007) Preparation and characterization of water self-dispersible poly(3-hexylthiophene) particles. Polym J 40:33–36

Soles CL, Ding Y (2008) Nanoscale polymer processing. Science 322:689–690

Stoclet G, Seguela R, Lefebvre JM, Elkoun S, Vanmansart C (2010a) Strain-induced molecular ordering in polylactide upon uniaxial stretching. Macromolecules 43:1488–1498

Stoclet G, Seguela R, Lefebvre JM, Rochas C (2010b) New Insights on the strain-induced mesophase of poly(d, l-lactide): in situ WAXS and DSC study of the thermo-mechanical stability. Macromolecules 43:7228–7237

Strobl G (2006) Crystallization and melting of bulk polymers: new observations, conclusions and a thermodynamic scheme. Prog Polym Sci 31:398–442

Svorcik V, Slepicka P, Siegel J, Reznickova A, Lyutakov O, Kvitek O, Hubacek T, Slepickova Kasalkova N, Kolska Z (2013) Nanostructuring of solid surfaces. In: Dong Y (ed) Nanostructures: properties, production methods and applications. Nova Science Publishers, Hauppauge, NY

Tong L, Cheng BW, Liu ZS, Wang Y (2011) Fabrication, structural characterization and sensing properties of polydiacetylene nanofibers templated from anodized aluminum oxide. Sens Actuat B Chem 155:584–591

Tsui OKC, Zhang HF (2001) Effects of chain ends and chain entanglement on the glass transition temperature of polymer thin films. Macromolecules 34:9139

Vanroy B, Wübbenhorst M, Napolitano S (2013) Crystallization of thin polymer layers confined between two adsorbing walls. ACS Macro Lett 2:168

Wang Y, Chan C-M, Ng K-M, Li L (2008) What controls the lamellar orientation at the surface of polymer films during crystallization? Macromolecules 41:2548–2553

Wang Y, Li M, Wang K, Hao C, Li Q, Shen C (2014) Unusual structural evolution of poly(lactic acid) upon annealing in the presence of an initially oriented mesophase. Soft Matter 10:1512–1518

Wasanasuk K, Tashiro K (2011) Structural regularization in the crystallization process from the glass or melt of poly(l-lactic acid) viewed from the temperature-dependent and time-resolved measurements of FTIR and wide-angle/small-angle X-ray scatterings. Macromolecules 44:9650–9660

Wasanasuk K, Tashiro K, Hanesaka M, Ohhara T, Kurihara K, Kuroki R, Tamada T, Ozeki T, Kanamoto T (2011) Crystal structure analysis of poly(l-lactic acid) α form on the basis of the 2-dimensional wide-angle synchrotron X-ray and neutron diffraction measurements. Macromolecules 44:6441–6452

Wei X-F, Bao R-Y, Cao Z-Q, Zhang L-Q, Liu Z-Y, Yang W, Xie B-H, Yang M-B (2014) Greatly accelerated crystallization of poly(lactic acid): cooperative effect of stereocomplex crystallites and polyethylene glycol. Colloid Polym Sci 292:163–172

Wübbenhorst M, Lupascu V (2005) Glass transition effects in ultra-thin polymer films studied by dielectric spectroscopy—chain confinement vs. finite size effects. In: 12th international symposium on electrets, p 87

Wunderlich B (1973) Macromolecular physics, vol 1: crystal structure, morphology, defects. Academic, New York

Xiong Z, Liu G, Zhang X, Wen T, DE Vos S, Joziasse C, Wang D (2013) Temperature dependence of crystalline transition of highly-oriented poly(l-lactide)/poly(d-lactide) blend: in-situ synchrotron X-ray scattering study. Polymer 54:964–971

Xu J, Guo B-H, Zhou J-J, Li L, Wu J, Kowalczuk M (2005) Observation of banded spherulites in pure poly(l-lactide) and its miscible blends with amorphous polymers. Polymer 46:9176–9185

Yu DG, Branford-White C, Williams GR, Bligh SWA, White K, Zhu LM, Chatterton NP (2011) Self-assembled liposomes from amphiphilic electrospun nanofibers. Soft Matter 7:8239–8247

Yuan K, Li F, Chen YW, Wang XF, Chen L (2011) In situ growth nanocomposites composed of rodlike ZnO nanocrystals arranged by nanoparticles in a self-assembling diblock copolymer for heterojunction optoelectronics. J Mater Chem 21:11886–11894

Zhang J, Tashiro K, Tsuji H, Domb AJ (2008) Disorder-to-order phase transition and multiple melting behavior of poly(l-lactide) investigated by simultaneous measurements of WAXD and DSC. Macromolecules 41:1352–1357

Zhang J, Duan Y, Domb AJ, Ozaki Y (2010) PLLA mesophase and its phase transition behavior in the PLLA–PEG–PLLA copolymer as revealed by infrared spectroscopy. Macromolecules 43:4240–4246

Chapter 7
Controlling Morphology in 3D Printing

Ana Tojeira, Sara S. Biscaia, Tânia Q. Viana, Inês S. Sousa, and Geoffrey R. Mitchell

7.1 Introduction

3D printing is a term widely used to refer to Additive Manufacturing (AM) or Direct Digital Manufacturing (DDM), which is a technology used to fabricate objects based on a layer-by-layer process. This technology has been revolutionising manufacturing industry by overcoming some of the limitations of traditional manufacturing processes. The concept of additive manufacturing first appeared in the mid 1980s with the first stereolithography and Selective Laser Sintering machines. In the following decade, other innovative additive processes were developed, responding positively to other challenges which had been identified, significantly impacting on both commercial and academic accomplishments (Bourel et al. 2014).

Even though 3D printing has been subject to intense development, there remains a doubt regarding its suitability and the value added to industries like biomedical, tools and plastics. Traditional and mass production companies may not directly benefit compared with industries of high value/low volume productions, but they can profit from additive technologies when associated with Rapid Tooling (Conner et al. 2014). There are several pre-production steps required before an object can be manufactured (Fig. 7.1). These dictate the quality and functionality of the final product.

Generically, additive manufacturing begins with a conceptual digital design which is converted into a parametric design using computer-aided design (CAD) software. This initial model can be analysed and optimised through computer-aided

---

A. Tojeira (✉) • S.S. Biscaia • T.Q. Viana • I.S. Sousa • G.R. Mitchell
Centre for Rapid and Sustainable Product Development, Institute Polytechnic of Leiria, Marinha Grande, Portugal
e-mail: ana.tojeira@ipleiria.pt

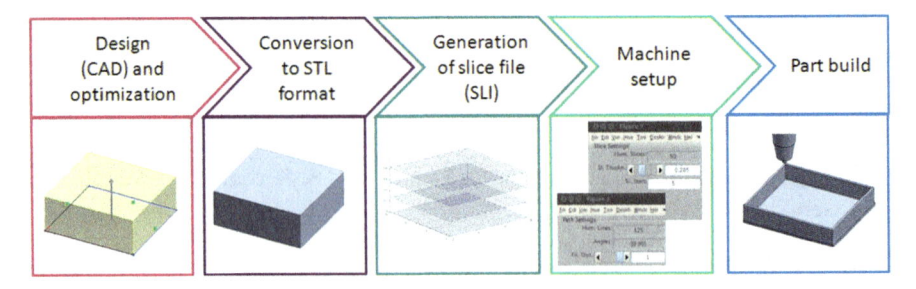

Fig. 7.1 Fused deposition modelling process (adapted from Gibson et al. 2010a)

engineering (CAE) software which mathematically predicts the properties of the part and facilitates the redesign of the part (Wong and Hernadez 2012). The next step involves the conversion of the digital design to an STL file format (ASCII or binary) which consists of a model of the part generated by a combination of the vertices and the normals of triangles defining the boundaries of the part.

Subsequently, the STL file is imported in a pre-processing software package such as Magic RP®, Viscam® and Netfabb® to define the orientation of the part on the build platform, the support structures and the layer densities. The slice thicknesses and tool path are both defined and saved in an SLI file (slice). Finally, the file is sent digitally to the selected printer for construction (Petrovic et al. 2011). The manufacture of 3D printed parts is often followed by operations such as removal of the support structures (simple hand separation or dissolution baths), clean-up and post-processing. The latter is designed to reduce surface roughness (see Sect. 7.3.1).

Developments in additive technologies provide a number of advantages including (Petrovic et al. 2011):

- Product customisation with very low constraints in the geometric complexity of the part (freedom of design)
- Capability of print on demand due to high speed of small series of parts
- There is no need for extra tools or moulds (any support necessary to production will be created in each slice)
- Parts with variety of density can be manufactured (depending on the layers fill settings)
- Possibility of usage of recyclable materials
- Near to zero material waste

However, 3D printing technology still presents some challenges to overcome:

- There is increased production costs for larger series
- Lower mechanical and thermal performance of printed parts compared to conventional manufacturing techniques
- Limited choice of materials for additive technologies

7 Controlling Morphology in 3D Printing

- The staircase effect due to layered construction which impacts the surface finish and dimensional accuracy
- No recovery of support materials

7.1.1 Light-Based Technologies

The earliest AM systems employed high intensity laser light beams to process resins or powders.

The laser curing process causes the liquid resin to solidify using frequency-specific light that moves point-by-point or layer-by-layer across the part area. Stereolithography (SLA) was first introduced to the market in 1986 by the 3D Systems® company and has been continuously used since then due to its accuracy and resolution of ~20 μm (Melchels et al. 2010). At the present time, stereolithographic technologies use either UV or infra-red lights or a combination of both (Bártolo and Mitchell 2003). Some examples of SLA-based 3D printers are detailed in Table 7.1.

In Selective Laser Sintering (SLS) technology, the light causes powder to controllably melt and fuse with neighbouring particles. Upon removal of the laser, the molten powder rapidly solidifies producing a consistent layer (Gibson et al. 2010b). This technology was originally developed at the University of Texas and commercialised by the DTM Corporation (Beaman et al. 1997; Hon and Gill 2003). Table 7.2 shows the most recent commercial SLS 3D printers.

Table 7.1 Stereolithography-based 3D printers

3D printer	Print volume (cm^3)	Laser wavelength (Nm)	Layer thickness (μm)	References
Form 1+	~2580	405	25, 50, 100	Formlabs (2014)
Titan 1	~5040	400	–	Kudo 3D (2014)
Pegasus touch	~7145	405	25–100	Kickstarter, Inc. (2014)

Table 7.2 Selective laser sintering-based 3D printers

3D printer	Print volume (cm^3)	Laser power (W)	Layer thicknesses (mm)	References
Ice 1	10,000	10	0.1–0.15	Norge Ltd (2014a)
Ice 9	40,500	40		Norge Ltd (2014b)
Sintratec	~2200	2.3	–	Sintratec (2015)
FORMIGA P 110	16,500	30	0.06–0.12	EOS (2015)
Elite P3200	61,440	60	0.13–0.18	3D Printing Systems (2014)
Elite P3600	77,760	80	0.13–0.18	
Elite P5500	181,500	100	0.15–0.1	

7.1.2 Print-Based Technologies

Printing-based additive technologies can be divided in two categories: tri-dimensional printing (3DP) and Inkjet technology. The former is the most primitive printing-based technology, first reported in 1989 by Sachs and co-workers. As the construction of the part proceeds, the system deposits successive individual powder layers that are bonded with a binding liquid which is selectively delivered by an inkjet print head, forming a pattern. The excess non-bonded powder is removed by an air jet and used on the subsequent layers (Pfister et al. 2004; Butscher et al. 2011).

As an alternative to the powder-based technologies, Inkjet Printing has been developed as a versatile process using liquid viscous materials. Inkjet printers can be further subdivided accordingly to the method of the delivering the material onto a moving build platform: drop-on-demand and continuous jet.

7.1.3 Extrusion-Based Technologies: Melt or Solution

Melt-based techniques are associated with high temperature material processes such as Fused Deposition Modelling (FDM), Fused Filament Fabrication (FFF), 3D fibre deposition, Precise Extrusion Manufacturing (PEM) and Multiphase Jet Solidification (MJS) which deposits a molten material onto the build platform. In the case of natural-derived materials, high temperatures compromise the product integrity, and it is necessary to resort to solution-based room temperature technologies. Regarding each type of extrusion system, four types of nozzles have been used in several studies: pressure-actuated, solenoid-actuated, piezoelectric and volume-actuated nozzles (Giannatsis and Dedoussis 2009).

7.1.3.1 Fused Deposition Modelling/Fused Filament Fabrication

Fused deposition modelling (FDM), first developed by Stratasys® and commercialised in 1992, uses a thermoplastic filament and a filament support material that are heated above their melting temperature in the nozzle where it is further extruded forming again a filament in a predefined position on the platform. The first layer, the so-called raft, is deposited onto a descending platform to form the compact and even base for the construction of the part. The molten material quickly solidifies and provides enough mechanical support to the part during printing. After each layer is built, the platform is lowered to enable a new layer deposition with user-defined build parameters such as writing speed, layer density and temperature, until the part is completed. The system is shown schematically in Fig. 7.2 (Chang and Huang 2011).

Fig. 7.2 Fused deposition modelling machine layout

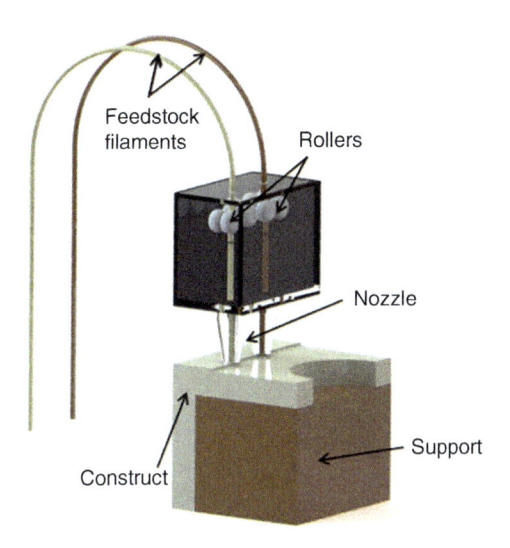

7.2 FDM/FFF Feedstock Materials

Fused deposition modelling and fused filament fabrication, as outlined above, is a melt-based process which uses heat to transform the material in to a more pliable manner. Tri-dimensional constructs require both building and support materials generally made from acrylonitrile-butadiene-styrene (ABS), polycarbonate (PC), blends of PC/ABS or polyphenylenesulfone (PPSU) (Wendel et al. 2008).

As the end use of AM parts has evolved from prototypes to marketable and functional components, developments have focused efforts on application-specific demands. Recent studies have reported usage of other thermoplastics and thermoplastic-based composites to obtain components with improved thermal or mechanical properties or even enhanced biodegradability (Table 7.3).

Polymer–metal compounds have been developed specifically for direct rapid-tooling applications as proposed by Masood and Song (2004). In that study, the composite filament was prepared by adding iron powder to a polyamide thermoplastic (P301) in order to increase the strength, the toughness and the conductivity of 3D printed injection moulding inserts. It was observed that the use of the FDM polymer/metal inserts reduced the injection cycle time and pressure of ABS parts compared to the accurate clear epoxy solid (ACES) injection moulding obtained from SLA.

Shofner et al. (2003) studied the effect of dispersion of vapour-grown carbon fibres (VGCF) in an ABS copolymer matrix on the mechanical properties of specimens printed in a Stratasys FDM 1600 Modeler. They observed good dispersion and alignment of the fillers and a lower porosity.

3D printing has also been used to produce medical implants and tissue engineering constructs involving biocompatible and bioresorbable scaffolds. Skowyra

Table 7.3 Fused deposition modelling-based 3D printers

3D printer	Print volume (cm^3)	Printable materials	Heated bed	Print speed (mm/s)	Print resolution (μm)	Nozzle size (mm)	Layer thickness (μm)	References
Printrbot Plus	~16,400	ABS, PLA	Yes	80 (max.)	100	–	–	Printrbot (2015)
Bukito	~3850	ABS, PLA, nylon	No	>200	–	0.5	–	Deezmaker Bukobot (2014a)
Ultimaker	~10,600	ABS, PLA	No	30–300	Z: up to 20	0.4	–	Ultimaker BV (2015)
Up Plus 2	~2650	ABS, PLA	Yes	–	–	–	150–400	PP3DP (2012a)
Cube 3	~3550	ABS, PLA	No	–	–	–	70–200	3D Systems, Inc (2015a)
Up Mini	~1730	ABS, PLA	Yes	–	–	–	200–350	PP3DP (2012b)
Felix 3.0	~13,070	PLA, ABS, Nylon	Yes	–	XY: 13 Z: 0.4	0.35	50–300	Felixrobotics BV (2014)
LulzbotTaz4	~20,500	ABS, PLA, PVA, HIPS, Laywood	Yes	200 (max)	–	–	75–350	Aleph Objects, Inc (2015)
Bukobot 8v2	8000	ABS, PLA, nylon, PC, PVA, HIPS, Laywood, Laybrick, MABS, PETT, TPE	Yes	200	–	–	350	Deezmaker Bukobot (2014b)
Sharebot NG—next generation	10,000	ABS, PLA, Nylon, PU, PS, PET	Yes	4.2	XY: 0.06 Z: 0.0025	0.35	50–350	Sharebot (2015)
3D touch	~16,000	ABS, PLA	No	15	Z: 125	–	–	3D Systems, Inc (2011)
uPrint SE	~4700	ABS	–	–	–	–	254	Stratasys Inc (2013)
CubePro	~17,700	PLA, ABS	No	15	Z: 0.100	–	70–300	3D Systems, Inc (2015b)

et al. (2015) reported on a study involved patient-personalised extended-release prednisolone tablets fabricated via FDM. In that study, polyvinyl alcohol (PVA) tablets were prepared in a modified MakerBot Replicator 3D Printer in order to control drug dose via printed volume. In a similar study, Teo et al. (2011) used FDM technology to fabricate poly(caprolactone)-based scaffolds for delivery of gentamicin sulphate that acts as an antibacterial agent.

7.3 Parameters in FDM/FFF 3D Printers

A number of reports show how each parameter in the FDM process contributes to the quality (less distortion) and functionality of printed parts. The complexity of the 3D printing process arises from the number of variables which are used to define the operation of the 3D printer machine and the material. Bearing in mind the possibility of producing solid or shell objects, the correct selection of the manufacturing parameters contributes to the improvement of the properties, such as density, porosity, layer thickness, and material properties (Villalpando et al. 2014). The intrinsic layer parameters such as printing and feed roller speeds and, extrusion temperature should not be ignored in any optimisation process (Fig. 7.3).

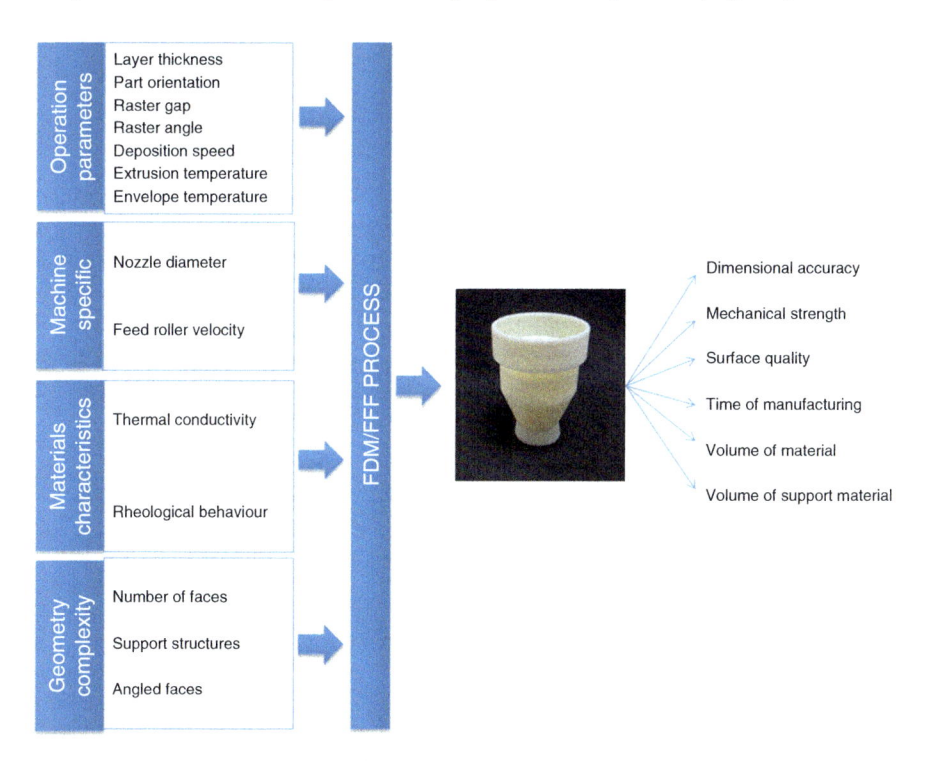

Fig. 7.3 3D printing process parameters

7.3.1 Surface

Central to AM processes is the layer-by-layer construction which produces parts with a variety of surface roughness. 3D printers based on FDM/FFF technologies operate accordingly to the STL derived from the CAD model. Galantucci et al. (2009) pointed out the importance of the STL refinement in order to obtain smooth and model-like surfaces. The staircase-like effect is inevitable and results in angled surfaces (Ahn et al. 2009a, b; Pandey et al. 2003).

7.3.1.1 Layer Thickness

It is generally understood that thinner layers lead to smoother surfaces and more dimensionally accurate parts. The thickness of each layer is defined in the third step (slicing) of 3D printing process when the construction paths are established. Therefore, the slice thickness determines the Z-axis displacement relative to the subsequent layer during production. The definition of this parameter is considered one of the most important parameters that determines surface quality (Fig. 7.4).

Bakar et al. (2010) investigated the effect of layer thickness on the surface roughness of cylindrical parts using a surface tester. From this study, they determined that horizontal surfaces were smoother compared to vertical ones (Fig. 7.5).

7.3.1.2 Part Orientation

In the FDM manufacturing process, an important factor for the quality of printed objects is how the part is orientated on the build platform. It has been established that this is often incorrectly chosen by the user. Not only does the orientation of the part affects the mechanical properties but may also contribute to extended production times, excessive support structures, decreased dimensional accuracy, increased surface roughness and higher cost of the production (Thrimurthulu et al. 2004). Lee et al. (2007) demonstrated the impact of axial and transversal build directions of

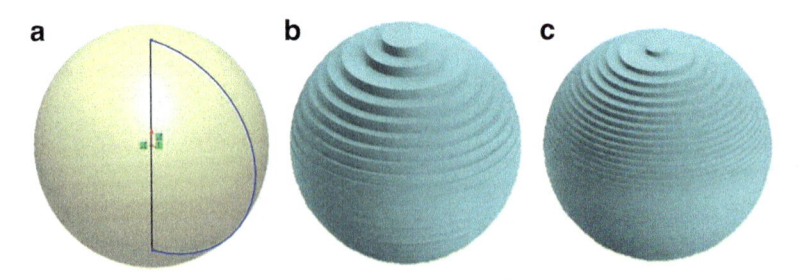

Fig. 7.4 (**a**) CAD model, (**b**) thick layers (rough surface) and (**c**) thin layers (smoother surface)

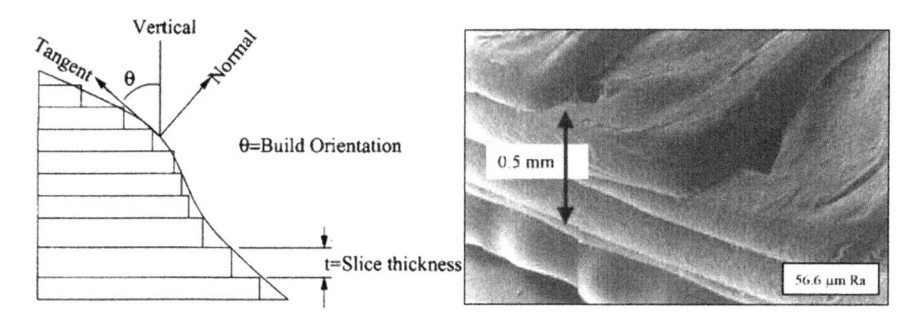

Fig. 7.5 Staircase effect (*left*) Reprinted from International Journal of Machine Tools and Manufacture, Vol. 44, K. Thrimurthulu, Pulak M. Pandey, N. Venkata Reddy, Optimum part deposition orientation in fused deposition modelling, 585–594, Copyright (2004) with permission from Elsevier; and micrograph of 3D printed part with layer thickness of 0.25 mm (*right*) Reprinted from Computer-Aided Design, Vol. 34, R. I. Campbell, M. Martorelli, H.S. Lee, Surface roughness visualisation for rapid prototyping models, 717–725, Copyright (2002) with permission from Elsevier

FDM parts on mechanical properties. Samples with 25.4 mm in length and 12.7 mm in diameter were made from ABS with a filling pattern of 45°/45° angles. From this, it was concluded that the axial samples presented 11.6 % more compressive strength compared to the transversal samples.

7.3.2 Layer and Cross-Section

Depending on the function of the 3D printed parts, the operator may define a different fraction of the layer infill with variable raster gaps and different orientation of the filaments which produces parts with different densities and localised mechanical properties.

7.3.2.1 Gap Between Filaments

The density and mechanical behaviour of FDM parts are greatly affected by the contour and the filling defined during the process. Depending on the final application of the part, the layer density can be tuned by optimisation of the distance between two neighbouring filaments. Low density parts may be produced with larger gaps without requiring supports. Figure 7.6 demonstrates ABS parts in the software package uPrint® with different layer patterns printed.

Smaller or no gaps generate solid layers in which filaments are thermally bonded together. Bond formation is solely driven by thermal energy of the molten material. The heat transfer necessary to join two adjacent filaments is influenced by thermal properties of the material as well as the liquefier and envelope temperatures, nozzle

Fig. 7.6 uPrint ABS parts with different infill patterns: (**a**) solid and (**b**) sparse

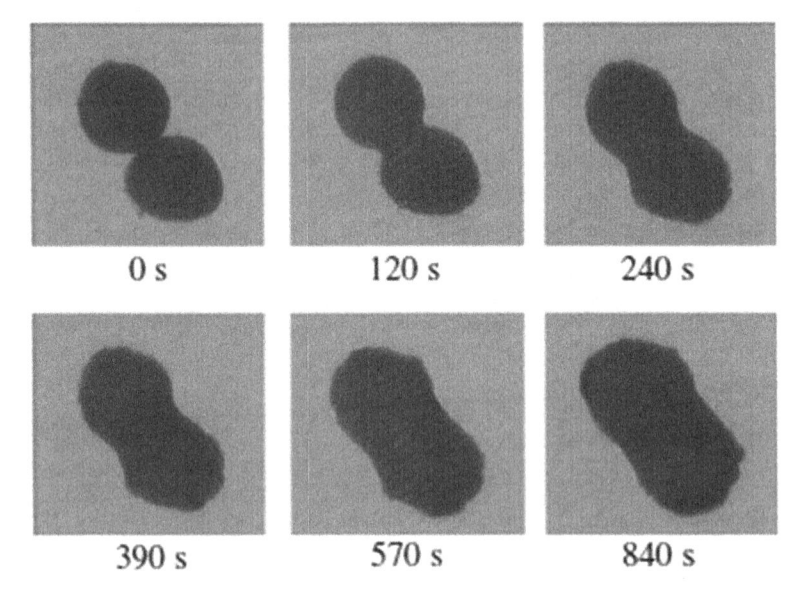

Fig. 7.7 Neck growth between adjacent filament at 200 °C (Reprinted from Journal of Manufacturing Processes, Vol. 6, Céline Bellehumeur, Longmei Li, Qian Sun, Peihua Gu, Modeling of Bond Formation Between Polymer Filaments in the Fused Deposition Modeling Process, 170–178, Copyright (2004), with permission from Elsevier

diameter and the flow and cooling rates. The bonding process occurs into three different stages: interfacial contact, neck formation and growth and molecular diffusion where polymer chains migrate across the interfacial zone.

Bellehumeur et al. (2004) studied the bonding quality of ABS P400 filaments. Figure 7.7 demonstrates the neck growth between adjacent filaments with no gap at a high envelope temperature with time. In a similar study, Reddy et al. (2007) reported the effect of small gaps on the inter-filament bond strength.

Too et al. (2002) conducted a study relating the layer microstructure obtained from gaps between the filaments ranging from 0 to 0.5 mm. This study allowed the development of an understanding of the versatility of the FDM process to obtain matrix-like structures with a consistent pore distribution. Compressive tests were performed to determine how the gap between the filaments affects the mechanical behaviour of ABS P400 printed parts. It was found that by setting the gap to 0.2 mm which yielded a measured porosity of ~59 % caused a decrease of the compressive strength by half compared with no gap samples.

7.3.2.2 Layer Patterning

An important operation parameter which directly affects the layer anisotropy is the raster angle that is defined by the direction of the deposited filaments on the plane of construction.

Ziemian et al. (2012) studied the impact of four different raster angles on the mechanical properties of ABS printed specimens: 0° (aligned with the longitudinal axis of the part), 45° (diagonal), 90° (transversal) and +45°/−45°. In order to validate the 3D printed parts, identical mechanical tests were conducted on the injected parts. From the results, they concluded that tensile properties were higher with the raster angle set at 0° and weaker with a raster angle of 90°.

Failure profiles were analysed, leading to the conclusion that the 0° raster angle causes localised delamination at the primary stages which develop in to monofilament deformations whereas 90° raster angles demonstrates a failure due to a weak filament bonding leading to instantaneous delamination. In contrast, the transversal orientation demonstrated higher compressive strength.

7.3.2.3 Materials Properties

In melt-based 3D printing, amorphous and semi-crystalline materials have been studied and optimised in order to obtain high quality parts. Ippolito et al. (1995) conducted a comparative study of SLA, SLS and FDM technologies and materials, where for the melt-based printing, polyolefin and polyamide resins were used. Through micrographs taken of the surface of the printed parts, they observed significant differences between the layer morphologies (Fig. 7.8) and the mechanical properties.

More recently, reinforced feedstock materials have been studied and are now commercially available to fabricate application specific parts involving mechanical, thermal, physical and biological properties.

An example of the development of materials used in FDM/FFF technologies is that conducted by Tekinalp et al. (2014) where they studied the benefits of short carbon fibre reinforcement on ABS load-bearing 3D printed parts. As observed in Fig. 7.9, carbon fibres were found to exhibit near-perfect alignment with the writing

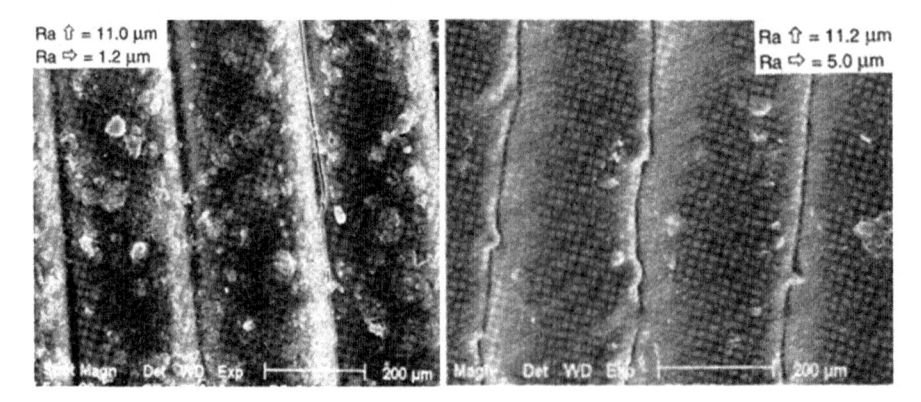

Fig. 7.8 Layer micrographs of 3D printed parts using P300 resin (*left*) and P200 resin (*right*) (Reprinted from CIRP Annals–Manufacturing Technology, Vol. 44, R. Ippolito, L. Iuliano, A. Gatto, Benchmarking of Rapid Prototyping Techniques in Terms of Dimensional Accuracy and Surface Finish, 157–160, Copyright (1995), with permission from Elsevier)

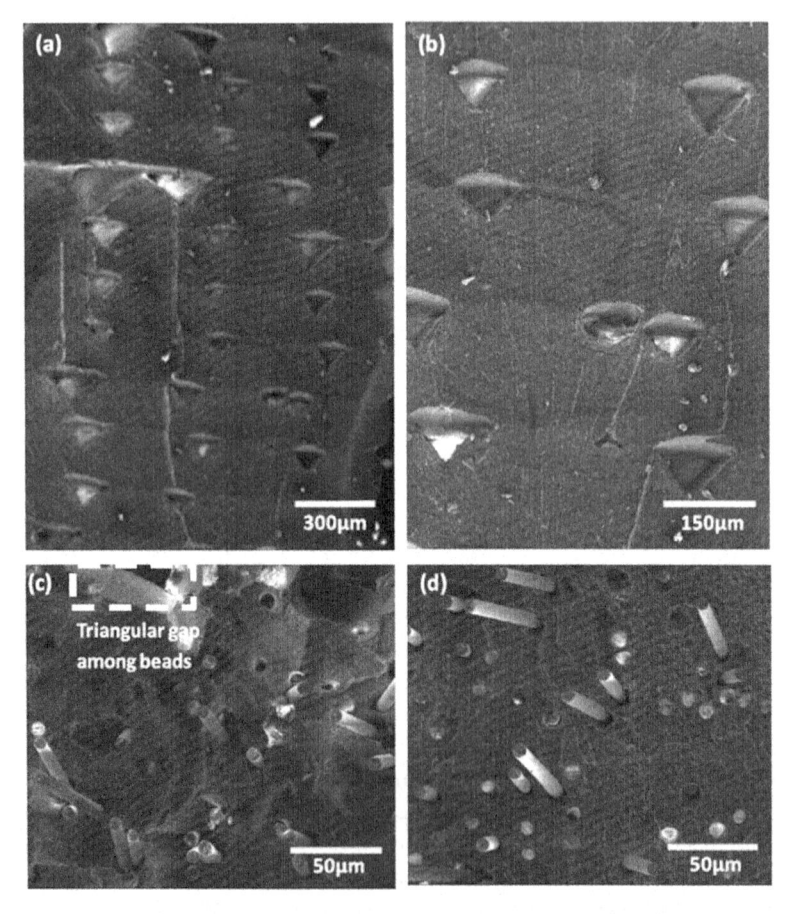

Fig. 7.9 Cross-sectional SEM micrographs: (**a**, **b**) neat ABS, and (**c**) ABS/10 wt% carbon fibre 3D printed parts; and (**d**) compression-moulded ABS/10 wt% carbon fibre (Reprinted from Composites Science and Technology, Vol. 105, Halil L. Tekinalp, Vlastimil Kunc, Gregorio M. Velez-Garcia, Chad E. Duty, Lonnie J. Love, Amit K. Naskar, Craig A. Blue, Soydan Ozcan, Highly oriented carbon-fiber-polymer composites via additive manufacturing, 144–150, Copyright (2014), with permission from Elsevier

direction and well dispersed within the polymer matrix. Furthermore, they demonstrated an upwards trend of the modulus and the tensile strength with the content of reinforcing agent.

7.3.3 Filaments

Several operating parameters affect the morphology of printed objects as they determine how the filaments solidify and subsequently respond to mechanical load. Aside from the geometrical parameters of printed parts, the microstructure obtained during material deposition is expected to be affected by the materials properties such as thermal expansion and viscosity and possibly by operational parameters such as deposition speed, feed roller velocity and extrusion temperature.

The authors of this work have undertaken a series of small angle X-ray scattering experiments (Fig. 7.10) in order to determine how each operating variable impacts the individual filament morphology. Individual strands, prepared using either poly (acid lactic) (PLA) or poly(ε-caprolactone) (PCL), were extruded under different conditions of deposition speed, feed roller velocity and temperature. The PLA samples were prepared using two different 3D commercial printers: 3D Touch® of 3D Systems, Inc. and BOX Basic® from AMcubed, Ltd.

Filaments patterns were collected using a SAXS Quantum 210r CCD detector from ADSC at the Non Crystalline Diffraction beamline at the ALBA Synchrotron with a sample-to-detector distance of ~6.1 m and a wavelength of ~1 Å. Silver behenate sample was used for calibration.

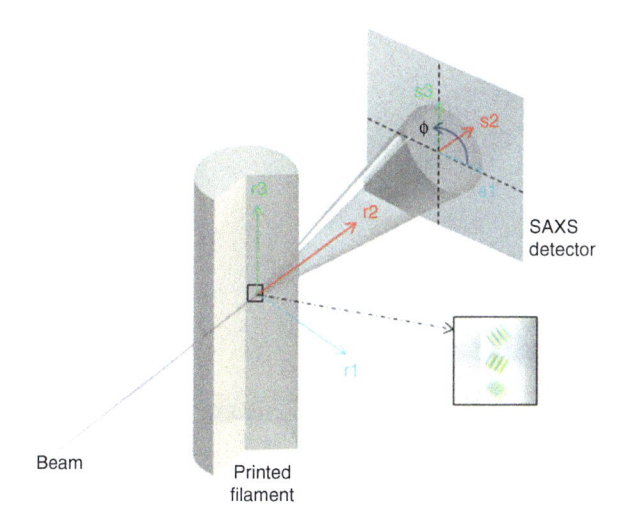

Fig. 7.10 Scattering experiments on 3D printed filaments, where $r1$, $r2$ and $r3$ are the cross-sectional planes of the filament, Φ is the azimuthal angle, $s1$ and $s3$ are the momentum transfer in equatorial and meridional directions ($s = 4\pi \sin(\theta)/\lambda$, where 2θ is the angle between the incident and scattered beam and λ is the X-ray wavelength) and $s2$ is the intensity of the scattering

7.4 Case Study 1: Feedstock Material

Typically, crystallisation from a quiescent melt tends to form a spherulitic morphology. A spherulite is a spherical organisation of the chain folded lamella crystals. However, during 3D printing, the flow stress applied to the liquefied polymer as it passes through the nozzle induces oriented crystallisation due to the formation of row nuclei (Tojeira et al. 2014). The addition of a small fraction of nucleating agents, in this case, graphene nanoflakes, alters the crystallisation mechanism and consequently the morphology of semi-crystalline filaments.

2D SAXS patterns were taken from vertically aligned PCL and PCL/graphene-printed filaments. The addition of different contents of filler induced significant modification in the PCL scattering pattern in the equatorial axis. The scattering patterns shown as inserts in Fig. 7.11 are typical of scattering patterns from semi-crystalline polymers which contain chain folded lamellar crystals. Azimuthal integration of the scattering pattern leads to the calculation of the orientation parameters which described the level of preferred orientation of the crystals with respect to the filament axis. Figure 7.11, demonstrates the variation of the orientation parameter $\langle P_2 \rangle$ determined based on the method described in Chap. 2, Sect. 2.4.1.2. $\langle P_2 \rangle$ values of PCL/graphene filaments indicate a preferential orientation of the crystals c-axis parallel with the filament axis. However, by doubling the amount of the filler no significant alteration was registered.

Analysis of the intensity profiles along the equatorial direction (Fig. 7.12a) shown an obvious peak at approximately 0.047 Å^{-1} which relates to a two-phase

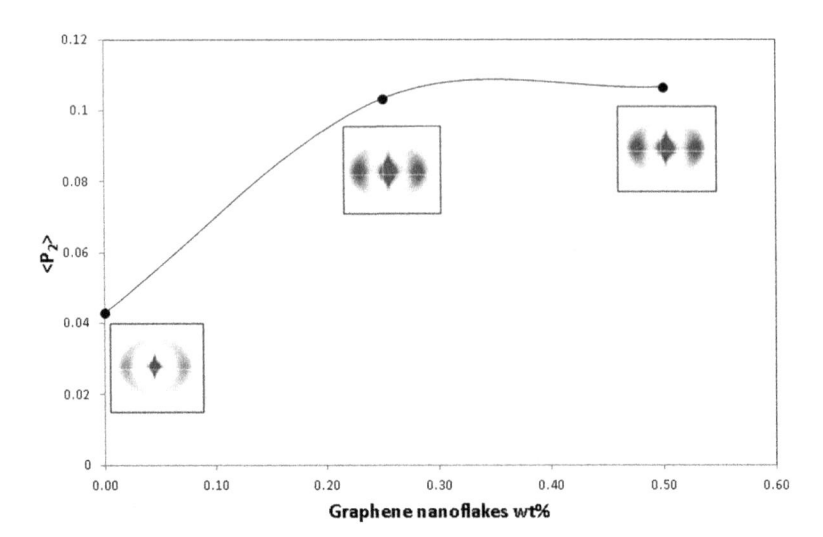

Fig. 7.11 2D SAXS pattern and orientation parameter ($\langle P_2 \rangle$) of vertically positioned PCL filaments with different contents on graphene nanoflakes: 0 wt%, 0.25 wt% and 0.50 wt%

7 Controlling Morphology in 3D Printing

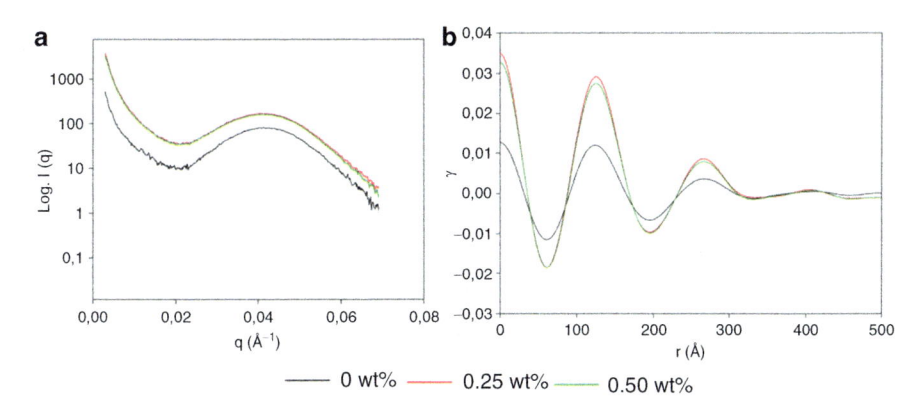

Fig. 7.12 (a) Scattering curves of 2D SAXS patterns of neat PCL filaments and graphene nanoflakes reinforced PCL filaments at different percentages and (b) correlation function obtained from scattering curves

structure (stacks of lamellae). The peaks position remains unchanged with the addition of graphene nanoflakes.

The long period, the crystalline and amorphous layer thicknesses and the bulk volume crystallinity were obtained from an analysis of the correlation functions, an example is shown in Fig. 7.12b. The correlation function is defined as:

$$\gamma(R) = \frac{1}{Q_s} \int_0^\infty I(q)q^2 \cos(qR) dq \tag{7.1}$$

where Q_s represents the relative invariant and which is defined by (7.2):

$$Q_s(t) = \int_0^\infty q^2 I(q) dq \approx \int_{q_1}^{q_2} q^2 I(q) dq \tag{7.2}$$

For correlation function calculation, data were smoothed using Hermite splines and extrapolated to low q values by fitting a linear combination of Gaussian and Cauchy distribution. Extrapolation to high q values was performed using the function in (7.3) considering constant level of background (B) throughout all the experiments (Rabiej and Rabiej 2011).

$$I(s) = \frac{K}{s^4} \cdot e^{-(2\pi\sigma s)^2} + B \tag{7.3}$$

Figure 7.13 shows the variation of long period and crystalline/amorphous layer thicknesses with increase content of graphene nanoflakes. From this, is it possible to observe that the graphene induces a thickening of the soft layer (Ab), leading to an increase of the long period, L_p and a reduction of bulk crystallinity.

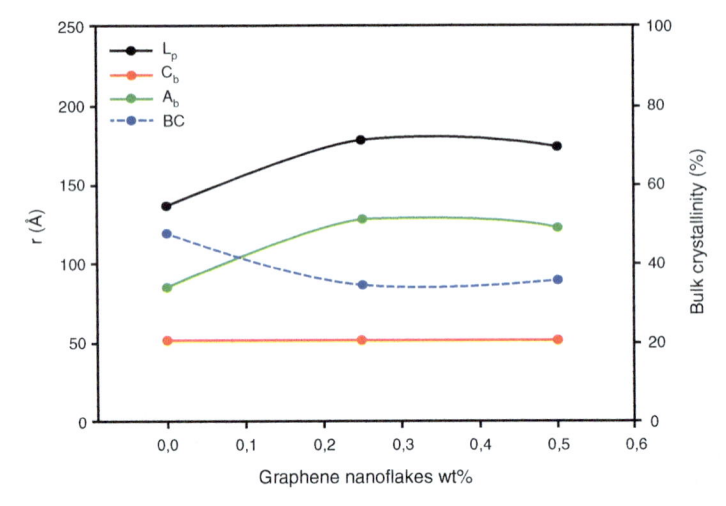

Fig. 7.13 A plot of the variation in the values of long period (L_p), crystalline (C_b) and amorphous layer (A_b) thicknesses and bulk crystallinity (BC) extracted from correlation functions with graphene nanoflake content

7.5 Case Study 2: Deposition Speed

Deposition velocities define how fast the filaments are drawn onto the platform. Very few studies have addressed how this parameter impacts the overall quality of the parts, but Anitha et al. (2001) determined that this variable can contribute ~15 % to the surface roughness of the part. However, within the microstructure of the printed parts, it is possible to observe the effect of varying such parameter and how it modifies mechanical properties.

Glassy polymer filaments do not exhibit hard segments as consequence of the thermal history or randomness along the polymer chains (Tojeira et al. 2014). Figure 7.14 shows radially integrated scattering curves of PLA filaments drawn at different deposition speeds using two different printers. We attribute the increase of the intensity of the scattering pattern at a smaller q to the increase of nanovoids in the PLA filaments.

Although there is no other visible changes on the filament structure, nanovoids formation generally leads to lower mechanical properties due to inter-lamellar slip, as verified in Fig. 7.15.

Semi-crystalline samples, on the other hand, deposited at high speeds produce thinner filaments due to the inherent stretching of the crystallising polymer at the tip of the nozzle. However, in this case study (Fig. 7.16), where the variation of speed was relatively low, there were no significant variations in the crystals orientation (i.e. $\langle P_2 \rangle_{10\,mm/s} = 0.04$ to $\langle P_2 \rangle_{30\,mm/s} = 0.057$) in the solidified filament.

7 Controlling Morphology in 3D Printing

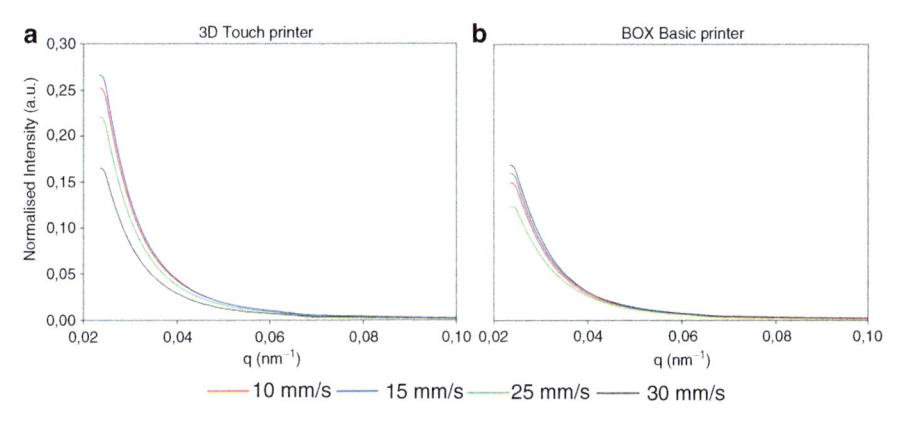

Fig. 7.14 Scattering curves of vertically positioned PLA filaments printed in: (**a**) 3D Touch® and (**b**) BOX Basic®, varying the deposition speed

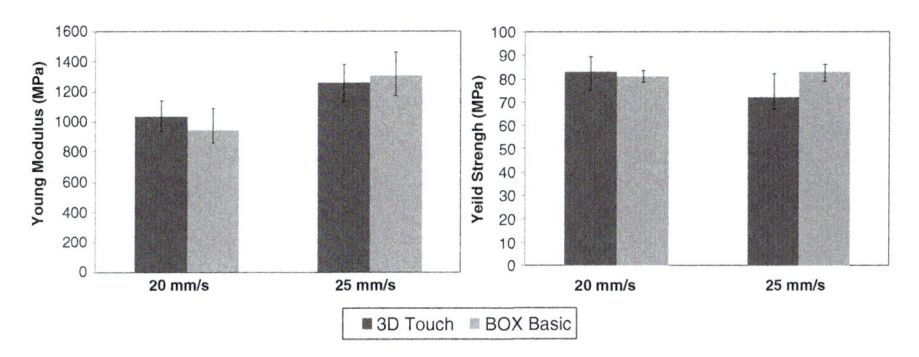

Fig. 7.15 Tensile properties of PLA filaments drawn at different speeds

Similarly to case study 1, PCL filaments drawn at different speeds did not exhibit a variation of the peaks position indicating a very minor impact in this range of velocities (Fig. 7.17a).

From an analysis of the correlation functions of the equatorial scattering curves (Fig. 7.17b), largely constant values regarding the bulk crystallinity and the amorphous and crystalline phases under these processing conditions were observed as shown in Fig. 7.18.

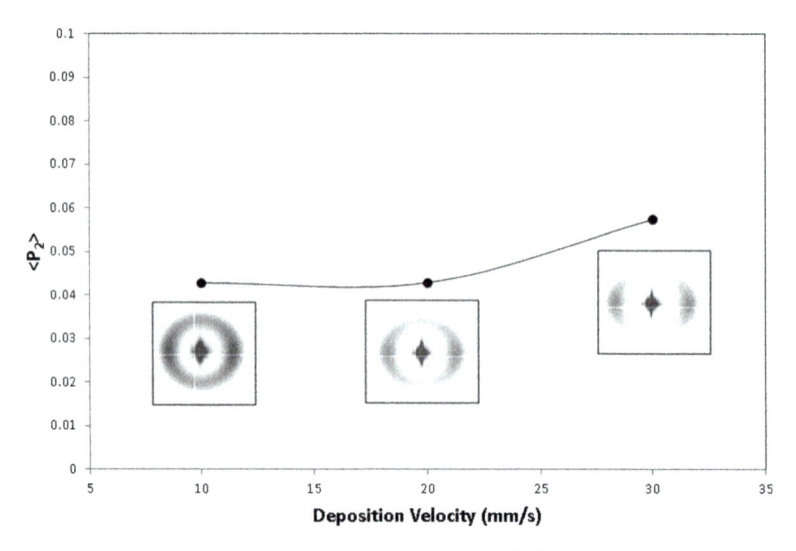

Fig. 7.16 2D SAXS patterns and orientation parameter ($\langle P_2 \rangle$) of vertically positioned PCL filaments drawn at different deposition speeds: 10 mm/s, 20 mm/s and 30 mm/s

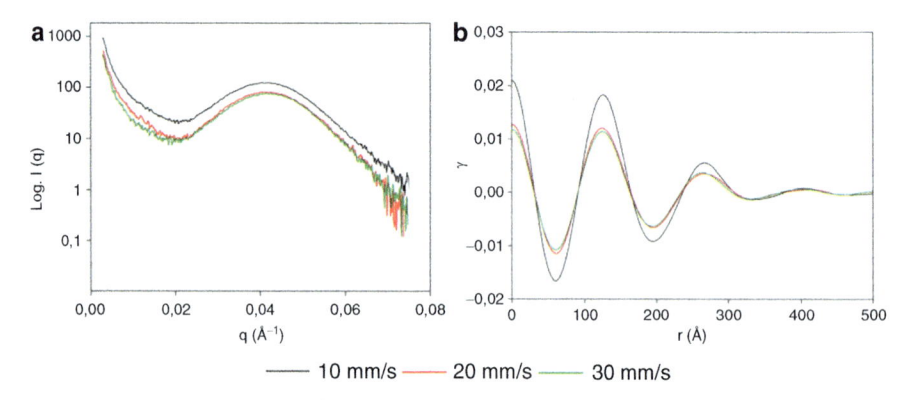

Fig. 7.17 (**a**) Scattering curves of 2D SAXS patterns of neat PCL filaments drawn at different deposition speed and (**b**) correlation function obtained from scattering curves

7.6 Case Study 3: Feed Roller Velocity/Screw Rotation Velocity

Extrusion-based 3D printers are generally efficient with feed roller mechanisms that push the feedstock wire downwards in to the heating chamber. Although smaller in number, pelleted-base printers use a screw-type mechanism to delivery molten material to the nozzle. The most recent example is the David 3D Printing by Sculptify (from Kickstarter®).

7 Controlling Morphology in 3D Printing

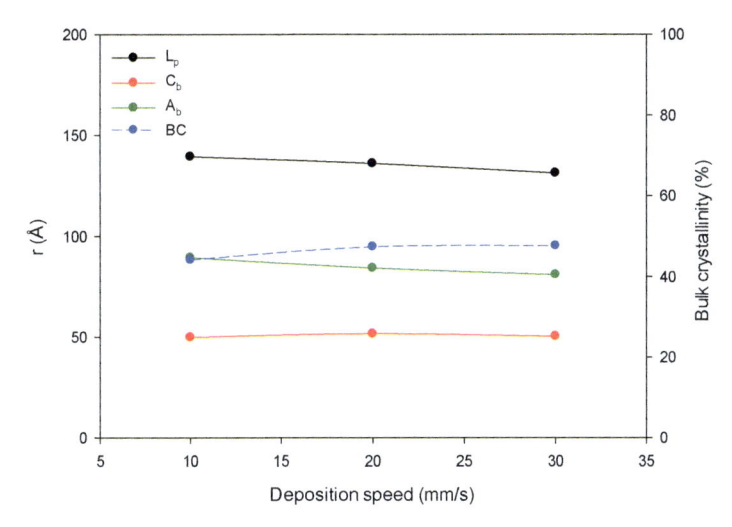

Fig. 7.18 Long period (L_p), crystalline (C_b) and amorphous layer (A_b) thicknesses and bulk crystallinity (BC) extracted from the correlation functions

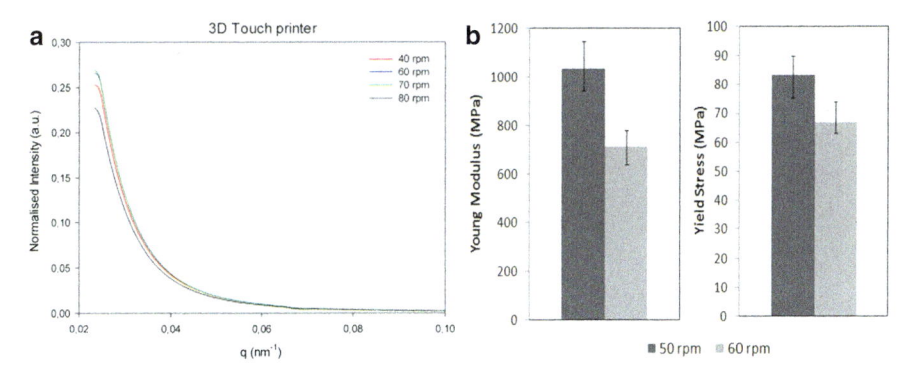

Fig. 7.19 (**a**) Scattering curves of PLA filaments printed in 3D Touch® varying feed roller speed and (**b**) their impact in mechanical properties

The former mechanism feeds the wire into the heated extrusion head using a pinch rollers connected to stepper motors (Turner et al. 2014).

In both printers, the PLA feedstock is compressed between the rollers and pushed against the aperture of the nozzle causing a critical increase of pressure at the tip of the nozzle.

2D SAXS patterns were taken from filaments extruded at different roller angular velocities (Fig. 7.19a). Significant differences in modulus and yield stress were observed between two feed roller speeds in the same 3D printer, as shown in Fig. 7.19b.

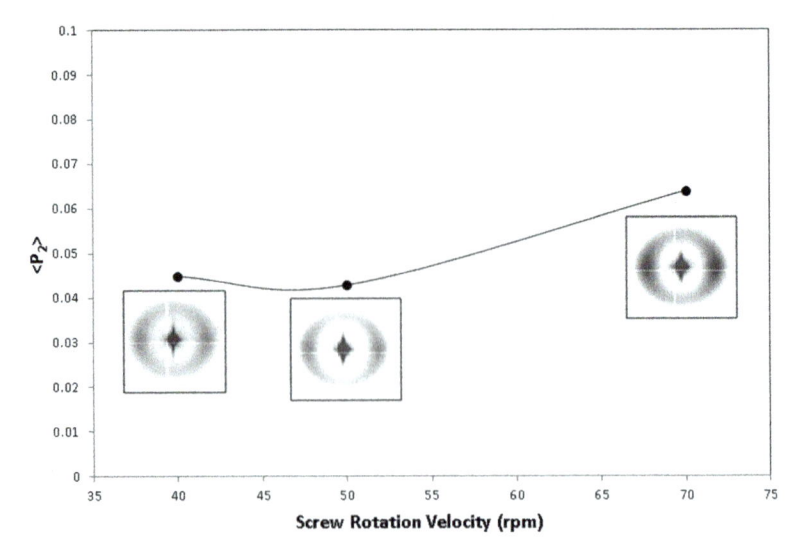

Fig. 7.20 2D scattering patterns and orientation parameter ($\langle P_2 \rangle$) of vertically positioned PCL filaments prepared using different screw rotation velocities: (**a**) 40 rpm, (**b**) 50 rpm and (**c**) 70 rpm

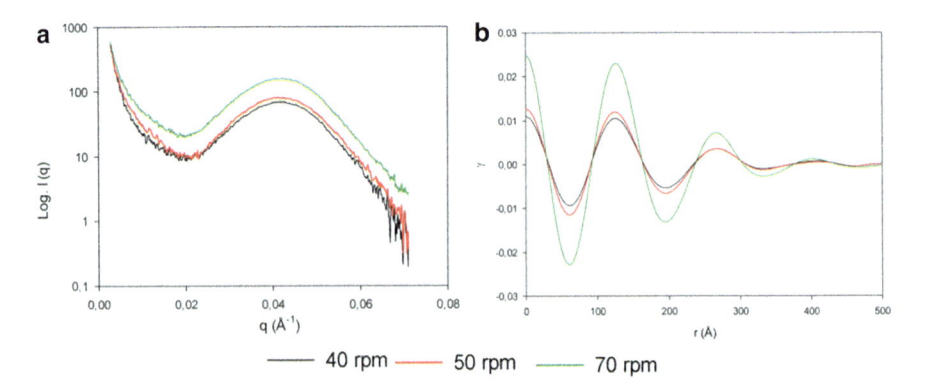

Fig. 7.21 (**a**) Scattering curves of 2D SAXS patterns of PCL filaments drawn at different screw rotation velocities and (**b**) correlation function obtained from the scattering curves

Semi-crystalline filaments, produced in the Bioextruder equipped with a screw-type feeding mechanism, demonstrated a minor increase of preferred orientation of the crystals under the same range of screw rotation velocity (Fig. 7.20).

Further analysis of the scattering curves shown in Fig. 7.21a, b revealed that the feed velocity did affect the thicknesses of the crystalline and amorphous blocks (Fig. 7.22).

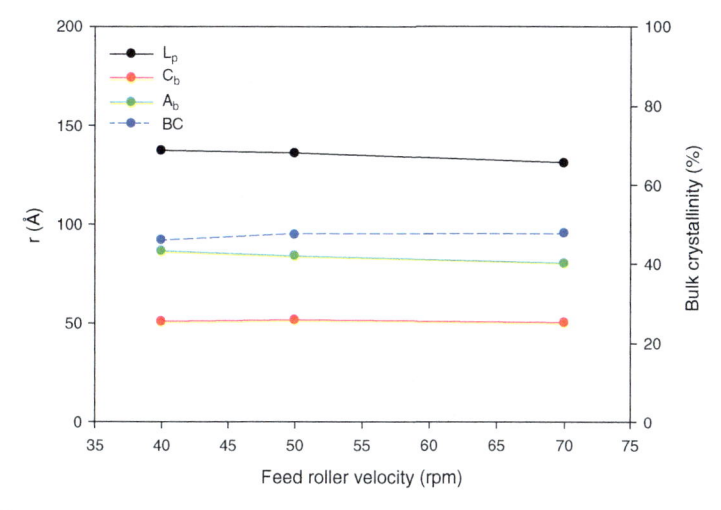

Fig. 7.22 Long period (Lp), crystalline (C$_b$) and amorphous layer (A$_b$) thicknesses and bulk crystallinity (BC) extracted from correlation functions

7.7 Case Study 4: Extrusion Temperature

The dynamics of the molten material in the extrusion head is a quite complex process involving the melting point of the materials coupled with flow processes and stresses which are in part related to the viscosity and surface energy as well as the design of the extruder. Most 3D printers use shear thinning materials that follow power-law viscosity models (7.4):

$$\eta = K(\dot{\gamma})^{n-1} \qquad (7.4)$$

where, η is the viscosity, $\dot{\gamma}$ is the shear rate and K and n are power-law fit parameters. It is important to consider that viscosity varies due to the non-isothermal flow of material across the extrusion head (Turner et al. 2014).

Similarly to case study 2, Fig. 7.23 demonstrates shows the variations of the elastics modulus and yield stress of PLA filaments produced using different temperatures in the extruder using different 3D printers. By fixing the extrusion parameters such as deposition velocity and screw rotation velocity, it is possible to observe that the different temperatures of the extruder nozzle produce strands with different mechanical properties.

We expect that the flow stresses on the liquid polymer in the extruder result in some deformation of the polymer chains in the melt. Before solidification, relaxation may take place on a timescale which is exponentially related to the temperature of the melt. At higher temperatures, relaxation will take place more quickly than at lower temperatures. Clearly, relaxation processes will cease on crystallisation in the case of crystallisable polymers. In this case study, the increase

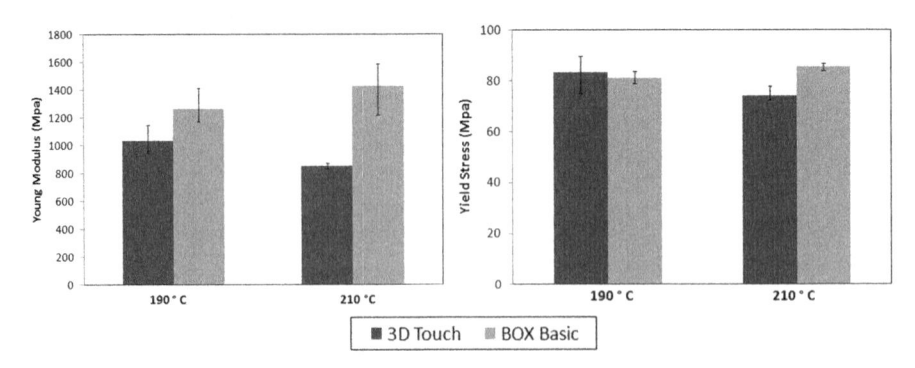

Fig. 7.23 Tensile properties of PLA filaments obtained from two different 3D printers and at varying the nozzle temperature

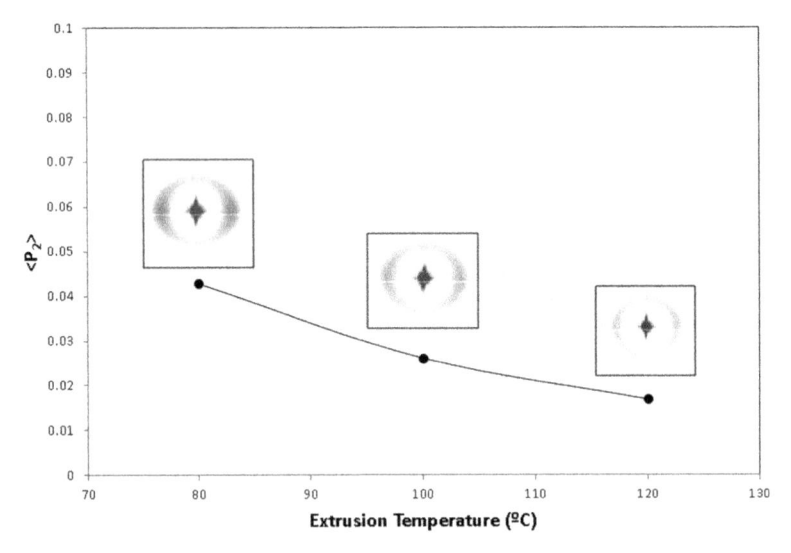

Fig. 7.24 2D scattering patterns and orientation parameter ($\langle P_2 \rangle$) of vertically positioned PCL filaments drawn at different extrusion temperatures: 80 °C, 100 °C and 120 °C

of the extrusion temperature leads to a decrease in the level of preferred orientation of the crystals as $\langle P_2 \rangle$ changed from 0.043 to 0.017 from 80 to 120 °C. The SAXS patterns show a change from anisotropic pattern to an almost isotropic pattern (Fig. 7.24).

Equatorial scattering curves were extracted from the 2D SAXS patterns and the correlation functions were obtained, as displayed in Fig. 7.25. Clearly the processing of PCL at higher extruder temperatures did not induce significant changes in the crystalline and amorphous layer thicknesses, as shown in Fig. 7.26. We attribute this to the fact that the crystallisation takes place at a

7 Controlling Morphology in 3D Printing

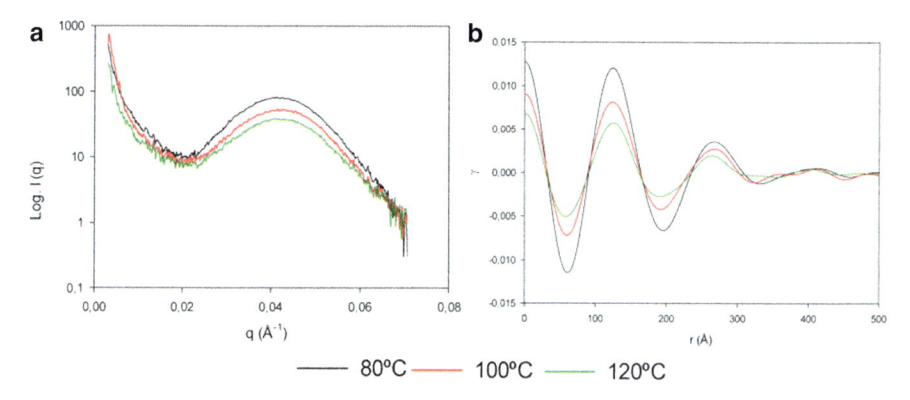

Fig. 7.25 (a) Scattering curves of 2D SAXS patterns of PCL filaments deposited at different temperatures and (b) correlation function of the scattering curves

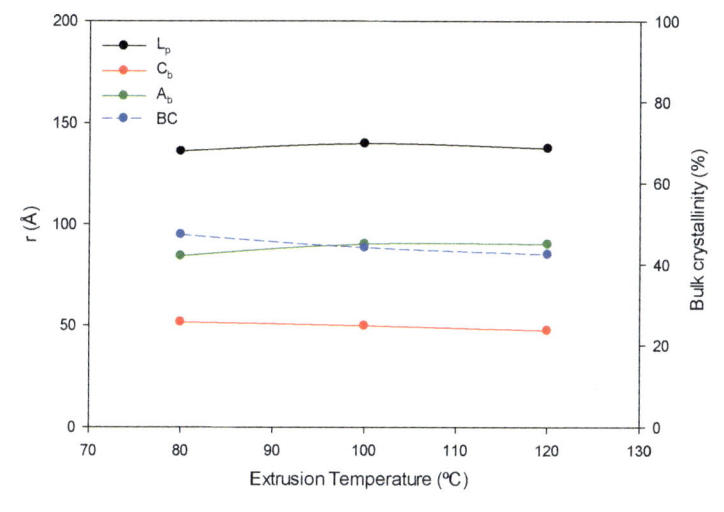

Fig. 7.26 Long period (Lp), crystalline (C_b) and amorphous layer (A_b) thicknesses and bulk crystallinity (BC) extracted from correlation functions

more or less constant temperature defined by the environment in the printer and the build platform. The changes observed in Fig. 7.24 are related to the rate of cooling of the molten filament. If crystallisation takes place from an unrelaxed and deformed melt any extended chains will direct the crystallisation as discussed in Chap. 3. The ability to fine tune the level of anisotropy of the crystalline morphology provides an opportunity to optimise the mechanical properties of the printed objects, particularly important in the field of scaffolds for tissue engineering.

7.8 Summary

In this chapter, we show that additive manufacturing and 3D printing in particular is a multi-step process where each step significantly affects the properties of the object being manufactured. We have found that as with more traditionally polymer processing procedures discussed in other chapters that the precise details of the 3D printing parameters have an impact on the internal morphology of the filaments as well as the external finish quality. In particular, the preferred alignment of the crystals and thicknesses of the amorphous and crystalline layers are effected by the composition of the feedstock and the extruder jet temperature. Clearly, these results highlight the possibilities to use the 3D printing process to fine tune the properties of application-specific components and to estimate how each parameter will affect the final properties.

Acknowledgments The SAXS experiments were performed at the Non Crystalline Diffraction (NCD) beamline at the ALBA Synchrotron with the collaboration of ALBA staff, with special regard to the beamline responsible, Dr. Marc Malfois and beamline scientists Dr. Christina Kamma-Lorger and Dr. Juan Carlos Martínez for the help provided during Small Angle X-ray Scattering experiments. This research work was supported by the Portuguese Foundation for Science and Technology (FCT) through the Project reference UID/Multi/04044/2013.

References

3D Printing Systems (2014) TPM® elite selective laser sintering (SLS) range of 3D printers—specifications. http://3dprintingsystems.com/products/3d-printers/elite-sls-3d-printer-specifications. Accessed 9 Jan 2015

3D Systems, Inc. (2011) 3D TOUCH: make personal colour manufacturing a reality. 3D Systems, Inc. http://cubify.s3.amazonaws.com/public/bfb/0911_touch_uk_3ds.pdf. Accessed 9 Jan 2015

3D Systems, Inc. (2015a) CUBE 3D printer Tech specs. http://cubify.com/en/Cube/TechSpecs. Accessed 9 Jan 2015

3D Systms, Inc. (2015b) CubePro® 3D printing.Real.Pro. http://cubify.com/en/CubePro/TechSpecs. Accessed 9 Jan 2015

Ahn D, Kweon J-H, Kwon S, Song J, Lee S (2009a) Representation of surface roughness in fused deposition modeling. J Mater Process Technol 209(15–16):5593–5600. doi:10.1016/j.jmatprotec.2009.05.016

Ahn D, Kim H, Lee S (2009b) Surface roughness prediction using measured data and interpolation un layered manufacturing. J Mater Process Technol 209(2):664–671. doi:10.1016/j.jmatprotec.2008.02.050

Aleph Objects, Inc. (2015) LulzBot TAZ 4 3D printer. https://www.lulzbot.com/products/lulzbot-taz-4-3d-printer. Accessed 9 Jan 2015

Anitha R, Arunachalam S, Radhakrishnan P (2001) Critical parameters influencing the quality of prototypes in fused deposition modelling. J Mater Process Technol 118(1–3):385–388

Bakar NSA, Alkahari MR, Boejang H (2010) Analysis on fused deposition modelling performance. J Zhejiang Univ Sci A Appl Phys Eng 11(12):972–977

Bártolo PJ, Mitchell G (2003) Stereo-thermo-lithography: a new principle for rapid prototyping. Rapid Prototyping J 9(3):150–156, http://dx.doi.org/10.1108/13552540310477454

Beaman JJ, Barlow JW, Bourell DL, Crawford RH, Marcus HL, Mcalea KP (1997) Solid freeform fabrication—a new direction in manufacturing. Kluwer Academic, Boston

Bellehumeur C, Li L, Sun Q, Gu P (2004) Modeling of bond formation between polymer filaments in the fused deposition modeling process. J Manuf Process 6(2):170–178. doi:10.1016/S1526-6125(04)70071-7

Bourel DL, Rosen DW, Leu MC (2014) The roadmap for additive manufacturing and its impact. 3D Print Addit Manuf 1(1):6–9. doi:10.1089/3dp.2013.0002

Butscher A, Bohner M, Hofmann S, Gauckler L, Muller R (2011) Structural and material approaches to bone tissue engineering in powder-based three-dimensional printing. Acta Biomater 7(3):907–920. doi:10.1016/j.actbio.2010.09.039

Chang D-Y, Huang B-H (2011) Studies on profile error and extruding aperture for the RP parts using the fused deposition modeling process. J Adv Manuf Technol 53(9–12):1027–1037. doi:10.1007/s00170-010-2882-1

Conner BP, Monogharan GP, Martof AN, Rodomsky LM, Rodomsky CM, Jordan DC, Limperos JW (2014) Making sense of 3-D printing: creating a map of additive manufacturing products and services. Addit Manuf 1–4:64–76. doi:10.1016/j.addma.2014.08.005, Inaugural Issue

Deezmaker Bukobot (2014a) Bukito v1 (2013). http://bukobot.com/bukito. Accessed 9 Jan 2015

Deezmaker Bukobot (2014b) Bukobot 8 v2 Vanilla & Duo (2013). http://bukobot.com/bukobot-v2. Accessed 9 Jan 2015

EOS (2015) FORMIGA P 110: compact-class additive manufacturing system that offers a cost-efficient and highly productive entry into the world of additive manufacturing. http://www.eos.info/systems_solutions/plastic/systems_equipment/formiga_p_110. Accessed 9 Jan 2015

Felixrobotics BV (2014) User manual FELIX 3.0, 3d printer VERSION 8. FELIXroboticsBV. http://shop.felixprinters.com/downloads/instruction%20manuals/20140821%20-%20User%20Manual_FELIX_3_0_V10.pdf. Accessed 9 Jan 2015

Formlabs (2014) Form 1+ SLA 3D printer—Tech specs. http://formlabs.com/en/products/form-1-plus/tech-specs/. Accessed 9 Jan 2015

Galantucci LM, Lavecchia F, Percoco G (2009) Experimental study aiming to enhance the surface finish of fused deposition modelled parts. CIRP Ann Manuf Technol 58(1):189–192. doi:10.1016/j.cirp.2009.03.071

Giannatsis J, Dedoussis V (2009) Additive fabrication technologies applied to medicine and health care: a review. Int J Adv Manuf Technol 40(1–2):116–127. doi:10.1007/s00170-007-1308-1

Gibson I, Rosen DW, Stucker B (2010a) Development of additive manufacturing technology. In: Additive manufacturing technologies. Springer, pp 17–39. doi: 10.1007/978-1-4419-1120-9_3

Gibson I, Rosen DW, Stucker B (2010b) Development of additive manufacturing technology. In: Additive manufacturing technologies. Springer, pp 17–39. doi: 10.1007/978-1-4419-1120-9_2

Hon KKB, Gill TJ (2003) Selective laser sintering of SiC/polyamide composites. CIRP Ann Manuf Technol 52(1):173–176. doi:10.1016/S0007-8506(07)60558-7

Ippolito R, Iuliano L, Gatto A (1995) Benchmarking of rapid prototyping techniques in terms of dimensional accuracy and surface finish. CIRP Ann Manuf Technol 44(1):157–160. doi:10.1016/S0007-8506(07)62296-3

Kickstarter, Inc. (2014) Pegasus Touch Laser SLA 3D printer: low cost, high quality. https://www.kickstarter.com/projects/fsl/pegasus-touch-laser-sla-3d-printer-low-cost-high-q. Accessed 9 Jan 2015

Kudo 3D (2014) Titan 1 with orange diamond cover—product description. http://www.kudo3d.com/product/titan1-orange-diamond-package/. Accessed 9 Jan 2015

Lee CS, Kim SG, Kim HJ, Ahn SH (2007) Measurement of anisotropic compressive strength of rapid prototyping parts. J Mater Process Technol 187–188:627–630. doi:10.1016/j.jmatprotec.2006.11.095

Masood SH, Song WQ (2004) Development of new metal/polymer materials for rapid tooling using fused deposition modelling. Mater Des 25(7):587–594. doi:10.1016/j.matdes.2004.02.009

Melchels FPW, Feijen J, Grijpma DW (2010) A review on stereolithography and its application in biomedical engineering. Biomaterials 31(24):6121–6130. doi:10.1016/j.biomaterials.2010. 04.050

Norge Ltd (2014a) Ice1 the first desktop SLS 3D printer. http://www.norgesystems.com/ice-1-low-budget-desktop-sls-3d-printer. Accessed 9 Jan 2015

Norge Ltd (2014b) Ice9, the first low-budget* SLS 3D printer. http://www.norgesystems.com/ice-9-low-budget-full-size-sls-3d-printer. Accessed 9 Jan 2015

Pandey PM, Reddy NV, Dhande SG (2003) Real time adaptive slicing for fused deposition modelling. Int J Mach Tool Manuf 43(1):61–71. doi:10.1016/S0890-6955(02)00164-5

Petrovic V, Gonzalez JVH, Ferrando OJ, Gordillo JD, Puchades JRB, Griñan LP (2011) Additive layered manufacturing: sectors of industrial application shown through case studies. Int J Prod Res 49(4):1061–1079

Pfister A, Landers R, Laib A, Hubner U, Schmelzeisen R, Mulhaupt R (2004) Biofunctional rapid prototyping for tissue-engineering applications: 3D bioplotting versus 3D printing. J Polym Sci A Polym Chem 42(3):624–638. doi:10.1002/pola.10807

PP3DP (2012a) UP Plus 2. http://www.pp3dp.com/index.php?page=shop.product_details&product_id=10&flypage=flypage.tpl&pop=0&option=com_virtuemart&Itemid=37&vmcchk=1&Itemid=37. Accessed 9 Jan 2015

PP3DP (2012b) UP! Mini. http://www.pp3dp.com/index.php?page=shop.product_details&flypage=flypage.tpl&product_id=6&option=com_virtuemart&Itemid=37. Accessed 9 Jan 2015

Printrbot (2015) Assembled Metal Printrbot Plus. http://printrbot.com/shop/assembled-metal-printrbot-plus/. Accessed 9 Jan 2015

Rabiej S, Rabiej M (2011) Determination of the parameters of lamellar structure of semicrystalline polymers using a computer program SAXSDAT. Polimery 56(9):662–670

Reddy BV, Reddy NV, Ghosh A (2007) Fused deposition modelling using direct extrusion. Virtual Phys Prototyp 2(1):51–60. doi:10.1080/17452750701336486

Sharebot (2015) Sharebot NG—next generation. http://www.sharebot.it/index.php/sharebot-next-generation/?lang=en. Accessed 9 Jan 2015

Shofner ML, Lozano K, Rodríguez-Macías FJ, Barrera EV (2003) Nanofiber-reinforced polymers prepared by fused deposition modeling. J Appl Polym Sci 89(11):3081–3090. doi:10.1002/app. 12496

Sintratec (2015) SINTRATEC: World's first desktop laser sintering 3d printer. http://sintratec.com/. Accessed 9 Jan 2015

Skowyra J, Pietrzak K, Alhnan M (2015) Fabrication of extended-release patient-tailored prednisolone tablets via fused deposition modelling (FDM) 3D printing. Eur J Pharm Sci 68:11–17. doi:10.1016/j.ejps.2014.11.009

Stratasys Inc. (2013) uPrint SE™ Proven.Powerful.Professional.—building 3D models at your desk is as easy as clicking "print". StratasysInc. http://www.stratasys.com/~/media/Main/Secure/System_Spec_Sheets-SS/DimensionProductSpecs/uPrintSESellSheet-INTL-ENG-10-13%20WEB.pdf. Accessed 9 Jan 2015

Tekinalp HL, Kunc V, Velez-Garcia GM, Duty CE, Love LJ, Naskar AK, Blue CA, Ozcan S (2014) Highly oriented carbon fiber-polymer composites via additive manufacturing. Compos Sci Technol 105(10):144–150. doi:10.1016/j.compscitech.2014.10.009

Teo EY, Ong S-Y, Chong MSK, Zhang Z, Lu J, Moochhala S, Ho B, Teoh S-H (2011) Polycaprolactone-based fused deposition modeled mesh for delivery of antibacterial agents to infected wounds. Biomaterials 32(1):279–287. doi:10.1016/j.biomaterials.2010.08.089

Thrimurthulu K, Pandey PM, Reddy NV (2004) Optimum part deposition orientation in fused deposition modelling. Int J Mach Tools Manuf 44(6):585–594. doi:10.1016/j.ijmachtools. 2003.12.004

Tojeira A, Biscaia S, Viana T, Bártolo P, Mitchell G (2014) Structure development during additive manufacturing. In: Bártolo P, et al. (eds) High value manufacturing: advanced research in virtual and rapid prototyping, Proceedings of the 6th international conference on advanced research in virtual and rapid prototyping. Taylor & Francis Group, London, pp 221–226. ISBN 978-1-138-00137-4, doi: 10.1201/b15961-42

Too MH, Leong KF, Chua CK, Du ZH, Yang SF, Cheah CM, Ho SL (2002) Investigation of 3D non-random porous structures by fused deposition modelling. Int J Adv Manuf Technol 19 (3):217–232. doi:10.1007/s001700200016

Turner BN, Strong R, Gold SA (2014) A review of melt extrusion additive manufacturing processes: I. Process design and modeling. Rapid Prototyping J 20(3):192–204. doi:10.1108/RPJ-01-2013-0012

Ultimaker BV (2015) Ultimaker 2: makes easy even easier. https://ultimaker.com/en/products/ultimaker-2-family. Accessed 9 Jan 2015

Villalpando L, Eiliat H, Urbanic RJ (2014) An optimization approach for components built by fused deposition modeling with parametric internal structures. Proc CIRP 17:800–805. doi:10.1016/j.procir.2014.02.050

Wendel B, Rietzel D, Kuhnlein F, Feulner R, Hulder G, Schmachttenberg E (2008) Additive processing of polymers. Macromol Mater Eng 293(10):799–809. doi:10.1002/mame.200800121

Wong KV, Hernadez A (2012) A review of additive manufacturing. ISRN Mech Eng 2012, Article ID 208760, 10 pages. doi:10.5402/2012/208760

Ziemian C, Sharma M, Ziemian S (2012) Anisotropic mechanical properties of ABS parts fabricated by fused deposition modelling. In: Gokcek M (ed) Mechanical engineering. ISBN: 978-953-51-0505-3, doi: 10.5772/34233

Chapter 8
Electrically Conductive Polymer Nanocomposites

Thomas Gkourmpis

8.1 Introduction

Polymer nanocomposites combine the properties of the matrix with those of the filler additive, thus allowing for the creation of totally new classes of materials with improved mechanical, electrical, optical and thermal properties. This combination of properties offers immense versatility and design capabilities and as consequence research in nanocomposites has been ever-growing. Traditionally carbon black has been the filler of choice for applications where electrical conductivity was required. This was done mainly due to the simplicity and versatility of the carbon black particles in combination with the relatively low cost preparation methods available (Kuhner and Voll 1993). Over the years other conductive fillers of anisotropic dimensions (high aspect ratio) like metal nanowires, graphene and carbon nanotubes (CNTs) have been introduced leading to a revolution in polymer nanocomposites. The potential of these fillers to achieve high conductivity due to their unique geometry at low or very low concentrations has attracted enormous scientific and commercial attention.

In order for this potential to be realised the possibility of producing and controlling well-defined systems is dependent on the detailed understanding of the structure–property relations of these nanocomposites. In this section we will review systems where the filler introduced creates a level of electrical conductivity in the resulting composition, and we will discuss how different classes of fillers affect the overall properties of the polymer. Furthermore, we will discuss the key structure–property relationships highlighting the underlying mechanisms that give rise to them. Of course such a wide topic cannot be treated within the limitations of this work, so the interested reader is encouraged to explore a number of comprehensive

---

T. Gkourmpis
Innovation and Technology, Borealis AB, Stenungsund SE 444-86, Sweden
e-mail: thomas.gkourmpis@borealisgroup.com

© Springer International Publishing Switzerland 2016
G.R. Mitchell, A. Tojeira (eds.), *Controlling the Morphology of Polymers*,
DOI 10.1007/978-3-319-39322-3_8

studies on polymer nanocomposites available (Bauhofer and Kovacs 2009; Moniruzzaman and Winey 2006; Winey and Vaia 2007; Winey et al. 2007; Byrne and Gunko 2010; Mutiso and Winey 2012; Gkourmpis 2014).

8.2 Percolation Theory

The connectivity and spatial arrangement of objects within a network structure and the resulting macroscopic effects can be described by the percolation theory. In all its variations the percolation theory focuses on critical phenomena that originate from the spatial formation of a network and result in sharp transitions in the behaviour of the system of interest (Kirkpatrick 1973). Percolation models have been applied with various degrees of success to the description of the electrical behaviour of polymer nanocomposites. In these systems the insulating polymer matrix is loaded with conductive filler whose network formation leads to a sharp insulation-conductor transition (Lux 1993). Experimental work and theoretical predictions have established that the system's conductivity σ follows a power-law dependence in accordance with percolation theory

$$\sigma \approx \sigma_o (\phi - \phi_C)^t$$

with σ_o a pre-exponential factor that is dependent on the conductivity of the filler, the network topology and the types of contact resistance. The terms ϕ and ϕ_C correspond to the filler concentration and the critical concentration at the transition (also known as percolation threshold) (Foygel et al. 2005). The critical exponent t is also conductivity-dependant with universal values of $t \approx 1.33$ and $t \approx 2$ for two and three dimensions, respectively (Stauffer and Aharony 1987; Sahimi 1994). It must be noted however that a wide range of values as high as $t \approx 10$ for the critical exponent have been reported (Bauhofer and Kovacs 2009; Gkourmpis 2014). The reason for this deviation from universality is still not well understood with complex tunnelling transport phenomena (Balberg 1987, 2009; Grimaldi and Balberg 2006; Johner et al. 2008) and filler variability in terms of production, entanglement, interconnections and surface chemistry being suggested as possible culprits (Mutiso and Winey 2012). In Fig. 8.1, the percolative behaviour of a system of polycarbonate with single-walled carbon nanotubes can be seen (Ramasubramaniam et al. 2003). This behaviour of sharp conductivity increase at a specific filler concentration is typical of all polymer nanocomposites and can be understood within the framework of the percolation theory.

As discussed previously, there are a number of percolation models dealing with electrical conductivity. Initial approaches were treated with systems where the particles were confined in well-defined periodic positions (lattice models). For such systems analytical solutions can be used in the case of one and two dimensions (Fisher and Essam 1961) and numerical solutions based on Monte Carlo methods for three dimensions (Kirkpatrick 1973). Obviously such models despite their

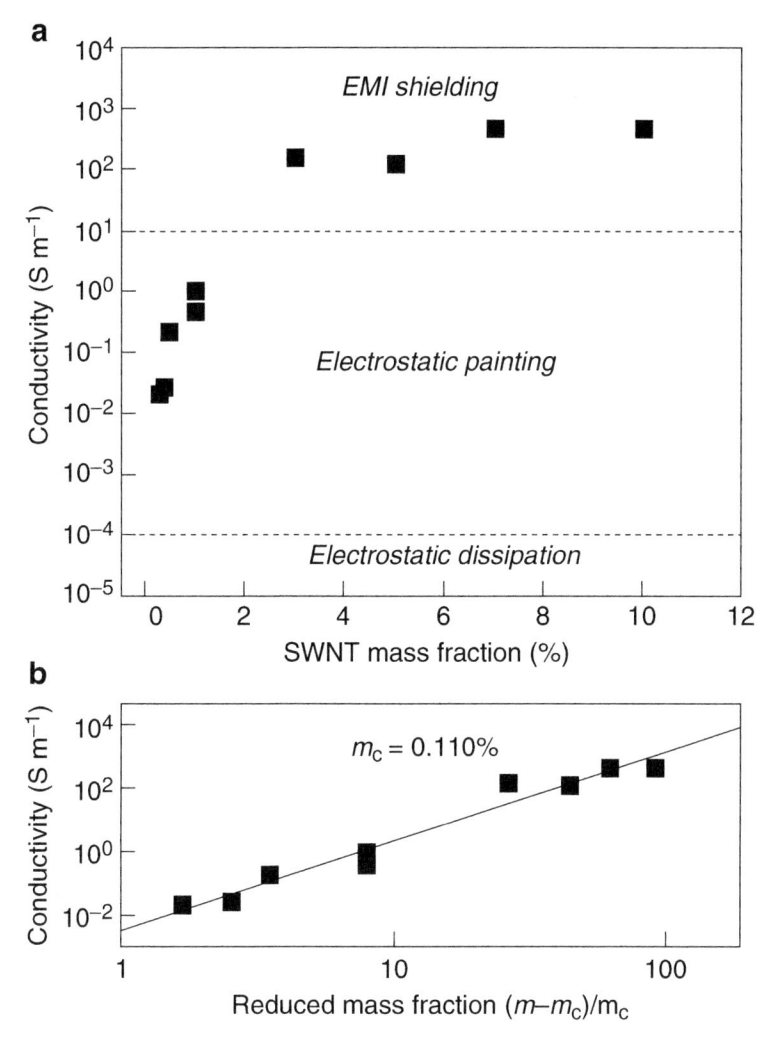

8 Electrically Conductive Polymer Nanocomposites

Fig. 8.1 (a) Electrical conductivity of polycarbonate/SWCNT composites as a function of the filler mass fraction. Different commercial applications are also noted in the graph. (b) Power-law application to experimental data for the calculation of the percolation threshold. Reprinted with permission from Ramasubramaniam et al. (2003)

usefulness are very poor representations of real systems, something that has led to the development of continuum (off-lattice) models that allow more realistic placement of the particles in space. The main types of continuum models can be classified as follows:

Soft-core systems: In this class of models the particles are assumed to be fully interpenetrating and are considered bonded when they overlap.

Hard-core systems: In this approach the particles are allowed to touch each other but not overlap.

Hard-core with soft shell: In this approach the electrical percolation depends not only on the geometry of the particles but on the tunnelling distance between them. In other words the particles are considered to have a hard non-penetrating core and a penetrable shell around the hard core.

For high aspect ratio systems the study gets computationally expensive due to the spatial orientation of the particles. One of the most common approaches to avoid complexity is to treat the particles with respect to their excluded volume (Hard-core) that can be significantly different from their true volume (Balberg et al. 1984). In such a representation the critical number of particles required for percolation N_C is inversely proportional to the excluded volume (V_{excl})

$$N_C \propto \frac{1}{V_{excl}}$$

In a similar manner, the percolation threshold ϕ_C can be expressed as the ratio of the true volume of the particle (V) over the excluded volume

$$\phi_C \propto \frac{V}{V_{excl}}$$

In the case of a cylinder the average excluded volume per rod can be written as (Balberg et al. 1984)

$$V_{excl} = \frac{32\pi}{3} R^3 \left[1 + \frac{3}{4}\left(\frac{L}{R}\right) + \frac{3}{8\pi}\langle \sin\theta \rangle \left(\frac{L}{R}\right)^2 \right]$$

with L and R the length and radius of the cylinder and θ the angle between two random rods. In a similar manner, the percolation threshold can be expressed as (White et al. 2009)

$$\phi_C = \frac{\frac{4}{3}\pi R^3 + \pi R^2 L}{\frac{32}{2}\pi R^3 + 8\pi R^2 L + \pi R L^2}$$

From these relations we can clearly see that the aspect ratio (L/R) will dominate the percolation threshold, and although in the limit $L/R \to \infty$ the above-mentioned equations require some alterations (Balberg et al. 1984; Berhan and Sastry 2007), the basic assumptions are still valid (Mutiso and Winey 2012). In the case of the hard-core with soft shell cylinders the equivalent relationship can be expressed as (Berhan and Sastry 2007; Wang and Ogale 1993)

8 Electrically Conductive Polymer Nanocomposites

Fig. 8.2 Schematic representation of a rod (*left*) and a sphere (*right*) indicating the impenetrable hard core and the penetrable soft shell

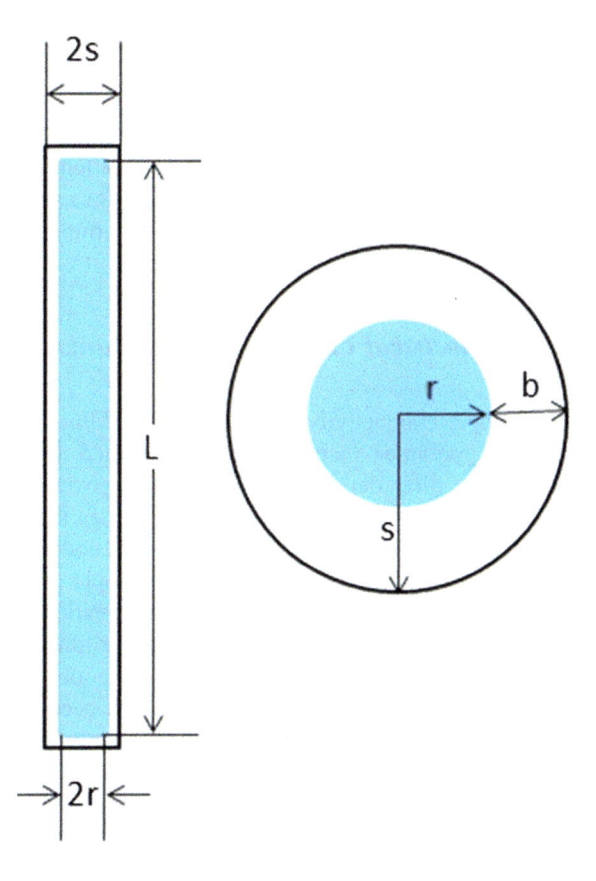

$$V_{excl} = \frac{32\pi}{3}R^3\left[(1-\kappa^3) + \frac{3}{4}\left(\frac{L}{R}\right)(1-\kappa^2) + \frac{3}{32}\left(\frac{L}{R}\right)^2(1-\kappa)\right]$$

with $\kappa = r/s$ being the core–shell ratio (see Fig. 8.2).

Most percolation studies rely on Monte Carlo simulations, whose properties of interest are critical cluster size, bond densities (i.e. number of bonds that lead to network per site), particle volume in the matrix, particle orientation and network geometry at the threshold. In most cases simulations treat the percolation in a statistical manner which means that the particles are randomly distributed in the matrix and network pathways are formed simply by increasing the particle volume fraction in the composite. Although such an approach has merit and provides valuable insight on the network formation, it is far away from reality, especially when polymers are concerned. Addition of particles in a polymer matrix is mainly performed via solution or melt mixing, which means that both particles and polymer chains are in motion and interact with each other. More advanced theoretical approaches do take into consideration the thermodynamic interactions between the composite constituents (particle–particle, particle–polymer and polymer–polymer) as well as any external stimulus such as shear forces. The need to account for the particle mobility and interactions manifests itself in cases where strongly interacting particles with a tendency to cluster after mixing is considered. A typical example of such system is epoxy/CNT composites whose very low reported and experimentally observed percolation ($\phi_C \ll 0.1$ wt%) is attributed to the strong attractive interactions between the carbon nanotubes (Bauhofer and Kovacs 2009; Mutiso and Winey 2012; Gkourmpis 2014; Bryning et al. 2005).

8.2.1 Electrical Conductivity Mechanisms

Macroscopic conductivity is achieved by addition of conductive particles in an insulating polymer matrix and is controlled by percolative behaviour. The conductivity-filler dependence is typically governed by an s-shape curve with three distinct regimes: (1) at very low loadings the filler amount in the matrix is low and the individual particle density is not enough to create an adequate network for current flow. In this regime the macroscopic (observed) conductivity is dominated by the matrix, (2) close to the percolation threshold the filler concentration is such that an adequate network capable of facilitating current flow is created, and this is followed by a sharp increase in the observed conductivity value and, (3) above the threshold the particle network becomes extensive, thus capable to facilitate current flow through multiple channels, and this is manifested by a high observed conductivity value. Within the framework of the percolation theory the electrical conductivity of a polymer nanocomposite is treated in geometrical terms in each of the three distinct regimes. In other words in order to have current flow through the matrix, the particles must be in contact in such numbers as to have a clear intact pathway (network) whose number or density determines the overall value of the conductivity. In the case where the particles are far apart with enough polymers occupying the space between them, no current flow is possible and the system is not conductive. A schematic representation of these ideas can be seen in Fig. 8.3.

Here we must note that even in the insulating region there is a certain level of conductivity despite the lack of adequate conductive network. The conductivity in this region is dominated by trapped charge carriers and polarisation effects and is mainly influenced by the morphology, polarity and electronic band structure of the polymer. Furthermore due to the frequency dependence of conductivity and permittivity of the matrix, the dielectric response of the system will be different for direct and alternating current. Nevertheless, this type of conductivity of essentially insulating materials is beyond the scope of this work but the interested reader is encouraged to consult recently published reviews and the references wherein (Glowacki et al. 2012).

It has been reported that filler particles can be separated by polymer at the high conductive region as well as in the insulating region (Balberg 2009). For electrical conductivity to exist in this case it has been suggested that the electric transport between the conductive particles is facilitated by quantum tunnelling and electronic

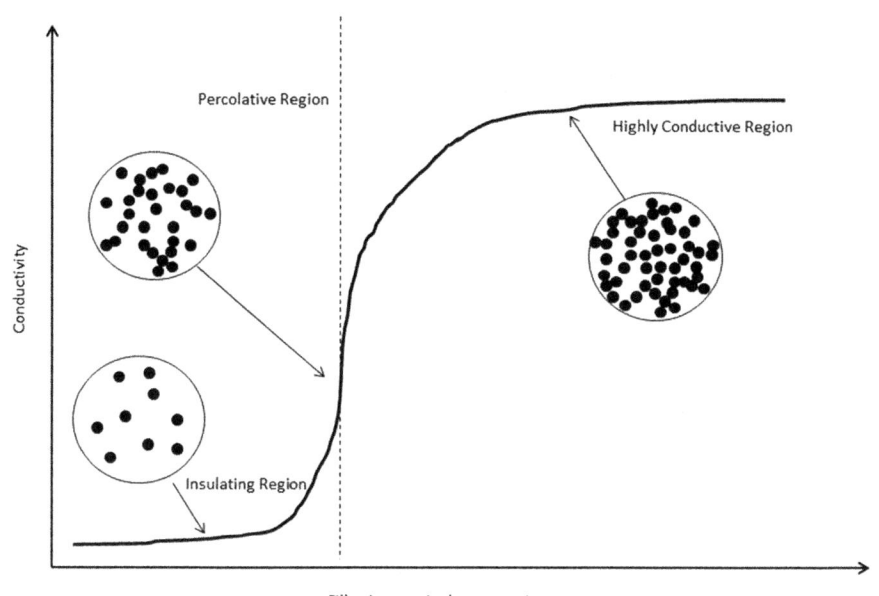

8 Electrically Conductive Polymer Nanocomposites

Fig. 8.3 Schematic representation of the percolative behaviour of a nanocomposite, indicating the different regions as a function of the electrical conductivity and the total amount of filler present. A schematic representation of the network creation for artificial spherical particles can also be seen for the different conductivity regions

hoping mechanisms (Balberg 1987, 2009, 2012; Glowacki et al. 2012; Strumpler and Glatz-Reichenbach 1999; Balberg et al. 2004; Sherman et al. 1983). Here we must note that while the percolation theory requires a sharp cut-off as onset to conductivity (i.e. the particles will either touch and conduct or be apart and not conduct) the tunnelling approximation is a continuous function of the interparticle distances (Stauffer and Aharony 1987; Sahimi 1994). Consequently, the tunnelling conductance decays exponentially with the interparticle distance, thus leading to a non-sharp cut-off (Balberg 2009). According to this we can express the interparticle conductance g between two particles i and j as (Balberg 2012; Ambrosetti et al. 2010)

$$g_{ij} = g_o \exp\left(-\frac{2\delta_{ij}}{\xi}\right)$$

with ξ the wavefunction decay length of the electron outside the particle to which it belongs and δ_{ij} the minimal distance between two particles. In the case of spheres of diameter D the minimal distance can be written as $\delta_{ij} = r_{ij} - D$ with r_{ij} the distance between the two centres. In this case we have two extremes that will lead to qualitatively different conductivity behaviours. In the case of large particles $\xi/D \rightarrow 0$ it becomes obvious from the above relationship that non-zero conductance is possible only when the particles touch each other. Therefore removing particles from the random closed packed limit will be equivalent of removing tunnelling bonds in a similar manner to a standard percolation lattice system (Sahimi 1994; Balberg 2012). In such a case the system will exhibit conductivity that follows percolation-type behaviour with critical exponent $t \approx 2$ and φ_C that correspond to the appropriate system-dependant percolation threshold (Johner et al. 2008). In the other extreme case $D/\xi \to 0$ the sites are randomly distributed in the system and site density ρ variations have no effect on the particle connectivity, but affect the distance $\delta_{ij} = r_{ij}$ between sites (Balberg 2009). In such case it has been seen that the conductivity behaviour in the context of hoping approximation in amorphous semiconductors can be expressed as (Balberg 2012; Ambrosetti et al. 2010)

$$\sigma \propto \exp\left[-\frac{1.75}{(\xi\rho)^{1/3}}\right]$$

which is the low density limit of the first extreme case.

The conductivity behaviour as a function of the percolation threshold and the aspect ratio for spheres, oblate (sheet-like) and prolate (rod-like) spheroids has been calculated using random sequential addition (RSA) and Monte Carlo techniques (Mutiso and Winey 2012; Balberg 2012; Shklovskii and Efros 1984; Tye and Halperin 1989; Berman et al. 1986; Shante and Kirckpatrick 1971; Balberg and Binenbaum 1987; Hunt and Ewing 2009; Ambrosetti et al. 2010). In Fig. 8.4 the distributions of impenetrable spheres and spheroids of different aspect ratio a/b (with a and b the polar and equatorial semi-axes, respectively) can be seen. In this framework the conductivity has been seen to reduce strongly with decreasing volume fraction of the particles (see Fig. 8.5), something attributed to the increasing interparticle distances that lead to a reduction of the local tunnelling conductances (Ambrosetti et al. 2010). This reduction depends strongly on the shape of the filler particles, and with increased anisotropy the overall conductivity onset decreases for a fixed ξ (Fig. 8.5b). Analytical relations for the tunnelling conductivity of randomly distributed prolate, oblate and spherical particles, respectively, can be expressed as (Ambrosetti et al. 2010)

$$\sigma \approx \sigma_0 \exp\left[-\frac{2D}{\xi}\frac{\gamma(b/a)^2}{\phi}\right]$$

$$\sigma \approx \sigma_0 \exp\left\{-\frac{2D}{\xi}\left[\frac{0.15(a/b)}{\phi}\right]^{4/3}\right\}$$

$$\sigma \approx \sigma_0 \exp\left[-\frac{2D}{\xi}\frac{1.65(1-\phi)^3}{12\phi(2-\phi)}\right]$$

In the previous discussion we have seen that the hoping and tunnelling conductivity models are extreme cases of the same problem. Therefore for a system with

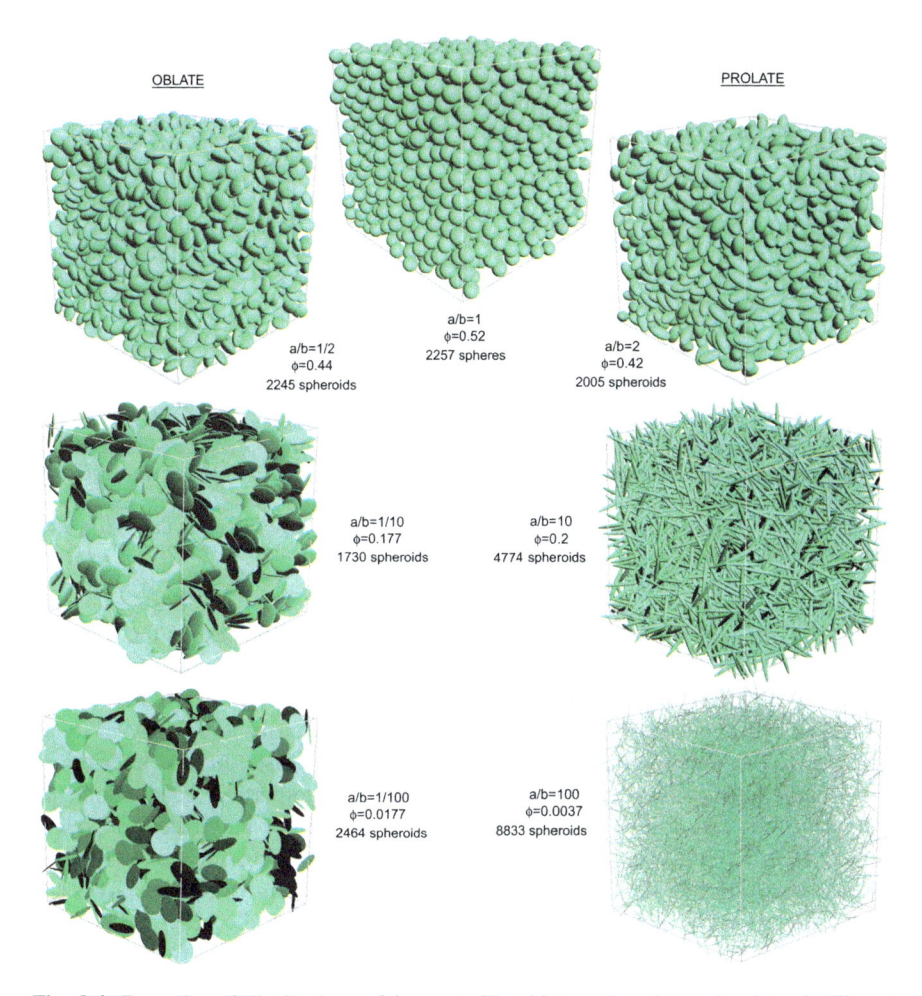

8 Electrically Conductive Polymer Nanocomposites

Fig. 8.4 Examples of distributions of impenetrable oblate and prolate spheroids of different aspect ratio (a/b) and volume fraction (φ) generated from computational techniques. Reprinted with permission from Ambrosetti et al. (2010)

interparticle conduction by tunnelling there must be a "well-defined" threshold. According to this and by assuming that the critical behaviour will depend on the distribution of conductance along the average local resistance of the system we can express the critical exponent as (Balberg 2012)

$$t - t_{\text{un}} = \frac{4(a-b)}{\xi} - 1$$

Although more accurate expressions of t are available (Johner et al. 2008), the above relationship is sufficient for this discussion. So far we have seen that both conductivity (under the hoping model) and the critical exponent (under the

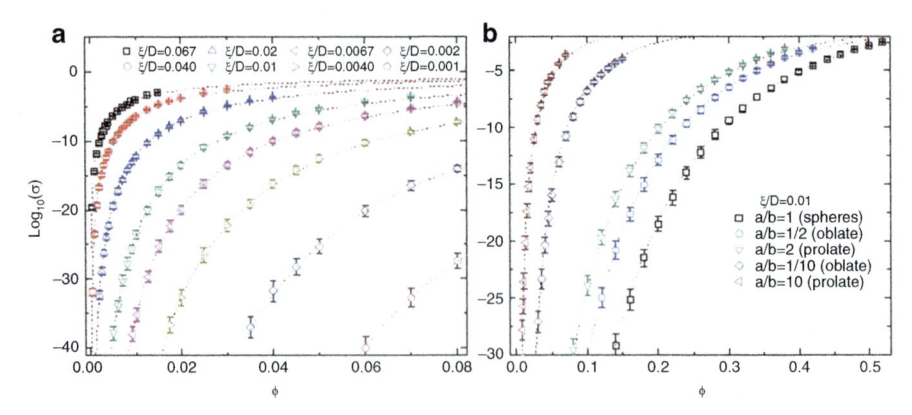

Fig. 8.5 Computational predictions indicating electrical conductivity dependence as a function of volume fraction on (**a**) tunnelling conductivity for hard prolate spheroids of aspect ratio $a/b = 10$ with different characteristic tunnelling distances and (**b**) tunnelling conductivity in a system of hard spheroids of different aspect ratios. *The dotted lines* are predictions based on the critical path approximation method. Reproduced with permission from Ambrosetti et al. (2010)

tunnelling model) are dependent on the interparticle distances in the system. Therefore we can see that when conductivity varies with the concentration of particles, the effect that will dominate is the variation of the local conductance instead of the connectivity of the system. This is obviously valid only in areas where an appropriate interparticle distance distribution exists. In this description, when tunnelling is considered as the mechanism of interparticle conduction, the conductivity function $\sigma(\varphi)$ leads to two distinct results for the critical exponent (Johner et al. 2008; Balberg 2012). Based on the previous discussion we can see that the two expressions of the critical exponent behaviour are equivalent (the physics is essentially the same in both approaches) and will merge in the $(2a - 2b)/\xi \gg 1$ limit. It is also possible to estimate the percolation threshold value for prolate

$$\phi_C \approx \frac{2D}{\xi} \frac{\gamma(b/a)^2}{\ln(\sigma_o/\sigma_m)}$$

and oblate particles as a function of the matrix conductivity σ_m

$$\phi_C \approx 0.15 \left(\frac{a}{b}\right) \left[\frac{2D}{\xi} \frac{1}{\ln(\sigma_o/\sigma_m)}\right]^{3/4}$$

The critical path approximation and the critical path distance δ_C are a very effective way of describing the tunnelling conductivity of polymer composites, since the geometrical connectivity problem of semi-penetrable particles in the continuum is of fundamental importance for the understanding of the filler dependencies (φ, D and a/b) on conductivity. From the above expressions for the percolation threshold (φ_C) we can see that it corresponds to the critical concentration required

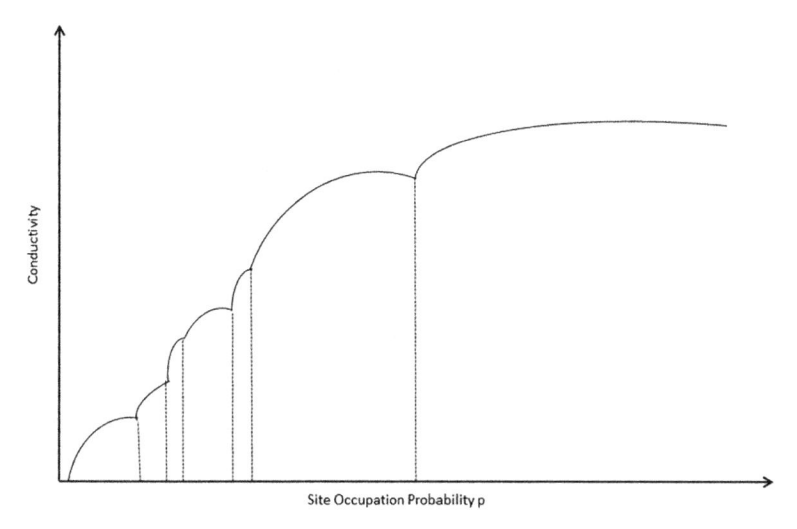

Fig. 8.6 Schematic representation of the electrical conductivity dependence on the site occupation probability when tunnelling is the main conduction mechanism. The *dotted lines* indicate the local percolation thresholds when all corresponding near neighbours are considered. The overall conductivity behaviour appears to have a staircase dependence of the occupation probability leading to a system of multiple thresholds (refer to the text for more details)

for percolation (onset of conductivity) by an ensemble of equal (or nearly equal) resistors (Balberg 2009, 2012; Balberg et al. 2004). Therefore in the tunnelling model more than just one onsets (percolations) are possible in a sequential manner, thus leading to a series of "steps" that are representative of the local conductances in the system (see Fig. 8.6). Thus in the case of a single or multiple mixed but not well-separated steps the data will be fitted with a single t in a manner similar to the standard percolation theory. In the case where multiple clearly defined steps are available multiple fits are required, leading to a larger value of t and a lower percolation threshold. In the latter case we must stress that the value of the percolation threshold does not have an intrinsic meaning.

We will close this discussion on conduction mechanisms by stressing that the tunnelling percolation model is in good agreement with experimental data and the geometrical percolation model. This is quite striking especially if we consider all the different factors that affect filler dispersion in a polymer matrix in comparison with the idealised models discussed previously. In reality fillers have non-uniform size, shape, aspect ratio and geometry distributions, and they are subject to orientation, entanglements, agglomeration and interaction with the polymer matrix. Furthermore, different mixing methods can have a huge impact on the dispersion level and structural integrity of the filler particles, especially in cases where high sheer and elongational forces are present (Gkourmpis 2014; Oxfall et al. 2015; Pötschke et al. 2004).

8.3 Filler Effects on Polymer Nanocomposites

In this section we are going to discuss the effect high aspect ratio fillers have in the polymer matrix and subsequent properties. We will discuss the effect the filler shape, size, distribution and orientation have on the polymer matrix focusing especially in the resulting morphology.

8.3.1 Aspect Ratio

High aspect ratio fillers percolate at low loadings in comparison with spherical or other low aspect ratio objects. This has been attributed to the efficiency of the network formation by large objects, due to the limited number of contacts required for the creation of a conductive cluster. This has resulted in percolation thresholds of the order of 0.1 wt% for CNT and graphene-based nanocomposites (Gkourmpis 2014; Stankovich et al. 2006; Ramasubramaniam et al. 2003; Kim and Macosko 2008; Kim et al. 2010). The theoretical percolation threshold of soft- and hard-core spheres in three dimensions is 0.29 and 0.16, respectively (Balberg 1986; Pike and Seager 1974). In the case of conventional fillers like carbon black the percolation threshold varies between 3 and 20 wt%, but these values are heavily dependent on the carbon black type, structure, surface purity and preparation method (Gkourmpis 2014). Recently, conducted theoretical studies on finite aspect ratio nanowires (Gelves et al. 2006, White et al. 2009) and randomly distributed graphite nanoplatelets (Li and Kim 2007) have highlighted the role of the aspect ratio in the percolation threshold. In both studies the filler diameter and thickness has been seen to have a direct impact on the predicted percolation threshold, and the overall aspect ratio increase leads to lower thresholds (see Fig. 8.7).

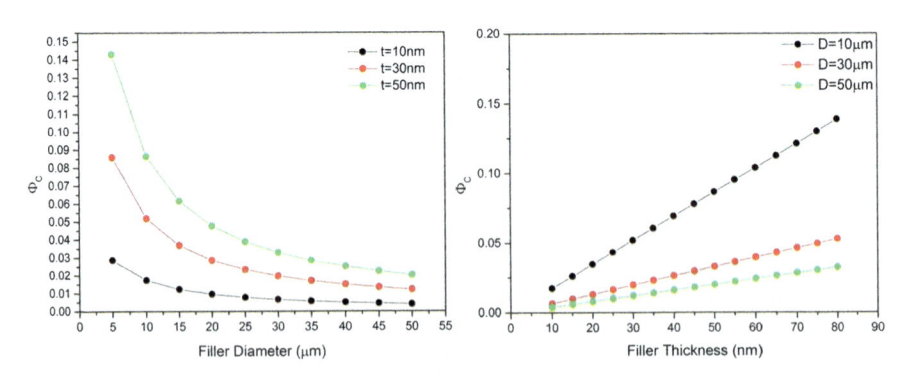

Fig. 8.7 Computational prediction on the effect of the diameter (*left*) and thickness (*right*) of randomly distributed graphene nanoplatelets of different aspect ratio (D/t) on the percolation threshold. Adopted from Li and Kim (2007)

8 Electrically Conductive Polymer Nanocomposites

Table 8.1 Effect of geometric shape factors and aspect ratio on the percolation threshold for graphene nanoplatelets with different matrix polymers

Polymer matrix	D (μm)	t (nm)	d/t	Percolation threshold (%vol)	Reference
High density polyethylene	6	10	600	4.46	Weng et al. (2004)
Epoxy	15	9.5	1579	1.13	Li and Kim (2007)
Poly(styrene-methyl methacrylate)	100	30	3333	0.878	Chen et al. (2001)
Polymethylmethacrylate	100	22	4545	0.529	Zheng and Wong (2003)
Polyisopropylene	50	10	5000	0.67	Chen et al. (2002)
Epoxy	46	4.5	10,222	0.5	Li et al. (2007)
Poly(ethylene vinyl-acetate)	25	6	4000	2.5	Gkourmpis (2014)
Poly(ethylene butyl-acrylate)	25	6	4000	6.9	Oxfall et al. (2015)

Here we must note that reported experimental results vary widely due to the different preparation methods that have a direct impact on the level of dispersion of the filler in the polymer matrix. All theoretical predictions are made assuming perfect or near-perfect dispersion, so a level of discrepancy between the theoretical and experimental threshold values is to be expected. These discrepancies become significant when melt mixing methods are employed. In Table 8.1 a small collection of experimental data for graphite platelets can be seen for a number of different polymers. These values are higher than the theoretical predictions seen in Fig. 8.7 indicating how important the filler geometry and preparation method is. A more comprehensive discussion on the different preparation methods and their effect on the nanocomposite properties has been presented elsewhere (Moniruzzaman and Winey 2006; Gkourmpis 2014; Kim and Macosko 2008; Kim et al. 2010).

8.3.2 Polydispersity

So far in the discussion we have assumed that all fillers introduced in the matrix are of the exact same size and shape. Such assumptions do not hold for real systems where the preparation methods of the filler and the nanocomposite have an effect on the geometry of the individual filler particle (Moniruzzaman and Winey 2006; Gkourmpis 2014; Kim and Macosko 2008; Kim et al. 2010). Carbon nanotubes especially are known to be very sensitive to their preparation methods that can have significant effects on their diameter, length and chirality. The latter is of particular importance since it has a direct influence on the electrical conductivity, resulting to metallic and semiconductive tubes (Moniruzzaman and Winey 2006). As discussed previously the aspect ratio has a strong influence on the percolation phenomena, therefore it is expected that any changes in the distribution of aspect ratio, length, thickness (or diameter) of the filler particles will have an effect on the overall electrical conductivity of the nanocomposite.

It has been reported that the percolation threshold exhibits an inversely proportional relationship to the weight average of the length distribution in the case of rods, an effect that is getting more pronounced with increasing width of the length distribution (Kyrylyuk and van der Schoot 2008). Analytical studies predict that the incorporation of small amounts of longer rods significantly reduces the percolation threshold. This effect is also more pronounced with increasing amounts of longer rods in the system (Otten and van der Schoot 2009). The inverse relation between percolation and amount of conductive filler indicates that the electrical conductivity of the nanocomposite is governed solely by the existence of conductive particles and not the polymer matrix. Although this appears self-evident in real nanocomposites, the situation is not exactly like that. In a real system the distribution of conductivities from the high aspect ratio fillers varies in an almost continuous manner taking into consideration the morphology and the possible defects, leading to a less profound effect on the percolation than predicted theoretically. An example of this is the case of CNT-based systems, where the individual conductivity of SWCNT varies significantly, sometimes to levels of several orders of magnitude (Ebbesen et al. 1996). The contact resistance between CNTs also differs with metallic–semiconducting junctions exhibiting significantly higher values than metallic–metallic or semiconducting–semiconducting, something that has been attributed to the Schottky and tunnelling barrier between different tubes (Fuhrer et al. 2000).

The addition of low aspect ratio particles in the high aspect ratio conductive network creates a hybrid mixture of particles of different sizes. This is a common practice in both research and industrial materials as it allows for unique versatility and balancing of the overall conductivity with other desired properties of interest (e.g. viscosity, mechanical reinforcement, crystallinity). For example it has been reported that the addition of (relatively) more expensive CNTs or GNPs in standard polymer/carbon black compositions offers significant improvements in conductivity and other properties of interest (Oxfall et al. 2015; Ma et al. 2009; Sun et al. 2009; Brigandi et al. 2014; Fan et al. 2009; Wei et al. 2010; Kostagiannakopoulou et al. 2012). For epoxy and poly(ethylene butyl acrylate) copolymers it has been reported that by substituting just 10 wt% of the high aspect ratio filler (graphene nanoplatelets in this case) with carbon black a reduction of the percolation threshold from 0.75 to 0.5 wt% and 6.9 to 4.6 vol.% was achieved for the two systems, respectively (Oxfall et al. 2015; Fan et al. 2009). In a similar manner the creation of hybrid blends of graphene, nanotubes and carbon black has been seen to offer improvements in mechanical properties (Wei et al. 2010; Kostagiannakopoulou et al. 2012). The reason for the improved conductivity of hybrids has been attributed to synergistic effects between the structural arrangements of the two particle networks. Due to thermodynamic reasons it is expected that each conductive particle type will create a network structure in the matrix. This structure will have active pathways capable of enhancing conductivity and the so-called dead ends which are parts of the network that do not contribute in the overall

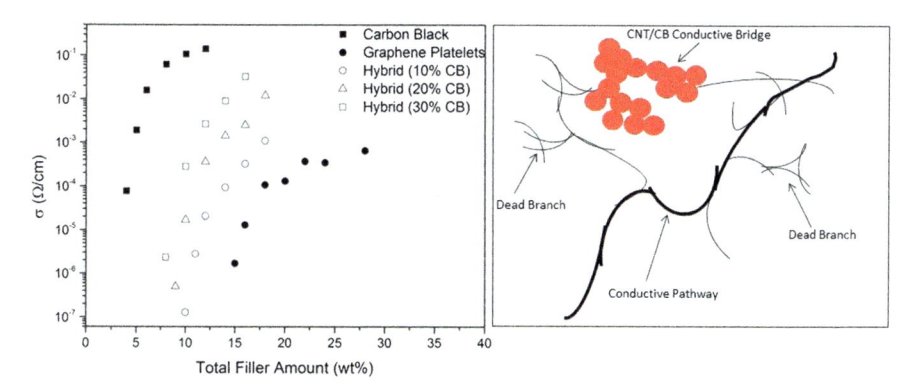

Fig. 8.8 (*Left*) Electrical conductivity as a function of filler content for EBA with carbon black, graphene platelets, and hybrid systems of mixtures between carbon black and graphene platelets. Adopted from Oxfall et al. (2015). (*Right*) schematic representation of a CB/CNT hybrid indicating the preferential localisation of the filler particles, the existence of active (conductive) and dead (non-conductive) network branches and the conductive bridges between the two filler networks existing in the matrix

conductivity. Due to size the network structure of high aspect ratio filler will be fairly extended, whereas the network of the low aspect ratio carbon black is expected to be fairly compact. Therefore, the low aspect ratio filler network can localise partially between branches of the extended high aspect ratio network creating conductive bridges (see schematic in Fig. 8.8b). By doing so the percolative behaviour of the overall composition changes, leading to a lower threshold while at the same time the overall conductivity follows a more complex expression than the traditional power law we discussed earlier. This can be seen in Fig. 8.8 where the percolative behaviour of a system of poly(ethylene butylacrylate) with carbon black and graphene nanoplatelets is presented. Both systems containing a single filler type exhibit the traditional power-law percolative behaviour, whereas the different hybrids (systems of graphene with carbon black added) deviate from it while at the same time exhibit lower percolation threshold values (Oxfall et al. 2015).

8.3.3 Orientation

All fillers are capable of orienting with external stimulus in the melt, a situation that is common in most polymer-based processes. High aspect ratio filler orientation has a far greater impact on conductivity, mainly due to the reduction of contact points between the individual particles. This contact reduction destroys the integrity of the filler network, leading to a substantial decrease in conductivity. The extent of the orientation can be described through the orientation distribution function $n(\theta, \varphi)$ and by assuming that the polar and azimuthal angles θ and ϕ are independent the axial and planar orientation parameters for the case of rods can be expressed as (Munson-McGee 1991)

$$f_p = 2\langle \cos^2\varphi \rangle - 1$$

$$f_a = \frac{1}{3}(3\langle \cos^2\theta \rangle - 1)$$

These parameters can have values between 0 and 1 that correspond to random and axial spatial arrangement, respectively, in a matter similar to the nematic order parameter in liquid crystal polymers (both expressed in the literature as S) (Mitchell et al. 1987; Pople and Mitchell 1997; Lacey et al. 1998; Andersen and Mitchell 2013). Theoretical predictions and experimental work on nanorods in two and three dimensions have indicated that the electrical percolation takes its lowest value in the case of isotropic ($S = 0$) or slightly isotropic ($S \sim 0.1$–0.2) system. As the level of anisotropy increases the percolation threshold takes ever higher values that are correlated with the level of alignment of the filler in the system (large values of S) (Mutiso and Winey 2012; White et al. 2009; Behnam et al. 2007).

Filler orientation has a significant effect on the overall electrical conductivity especially in melt-mixing methods. This is the result of the disruption of the filler network integrity due to the shear and elongational forces during mixing. In almost all cases where the system (matrix and filler) flows the polymer chains will align along the flow direction creating extended chain conformations. This chain alignment leads to further alignment of the filler particles along the flow direction. Obviously the alignment level of the polymer is strongly dependant to the chain conformation, but in principle all polymers do exhibit a level of orientation along the flow direction. As soon as the external stimulus stops the polymer chains revert to the "natural" random coil conformations, but the filler particles due to their size cannot do the same in a very short period of time. If now we consider that in most cases when the external stimulus ends the system undergoes a certain level of cooling (e.g. injection moulding, film blowing, fibre spinning) then we can see that the crystallisation process in the matrix makes the filler movement even more difficult. Large fillers like graphene and nanotubes tend to suffer from this effect more than more traditional ones like carbon black, although this depends heavily on the shear forces and the polymer/filler composition.

8.3.4 *Dispersion and Filler Localisation*

Filler incorporation in a polymer matrix is achieved by a number of different methods, but all generally fall under three main categories, solution, melt and in-situ polymerisation mixing. In solution mixing, both polymer and filler dispersion is facilitated by solvents that are removed after mixing. Melt mixing is by far the most common method used commercially and in this case the filler is added to the polymer melt and dispersion is facilitated by an external stimulus, usually in the form of a compounding or extruding machine. In the case of in-situ polymerisation mixing, the filler is added during the growth phase of the polymer. Each of the mixing methods has advantages and disadvantages, that have been discussed elsewhere (Gkourmpis 2014), but in the context of this work is sufficient to say that the level of dispersion is strongly influenced by the way the filler is incorporated in the matrix.

When filler is incorporated in a matrix, a wide range of interactions is available in the system that can have a direct link in the overall morphology, topology and properties of the nanocomposite. Polymer–filler interactions can extend over a range of distances and length-scales and can be weak or strong (Kyrylyuk and van der Schoot 2008). The interfacial energy between polymer and filler has been seen to affect the compatibility and efficiency of mixing, although this is strongly related to the chain conformation and filler structure (Alig et al. 2012). This property has been exploited in immiscible polymer blends where preferential localisation along the interface (Filippone et al. 2014; Huang et al. 2014; Gubbels et al. 1998) or in one of the phases allows for reduced percolation threshold (Gubbels et al. 1994, 1995; Gkourmpis et al. 2013). This preferential localisation enhances conductivity only in the host phase leaving part of the matrix insulating and filler free. Consequently, the overall conductivity of the nanocomposite will be determined by the conductivity of the phase that hosts the filler in a manner similar to the single polymer cases discussed previously. Obviously for the entire nanocomposite to be conductive a network is required, which means that the two polymer phases must be co-continuous. Depending on the polymers immiscible blends tend to form co-continuous phases at ranges of 30–70 %, but addition of fillers can lead to phase swelling, thus creating systems where phase continuity can be achieved at far lower amounts. An example of this is an immiscible blend of polyethylene (PE) and polystyrene (PS) that reaches co-continuity at a range of 30–45 wt% PE. The amount of PE required to create a co-continuous phase can drop to 5 wt% by addition of small amounts of carbon black (Gubbels et al. 1995). This effect is a simple manifestation of polymer thermodynamics, since the filler particle occupies a larger volume it means that the polymer phase containing it needs to take into consideration the volume increase. In terms of thermodynamics the criterion of the filler localisation and phase behaviour is the minimisation of the free energy of the system. Therefore if the free energy criterion is satisfied by the filler located in the polymer phase, then that phase will have to adjust to the volume change by swelling in a manner similar to block copolymers (Matsen 2002). If this is not possible, then the filler will localise to the larger phase that is capable of adjusting to the subsequent volume changes.

For the understanding of the percolative behaviour of conductive fillers in immiscible polymer blends, the details of the filler distribution in the different phases are important. The critical condition for achieving the all-important co-continuity of the conductive filler-rich phase has been suggested to be determined by the viscosity of the phases as (Paul and Barlow 1980)

$$\frac{\phi_A}{\phi_B} \times \frac{\eta_B}{\eta_A} = X$$

with φ and η the volume fraction and melt viscosity of the two phases (noted as A and B), respectively. For $X > 1$ phase A is continuous, $X \approx 1$ both phases are co-continuous and for $X < 1$ phase B is continuous (Jordhamo et al. 1986). Interfacial tension and wetting coefficient of the different mixture components have also been identified to influence the filler distribution and subsequent morphology of the blend. For carbon black particles in an immiscible blend it has been suggested that their distribution can be qualitatively predicted by the wetting coefficient ω. The interfacial free energy minimisation can lead to Young's equation (Sumita et al. 1991)

$$\omega_{AB} = \frac{\gamma_{CB-B} - \gamma_{CB-A}}{\gamma_{AB}}$$

with γ_{AB}, γ_{CB-A} and γ_{CB-B} the interfacial tension between polymer A and B, CB and polymer A and CB and polymer B, respectively. According to this the distribution of carbon black particles can be classified as: for $\omega_{AB} > 1$ the particles localise in phase A, for $-1 < \omega_{AB} < 1$ the particles localise along the interface and for $\omega_{AB} < -1$ the particles localise in phase B. The interfacial tension between the phases can be estimated by Wu's harmonic mean average (Wu 1982)

$$\gamma_{AB} = \gamma_A + \gamma_B - 4\left(\frac{\gamma_A^d \gamma_B^d}{\gamma_A^d + \gamma_B^d} + \frac{\gamma_A^p \gamma_B^p}{\gamma_A^p + \gamma_B^p}\right)$$

with γ_A and γ_B the interfacial tension between polymers A and B, γ_A^d and γ_B^d the dispersive components of the surface tensions γ_A and γ_B and γ_A^p and γ_B^p the equivalent polar components. According to this three spreading coefficients λ can be defined for a ternary blend (Hobbs et al. 1988)

$$\lambda_{ikj} = \gamma_{ij} - (\gamma_{ik} + \gamma_{jk})$$

Except for these thermodynamic considerations, kinetic factors, such as viscosity, chain conformation, mixing procedures (time, speed, sequence of incorporation, shear forces etc.) and possible affinity or reaction of the polymer to the filler have been seen to play a vital role in the particle localisation. Mixing in particular has been seen to have a big effect on the overall morphology of the system, as shear forces have a tendency to break large agglomerates into smaller aggregates. These smaller filler clusters can in turn erode into even smaller aggregates given enough mixing time, thus leading into a system with increased filler dispersion. Obviously such concepts are fairly simple in terms of small conductive fillers like carbon black, but the situation becomes more complicated in the case of larger nanosize and high aspect ratio fillers like graphene or nanotubes. In the latter case dispersion

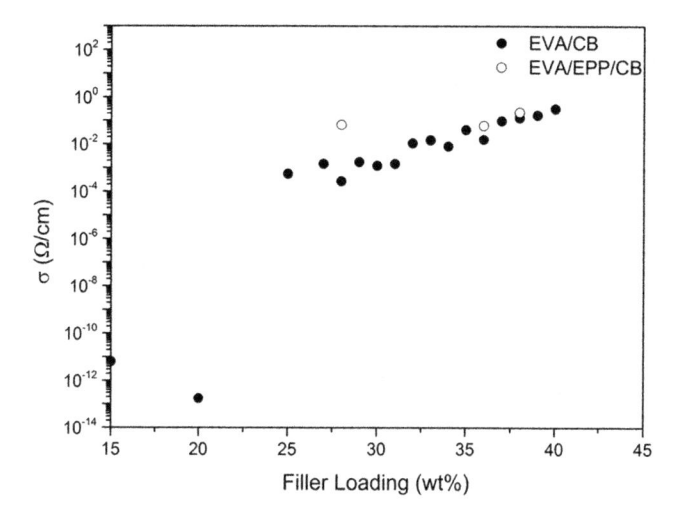

Fig. 8.9 Electrical conductivity as a function of filler loading for a system of EVA/CB and an immiscible blend of EVA/EPP/CB, indicating the effects of selective localisation. Adopted from Gkourmpis et al. (2013)

is even more difficult due to the filler size and the breakdown of the large agglomerates into smaller aggregates is truly challenging (Gkourmpis 2014).

The almost exclusive localisation of the filler in one phase of the blend has another very important effect as far as conductivity is concerned. Since the filler is the sole conductivity medium in the system (we can ignore the matrix conductivity for this discussion), its network formation will determine the amount of current flowing in the nanocomposite. As discussed previously if the system is co-continuous then the entire nanocomposite will be conductive, but its conductivity will arise *only* from the phase that contains the conductive filler. This concept is illustrated in Fig. 8.9 where the conductivity of a system of poly(ethylene vinyl-acetate)/poly(ethylene-co-propylene)/carbon black (EVA/EPP/CB) appears almost constant regardless of the filler loading (Gkourmpis et al. 2013), something that is achieved by keeping the ratio of the filler with respect to the EVA phase (where it localises) constant. In other words the percolative behaviour of the filler in the EVA phase is similar to the percolative behaviour of the same filler in the EVA phase in a single-component system.

8.3.5 Rheology and Mechanical Properties

The introduction of large conductive fillers into a polymer matrix has been seen to influence the mechanical and flow characteristics of the system. In almost all cases this addition leads to increase in storage modulus (G') and complex viscosity with increasing filling amounts (Brigandi et al. 2014). Storage modulus has been seen to increase sharply with adding filler amounts, leading to a situation where the system crosses over from a liquid-like response to a solid-like response. This transition is usually referred to as mechanical or rheological percolation (Penu et al. 2012) and a power-law relationship can be used to determine its threshold

$$G' \propto (m - m_C)^t$$

with m the mass fraction of the filler, m_C the threshold for percolation and t the critical exponent (Du et al. 2004).

Systems containing carbon black have been seen to exhibit significant shear thinning in comparison with single polymers or binary blends. This has been utilised in the case of poly(ethylene acrylic acid) (EAA) and polypropylene (PP) blends where by adding carbon black the rheological percolation has been reported to increase with increasing EAA content. This behaviour suggested that the rheological percolation was tightly coupled with the electrical percolation by forming network structures in both PP/CB and EAA/PP/CB composites (Chen et al. 2009).

The incorporation of conductive fillers has been reported to increase the strength and stiffness of the host matrix (Gkourmpis 2014). One of the most important factors affecting the overall mechanical properties of a nanocomposite is the interfacial interaction between filler particles and polymer. This interaction is of great importance as the level of adhesion of the filler particle to the polymer will affect the efficiency of the energy transfer between them under load. In cases where the adhesion is high the stress transfer between the filler particles and the host matrix will be efficient, thus leading to an overall reinforcement of the nanocomposite. Parameters like yield, strain and tensile strength can be modelled for different adhesion levels with relatively good comparison to experimental results (Yan et al. 2006).

The reinforcement effects, as seen through the tensile strength, discussed previously are large with the incorporation of small amount of carbonaceous fillers into the polymer matrix. In the case of high aspect ratio fillers these effects are even more pronounced. The reason for this behaviour can be traced in the level of dispersion of the filler particles in the polymer matrix. In Fig. 8.10, we can see that for a system of poly(vinyl alcohol) (PVA) with graphene nanosheets introducing 0.06 vol.% of graphene into the PVA matrix the tensile strength increases by 73 % whereas by increasing the filler level from 1.8 to 3 vol.% the tensile strength increase is negligible (from 42 to 43 MPa) with the elongation at break having similar decreasing behaviour (Zhao et al. 2010). This effect has been attributed to the rearrangement and further aggregation of graphene sheets as the overall amount increases due to van der Waals interactions. When this is happening the slippage of the aggregated sheets and energy transfer to the matrix will be less efficient in comparison with the case of single sheets during a tensile measurement. In Fig. 8.10, we can see a schematic representation of this with respect to the different dispersion and aggregation levels of the filler in the matrix. Case (1) is the initial

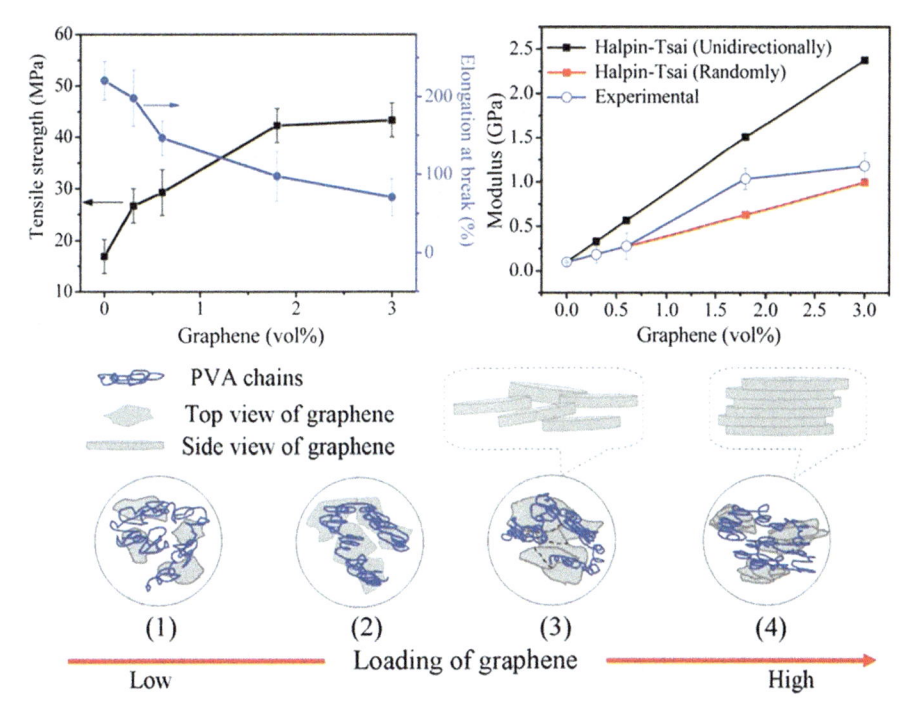

Fig. 8.10 Mechanical properties (tensile strength and elongation at break) of PVA/graphene nanocomposites as a function of graphene loading (*top left*). Comparison between experimental data of the Young's Modulus of PVA/graphene nanocomposites with theoretical predictions using the Halpin-Tsai models. Two cases of unidirectional and random distribution of the graphene in the PVA were considered for the theoretical estimations (*top right*). Schematic representation of the various levels of dispersion of graphene in the PVA. Refer to the text for a more detailed discussion of the different dispersion possibilities (*bottom*). Adopted and reprinted with permission from Zhao et al. (2010)

low loading situation where each filler particle is individually placed in the matrix, as more filler particles are introduced the spaces between the particles are filled and a network structure where the particles are in physical contact is created (case 2). Case (3) is an intermediate case where the particles are properly dispersed but a certain level of lose aggregation exists. In case (4) the individual particles are heavily aggregated creating larger structures that are detrimental to the efficiency of the reinforcement of the matrix. In a real system the spatial arrangement will move from case (1) to case (3) with increasing filler loading. A critical level of filler content called rheological (or mechanical) percolation exists on the interplay between the amount of filler and the efficiency of the mechanical reinforcement on the nanocomposite (Zhang et al. 2008). Below this content the filler particles are well dispersed and by increasing the loading a significant improvement on the mechanical performance can be realised. Above the percolation the filler particles tend to aggregate leading to inefficient structures that cannot transfer energy in an optimal manner leading to moderate mechanical improvement.

As discussed previously, a number of mathematical models for prediction of the mechanical reinforcement in polymer matrices exist. Halpin-Tsai (Halpin and Kardos 1976; Cadek et al. 2002) and Lewis-Nielsen (Lewis and Nielsen 1970) equations are based on simple approximations and provide reliable predictions of the modulus of reinforced matrices where the filler is unidirectional or randomly distributed. Therefore one can utilise these predictions to quantify the level of dispersion and the possible orientation of the filler particles in the matrix (see Fig. 8.10).

8.3.6 Crystallisation and Morphology

The introduction of conductive fillers in a polymer matrix except for the obvious effect on electrical conductivity which was discussed previously, has the possibility to affect the overall morphology of the nanocomposite. Here though we must note that the changes or their absence in the overall morphology are heavily dependent on the polymer, the type, regularity and size of the filler, the preparation methods as they affect the level of dispersion and the crystallisation conditions (Gkourmpis 2014).

For systems containing simple fillers like carbon black one expects a decrease in overall crystallinity, in comparison with the pure polymer, that is linked to the overall filler content (Gubbels et al. 1994, 1995, 1998; Gkourmpis et al. 2013). Crystallisation kinetics as seen through the melting and crystallisation temperatures are affected, something that can be attributed to the partial immobilisation of the macromolecular chains in the vicinity of the filler particles. In the case of binary blends where the filler localises in one of the two phases it is not uncommon to see morphological changes in both phases. This has been seen for a system of EVA/EPP where despite the apparent immiscibility of the two components mutual interactions between the two phases lead to the EPP phase to exhibit a split in its crystallisation peak and adopt two distinct phases. The introduction of carbon black in the system localises the particles in the EVA phase, further limiting the crystallisation of the host phase due to the reduced mobility of the polymer chains in the vicinity of the carbon black particles (Gkourmpis et al. 2013).

In the case where high aspect ratio fillers like graphene or nanotubes are introduced in the matrix the shape and geometry of the particles has been seen to have an even more direct effect on the overall crystallisation behaviour of the nanocomposite. Due to their sheer size and geometrical regularity these fillers can be very efficient nucleating agents, promoting or hindering crystallisation and leading to morphological changes (Gkourmpis 2014). CNTs have been reported to induce nanohybrid shish-kebab (NHSK) crystals for a number of polymers like PE (Laird and Li n.d.), PP (Xu et al. 2010) and poly(L-lactide) (PLLA) (Xu et al. 2011). The molecular origin of NHSK periodicity has been suggested to be closely related to the concentration gradient and the heat dissipation of the growth front of the lamellar structure. Once nucleation is initiated, the polymer concentration gradient at the crystal growth front adopts a periodic profile along the

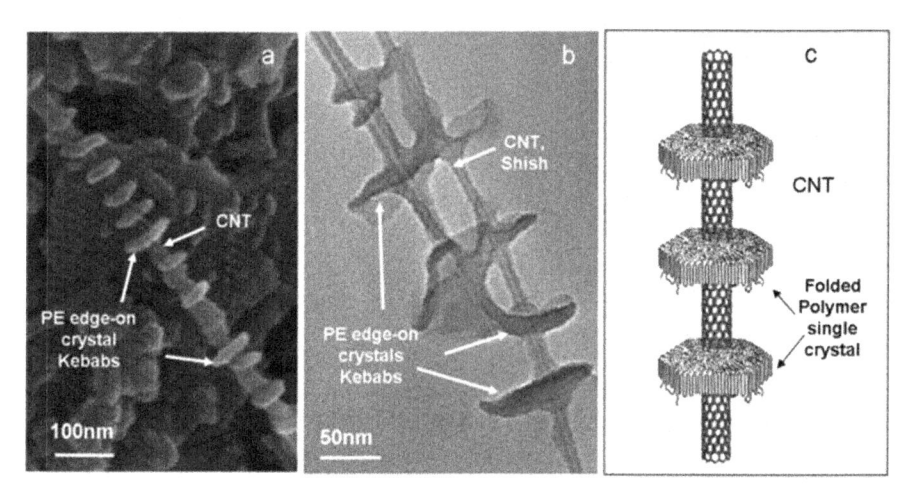

Fig. 8.11 NHSK structure of PE/MWCNT as seen by SEM (**a**), TEM (**b**) and a schematic representation indicating how lamellas grow from the CNT surface (**c**). Reprinted with permission from Li et al. (2005)

nanotube axis. This combined with the fact that most of the heat generated by the crystallisation process can be dissipated along the nanotube, leads to a further temperature gradient along the axis of the nanotube. Parameters like the crystal periodicity, the CNT diameter, the lamellar thickness and the nanotube chirality have been suggested to affect the NHSK formation and growth (Laird and Li n.d.; Li et al. 2009).

Control of the NHSK growth has been attributed to two factors, namely the epitaxial growth of the polymer on the CNT surface and the geometric confinement. Due to the small CNT dimensions the polymer chains exhibit preferential alignment along the tube axis regardless of the lattice matching between polymer and graphitic surface, leading to a soft epitaxy mechanism (Laird and Li n.d.). In the case of polyethylene the polymer behaves as being on a flat surface due to the significantly larger size of the nanotube, thus epitaxy is the main mechanism for crystal growth. As the chain begins crystallising on the surface geometric confinement, elements become important due to the comparable size of the crystal lamellar and the nanotube. Therefore the crystallisation will be preferential parallel to the nanotube axis without any regard for its chirality, leading to a situation where the orthogonal orientation of the lamellar is obtained (see Fig. 8.11) (Li et al. 2005). Computer simulations have suggested that CNT chirality is strongly influencing molecular crystallisation, something that has not so far been verified experimentally (Laird and Li n.d.).

As a final note we can say that the addition of CNTs in a semicrystalline polymer has the potential to impede, promote or have no effect on the nanocomposite crystallisation behaviour. This of course will be heavily dependent on the preparation method, the level of dispersion and the chemical configuration of both polymer and filler. It has also been suggested that the amount of CNTs in the system might have an effect of the overall crystallisation behaviour. Low levels of CNTs might promote crystallisation while higher loadings can actually impede the whole process (Laird and Li n.d.).

In the case of graphene the two-dimensional arrangement of the filler makes lattice matching the dominant form in the crystallisation process. This can be attributed to the sequential absorption of polymer chains on the graphitic surface that leads the polymer to undergo structural rearrangements in order to adopt the lowest energy. Therefore crystallisation will take longer time something also seen experimentally in the case of polypropylene (Xu et al. 2010). The flat graphitic surface can also allow the polymer crystals to grow into multiple nucleation points and orientations. It has been suggested that the multiple spatial opportunities for crystal growth might create interference from adjacent crystals leading to an overall suppression of the crystallisation kinetics process (Xu et al. 2010). Again as in the case of nanotubes the nanocomposite preparation method, the level of filler dispersion, the amount of filler, the polymer architecture and the filler's surface characterisation are critical on the overall crystallisation process. A good example of this are two independent studies on poly(ethylene vinyl acetate) filled with graphene nanoplatelets, that indicate a totally different behaviour with respect to the crystallisation process. In the case of a melt-mixed system the existence of the filler has been seen to reduce slightly the overall crystallinity, while crystallisation takes place away from the filler that has a minimal impact on the overall morphology of the system (Stalmann 2012) in a manner similar to the one expected from traditional fillers like carbon black (Gkourmpis et al. 2013). In the case of solution-based similar systems the existence of the filler has been reported to facilitate and promote crystallisation due to the high nucleating activity of the nanoplatelets (He et al. 2011; Pang et al. 2012).

8.4 Summary

Electrically conductive polymer nanocomposites are widely used especially due to their superior properties and competitive prices. It is expected that as the level of control of the overall morphology and associated properties increases we will see an even wider commercialisation on traditional and totally novel applications. In this section we have discussed the basic principles of the percolation theory and the different types of conduction mechanisms, outlined some of the critical parameters of controlling primarily the electrical performance and we have provided some indications on the effect such conductive fillers have on the overall morphology and crystallisation of the nanocomposite. The latter becomes even more critical if we take into consideration that modern nanosized fillers offer unique potential for superior properties at low loadings (low percolation thresholds) but have a more direct impact on the morphology of the system. Furthermore we have indicated that similar systems can have totally different behaviour as the preparation methods, the chain conformation and the surface chemistry of the fillers will have a massive impact on the resulting nanocomposite. Obviously such wide topic cannot be exhausted in a small chapter but we hope we have provided enough information to intrigue and assist the interested reader to explore this unique class of materials even further.

References

Alig I, Pötschke P, Lellinger D, Skipa T, Pegel S, Kasaliwal GR, Villmow T, Establishment T (2012) Morphology and properties of carbon nanotube networks in polymer melts. Polymer 53:4

Ambrosetti G, Grimaldi C, Balberg I, Maeder T, Danani A, Ryser P (2010) Solution of the tunneling-percolation problem in the nanocomposite regime. Phys Rev B 81:155434

E. M. Andersen, G. R. Mitchell (eds) (2013) Rheology: theory, properties and practical applications, Novapress, London 2013

Balberg I (1986) Excluded-volume explanation of Archie's law. Phys Rev B 33:3618

Balberg I (1987) Tunneling and nonuniversal conductivity in composite materials. Phys Rev Lett 59:1305

Balberg I (2009) Tunnelling and percolation in lattices and the continuum. J Phys D Appl Phys 42:064003

Balberg I, Binenbaum N (1987) Invariant properties of the percolation thresholds in the soft-core–hard-core transition. Phys Rev A 35:5174

Balberg I, Andreson C, Alexander S, Wagner N (1984) Excluded volume and its relation to the onset of percolation. Phys Rev B 30:3933

Balberg I, Yang X (eds) (2012) Semiconductive polymer composites: principles, morphologies, properties and applications. Willey-VCH Verlag, London, p 145

Balberg I, Azulay D, Toker D, Millo O (2004) Percolation and tunnelling in composite materials. Int J Mod Phys B 18:2091

Bauhofer W, kovacs JZ (2009) A review and analysis of electrical percolation in carbon nanotube polymer composites. Compos Sci Technol 69:1486

Behnam A, Guo J, Ural A (2007) Effects of nanotube alignment and measurement direction on percolation resistivity in single-walled carbon nanotube films. J Appl Phys 102:044313

Berhan L, Sastry SM (2007) Modeling percolation in high-aspect-ratio fiber systems. I. Soft-core versus hard-core models. Phys Rev E 75:041120

Berman D, Orr BG, Jaeger HM, Goldman AM (1986) Conductances of filled two-dimensional networks. Phys Rev B 33:4301

Brigandi PJ, Cogen JM, Pearson RA (2014) Electrically conductive multiphase polymer blend carbon-based composites. Polym Eng Sci 54:1

Bryning MB, Islam MF, Kikkawa JM, Yodh AG (2005) Very low conductivity threshold in bulk isotropic single-walled carbon nanotube–epoxy composites. Adv Mater 17:1186

Byrne MT, Gunko YK (2010) Recent advances in research on carbon nanotube-polymer composites. Adv Mater 22:1672

Cadek M, Coleman JN, Barron V, Hedicke K, Blau WJ (2002) Morphological and mechanical properties of carbon-nanotube-reinforced semicrystalline and amorphous polymer composites. Appl Phys Lett 81:5123

Chen GH, Wu DJ, Weng WG, Yan WL (2001) Preparation of polymer/graphite conducting nanocomposite by intercalation polymerization. J Appl Polym Sci 82:2506

Chen XM, Shen JW, Huang WY (2002) Novel electrically conductive polypropylene/graphene nanocomposites. J Mater Sci Lett 21:213

Chen G, Yang B, Guo S (2009) Ethylene–acrylic acid copolymer induced electrical conductivity improvements and dynamic rheological behavior changes of polypropylene/carbon black composites. J Polym Sci B Polym Phys 47:1762

Du F, Scogna RC, Zhou W, Brand S, Fischer JE, Winey KI (2004) Nanotube networks in polymer nanocomposites: rheology and electrical conductivity. Macromolecules 37:9048

Ebbesen TW, Lezec HJ, Hiura H, Bennett JW, Ghaemi HF, Thio T (1996) Electrical conductivity of individual carbon nanotubes. Nature 382:54

Fan Z, Zheng C, Wei T, Zhang Y, Luo G (2009) Effect of carbon black on electrical property of graphite nanoplatelets/epoxy resin composites. Polym Eng Sci 49:2041

Filippone G, Causa A, Filippone G, Causa A, de Luna MS, Sanguigno L, Acierno D (2014) Assembly of plate-like nanoparticles in immiscible polymer blends—effect of the presence of a preferred liquid–liquid interface. Soft Matter 10:3183

Fisher ME, Essam J (1961) Some cluster size and percolation problems. J Math Phys 2:609

Foygel M, Morris R, Anez D, French S, Sobolev VL (2005) Theoretical and computational studies of carbon nanotube composites and suspensions: electrical and thermal conductivity. Phys Rev B 71:104201

Fuhrer MS, Nygård J, Shih L, Forero M, Yoon Y-G, Mazzoni MSC, Choi HJ, Ihm J, Louie SG, Zettl A, McEuen PL (2000) Crossed nanotube junctions. Science 288:494

Gelves GA, Lin B, Sundararaj U, Haber JA (2006) Low electrical percolation threshold of silver and copper nanowires in polystyrene composites. Adv Funct Mater 16:2423

Gkourmpis T, Svanberg C, Kaliappan SK, Schaffer W, Obadal M, Kandioller G, Tranchida D (2013) Improved electrical and flow properties of conductive polyolefin blends: modification of poly(ethylene vinyl acetate) copolymer/carbon black with ethylene–propylene copolymer. Eur Polym J 49:1975

Gkourmpis T, Mercader GA, Haghi AK (eds) (2014) Nanoscience and computational chemistry: research progress. Apple Academic Press, Toronto, p 85

Glowacki I, Jung J, Ulanski J, Matyjaszewski K, Möller M (eds) (2012) Polymer science: a comprehensive reference, vol 2. Elsevier, London, p 847

Grimaldi C, Balberg I (2006) Tunneling and nonuniversality in continuum percolation systems. Phys Rev Lett 96:066602

Gubbels F, Jerome R, Teyssie P, Vanlathem E, Deltour R, Calderone A, Parente V, Bredas JL (1994) Selective localization of carbon black in immiscible polymer blends: a useful tool to design electrical conductive composites. Macromolecules 27:1972

Gubbels F, Blacher S, Vanlathem E, Jerome R, Deltour R, Brouers F, Teyssie P (1995) Design of electrical composites: determining the role of the morphology on the electrical properties of carbon black filled polymer blends. Macromolecules 28:1559

Gubbels F, Jerome R, Vanlathem E, Deltour R, Blacher S, Brouers F (1998) Kinetic and thermodynamic control of the selective localization of carbon black at the interface of immiscible polymer blends. Chem Mater 10:1227

Halpin JC, Kardos JL (1976) The Halpin-Tsai equations: a review. Polym Eng Sci 16:344

He F, Fan J, Lau S, Chan LH (2011) Preparation, crystallization behavior, and dynamic mechanical property of nanocomposites based on poly(vinylidene fluoride) and exfoliated graphite nanoplate. J Appl Polym Sci 119:1166

Hobbs SY, Dekkers MEJ, Watkins VH (1988) Effect of interfacial forces on polymer blend morphologies. Polymer 29:1598

Huang J, Mao C, Zhu Y, Jiang W, Yang X (2014) Control of carbon nanotubes at the interface of a co-continuous immiscible polymer blend to fabricate conductive composites with ultralow percolation thresholds. Carbon 73:267

Hunt A, Ewing R (2009) Percolation theory for flow in porous media. Springer, Berlin

Johner N, Grimaldi C, Balberg I, Ryser P (2008) Transport exponent in a three-dimensional continuum tunneling-percolation model. Phys Rev B 77:174204

Jordhamo GM, Manson JA, Sperling LH (1986) Phase continuity and inversion in polymer blends and simultaneous interpenetrating networks. Polym Sci Eng Sci 26:517

Kim H, Macosko CW (2008) Morphology and properties of polyester/exfoliated graphite nanocomposites. Macromolecules 41:3317

Kim H, Abdala AA, Macosko CW (2010) Graphene/polymer nanocomposites. Macromolecules 43:6515

Kirkpatrick S (1973) Percolation and conduction. Rev Mod Phys 45:574

Kostagiannakopoulou C, Maroutsos G, Sotiriadis G, Vavouliotis A, Kostopoulos V (2012). In: Third international conference on smart materials and nanotechnology in engineering, April 2012

Kuhner G, Voll M (1993) Manufacture of carbon black. In: Donnet J-B, Bansal RC, Wang M-J (eds) Carbon black science and technology. Taylor & Francis, London, p 1

Kyrylyuk AV, van der Schoot P (2008) Continuum percolation of carbon nanotubes in polymeric and colloidal media. Proc Nat Acad Sci USA 105:8221

Lacey D, Beattie HN, Mitchell GR, Pople JA (1998) Orientation effects in monodomain nematic liquid crystalline polysiloxane elastomers. J Mater Chem 8:53

Laird ED, Li CY (2013) Structure and morphology control in crystalline polymer–carbon nanotube nanocomposites. Macromolecules 46:2877

Lewis TB, Nielsen LE (1970) Dynamic mechanical properties of particulate-filled composites. J Appl Polym Sci 14:1449

Li J, Kim J-K (2007) Percolation threshold of conducting polymer composites containing 3D randomly distributed graphite nanoplatelets. Compos Sci Technol 67:2114

Li CY, Li L, Cai W, Kodjie SL, Tenneti KK (2005) Nanohybrid shish-kebabs: periodically functionalized carbon nanotubes. Adv Mater 17:1198

Li J, Kim JK, Sham ML, Marom G (2007) Morphology and properties of UV/ozone treated graphite nanoplatelet/epoxy nanocomposites. Compos Sci Technol 67:296

Li L, Li B, Hood MA, Li CY (2009) Carbon nanotube induced polymer crystallization: the formation of nanohybrid shish–kebabs. Polymer 50:953

Lux F (1993) Models proposed to explain the electrical conductivity of mixtures made of conductive and insulating materials. J Mater Sci 28:285

Ma PC, Liu MY, Zhang H, Wang SQ, Wang R, Wang K, Wong YK, Tang BZ, Hong SH, Paik KW, Kim JK (2009) Enhanced electrical conductivity of nanocomposites containing hybrid fillers of carbon nanotubes and carbon black. ACS Appl Mater Interfaces 1:1090

Matsen MW (2002) The standard Gaussian model for block copolymer melts. J Phys Condens Matter 14:R21

Mitchell GR, Davis FJ, Ashman A (1987) Structural studies of side-chain liquid crystal polymers and elastomers. Polymer 28:639

Moniruzzaman M, Winey KI (2006) Polymer nanocomposites containing carbon nanotubes. Macromolecules 39:5194

Munson-McGee SH (1991) Estimation of the critical concentration in an anisotropic percolation network. Phys Rev B 43:3331

Mutiso RM, Winey KI, Matyjaszewski K, Möller M (eds) (2012) Polymer science: a comprehensive reference, vol 7. Elsevier, London, p 327

Otten RHJ, van der Schoot P (2009) Continuum percolation of polydisperse nanofillers. Phys Rev Lett 103:225704

Oxfall H, Ariu G, Gkourmpis T, Rychwalski RW, Rigdhal M (2015) Effect of carbon black on electrical and rheological properties of graphite nanoplatelets/poly(ethylene-butyl acrylate) composites. eXPRESS Polym Lett 9:66

Pang H, Zhong G, Xu J, Yan D, Ji X, Li Z, Chen C (2012) Non-isothermal crystallization of ethylene-vinyl acetate copolymer containing a high weight fraction of graphene nanosheets and carbon nanotubes. Chin J Polym Sci 30:879

Paul DR, Barlow JW (1980) Polymer blends (or alloys). J Macromol Sci Rev Macromol Chem C18:109

Penu C, Hu G-H, Fernandez A, Marchal P, Choplin L (2012) Rheological and electrical percolation thresholds of carbon nanotube/polymer nanocomposites. Polym Eng Sci 52:2173

Pike GE, Seager CH (1974) Percolation and conductivity: a computer study. I. Phys Rev B 10:1421

Pople JA, Mitchell GR (1997) WAXS studies of global molecular orientation induced in nematic liquid crystals by simple shear flow. Liquid Crystals 23:467

Pötschke P, Abdel-Goad M, Alig I, Dudkin S, Lellinger D (2004) Rheological and dielectrical characterization of melt mixed polycarbonate-multiwalled carbon nanotube composites. Polymer 45:8863

Ramasubramaniam R, Chen J, Liu H (2003) Homogeneous carbon nanotube/polymer composites for electrical applications. Appl Phys Lett 83:2928

Sahimi M (1994) Applications of percolation theory. Taylor & Francis, London

Shante VKS, Kirckpatrick S (1971) An introduction to percolation theory. Adv Phys 30:325

Sherman RD, Middleman LM, Jacobs SM (1983) Electron transport processes in conductor-filled polymers. Polym Sci Eng 23:36

Shklovskii BI, Efros AL (1984) Electronic properties of doped semiconductors. Springer, Berlin

Stalmann G (2012) MSc Dissertation University of Marburg

Stankovich S, Dikin DA, Dommett GHB, Kohlhaas KM, Zimney EJ, Stach EA, Piner RD, Nguyen ST, Ruoff RS (2006) Graphene-based composite materials. Nature 442:282

Stauffer D, Aharony A (1987) Introduction to percolation theory. Taylor & Francis, London

Strumpler R, Glatz-Reichenbach J (1999) Conducting polymer composites. J Electroceram 3:329

Sumita M, Sakata K, Asai S, Miyasaka K, Nakagawa H (1991) Dispersion of fillers and the electrical conductivity of polymer blends filled with carbon black. Polym Bull 25:265

Sun Y, Bao H-D, Guo Z-X, Yu J (2009) Modeling of the electrical percolation of mixed carbon fillers in polymer-based composites. Macromolecules 42:459

Tye S, Halperin BI (1989) Random resistor network with an exponentially wide distribution of bond conductances. Phys Rev B 39:877

Wang SF, Ogale AA (1993) Simulation of percolation behavior of anisotropic short-fiber composites with a continuum model and non-cubic control geometry. Compos Sci Technol 46:389

Wei T, Song L, Zheng C, Wang K, Yan J, Shao B, Fan Z-J (2010) The synergy of a three filler combination in the conductivity of epoxy composites. Mater Lett 64:2376

Weng WG, Chen GH, Wu DJ, Yan WL (2004) HDPE/expanded graphite electrically conducting composite. Compos Interface 11:131

White SI, DiDonna BA, Mu M, Lubensky TC, Winey KI (2009) Simulations and electrical conductivity of percolated networks of finite rods with various degrees of axial alignment. Phys Rev B 79:024301

Winey KI, Vaia RA (2007) Polymer nanocomposites. MRS Bull 32:314

Winey KI, Kashiwagi T, Mu M (2007) Improving electrical conductivity and thermal properties of polymers by the addition of carbon nanotubes as fillers. MRS Bull 32:348

Wu S, Polymer (1982) interface and adhesion, M. Dekker, London

Xu JZ, Chen T, Yang CL, Li ZM, Mao YM, Zeng BQ, Hsiao BS (2010) Isothermal crystallization of poly(l-lactide) induced by graphene nanosheets and carbon nanotubes: a comparative study. Macromolecules 43:5000

Xu JZ, Chen T, Wang Y, Tang H, Li Z-M, Hsiao BS (2011) Graphene nanosheets and shear flow induced crystallization in isotactic polypropylene nanocomposites. Macromolecules 44:2808

Yan W, Lin RJT, Bhattacharyya D (2006) Particulate reinforced rotationally moulded polyethylene composites—mixing methods and mechanical properties. Compos Sci Technol 66:2080

Zhang QH, Fang F, Zhao X, Li YZ, Zhu MF, Chen DJ (2008) Use of dynamic rheological behavior to estimate the dispersion of carbon nanotubes in carbon nanotube/polymer composites. J Phys Chem B 112:12606

X. Zhao, Q. Zhang, D. Chen, P. Lu (2010). Enhanced mechanical properties of graphene-based poly(vinyl alcohol) composites. Macromolecules 43, 2357

Zheng W, Wong SC (2003) Electrical conductivity and dielectric properties of PMMA/expanded graphite composites. Compos Sci Technol 63:225

Chapter 9
Nanodielectrics: The Role of Structure in Determining Electrical Properties

Alun S. Vaughan

9.1 Introduction

One of the most pressing challenges faced by the current generation is how to meet the ever-increasing global demand for energy without suffering the consequences of major climate change. The situation in developed economies is exemplified by that in the UK where, in November 2012, the Department of Energy and Climate Change (DECC) published the UK's Energy Security Strategy document. While highlighting the generalities of energy security and climate change as two critical societal issues, it also illustrated the scale of the problem. In the UK:

- Electricity usage is predicted to increase by at least 30 % and potentially by 100 % by 2050, as heating and transportation become increasingly electrified.
- Massive investment will consequently be required in electricity infrastructure.
- Future electricity systems will take a very different form from the proven systems of the past, as decarbonisation targets will drive an ever greater reliance on renewable generation, coupled with interconnection of neighbouring power systems to increase resilience against the intermittent nature of many renewable energy sources.

Such changes will involve a paradigm shift in the ways in which we generate and transmit electricity.

A central element of all power systems is insulation and, consequently, the development and deployment of advanced material solutions that are capable of supporting the above plans is critical. In Western Europe, for example, a total of 1.5 Mton per annum of polymeric materials are currently used for electrical insulation, with an estimated value of ~£2bn; the 2013 UK Renewable Energy Roadmap

---

A.S. Vaughan
ECS, University of Southampton, Southampton, UK
e-mail: asv@ecs.soton.ac.uk

© Springer International Publishing Switzerland 2016
G.R. Mitchell, A. Tojeira (eds.), *Controlling the Morphology of Polymers*,
DOI 10.1007/978-3-319-39322-3_9

indicates that between January 2010 and September 2013, DECC recorded announcements worth £31bn of private sector investment in renewable electricity generation in the UK; globally, the 2014 IEA Renewables Report indicates that global investment in new renewable power capacity was around $250bn in 2014 alone. The technological potential and economic value of novel, high performance insulation systems are therefore enormous.

9.1.1 Nanotechnology

Although largely unnoticed at the time, the lecture *There's Plenty of Room at the Bottom* given in December 1959 at Caltech by Richard Feynman is now often cited as the birth of nanotechnology. However, in reality, mankind has unwittingly been exploiting nanostructured materials for very much longer, with notable examples being the Lycurgus Cup (~400 AD), Damascus steel and, perhaps rather more prosaically, carbon black as an additive in automotive tyres. Nevertheless, credit for the current interest in nanostructured polymer-based material systems containing nanoscale inclusions is often traced back to work conducted by Toyota in the late 1980s, which led to the concept of nanocomposites (Kojima et al. 1993).

Many different standards and regulatory bodies have sought to define a nanomaterial, but a convenient definition derived from the 2010 European Commission Joint Research Centre report, *Considerations on a Definition of Nanomaterial for Regulatory Purposes*, is as follows: An intentionally manufactured material with one or more external dimensions, or an internal structure, on a scale from 1 to 100 nm, which exhibits novel properties not observed for coarser structures. This definition is informative in that it specifies a size range, considers the necessary dimensionality and introduces the concept that nanomaterials behave in unique ways which, in the case of nanocomposites, is generally associated with their enormous specific internal surface area. Consider, for example, the case of spherical metal oxide particles; if the diameter is 20 µm, then the consequent specific surface area is about 1.2 m^2 g^{-1} while, for 20 nm nanoparticles, this figure increases to ~1200 m^2 g^{-1}. Practically, when considering nanoparticles, it is useful to consider both the quoted size and the specific surface area of the system; in the case of the MKnano silica nanopowder MKN-SiO_2-015P, the average particle size and specific surface area are specified as 15 nm and 650 m^2 g^{-1}, respectively. In this case, both figures are broadly consistent but, where this is not the case, it is likely that the system contains primary nanoparticles that are agglomerated into larger structures, which can have adverse consequences.

The current interest in nanostructured dielectrics is generally attributed to John Lewis' 1994 paper, *Nanometric Dielectrics* (Lewis 1994), which considers dimensional effects in a dielectric that is formulated from two distinct components. For illustration, in the case of a composite containing a 10 % volume fraction of spherical particles 20 nm in diameter, the interfacial area will be ~3×10^7 m^{-1} and the typical separation between nearest neighbour nanoparticles will be comparable to the nanoparticle diameter. Lewis discusses the consequences of this (Lewis 2004, Lewis 2005), where it is suggested that interfacial effects can be considered in terms of local interactions that determine the intensity, I, of any property, such that the "extent of the interface is then defined by the change in I and will depend on the property of interest". That is, the region(s) immediately adjacent to the nanoparticle surface—so-called interphase(s)—will differ from both the matrix and nanoparticle when these are evaluated in the form of the relevant bulk material. Considering again the case of a system composed of 10 % by volume of isolated spherical nanoparticles 20 nm in diameter embedded within a continuous matrix, assuming an interphase thickness of just 5 nm, this will equate to nearly 25 % of the volume of the complete system, whereupon, the overall macroscopic properties will be reflective of this.

9.1.2 Nanocomposites: Scales of Structure

As in the case of other material systems, the macroscopic properties of nanocomposites are driven by their micro-/nanoscopic structure. From an electrical insulation perspective, polyethylene (PE) and epoxy resins constitute two technologically important material systems, each of which embodies in very different ways, a great deal of structural complexity. In the case of PE, the constituent molecules are the result of the inherently statistical polymerisation process, which can ultimately result in the formation of a hierarchical morphology in which different molecular fractions become segregated to specific morphological locations. In an epoxy resin, the epoxy monomer chemistry, the hardener and the stoichiometry can all be varied, to affect the network structure that evolves. In the case of nanocomposites, another layer of structural hierarchy is then overlaid upon and interacts with the inherent characteristics of the host matrix.

Figure 9.1 represents schematically the structural complexity of a nanocomposite. First, at the level of the primary nanoparticle (Fig. 9.1b), relevant

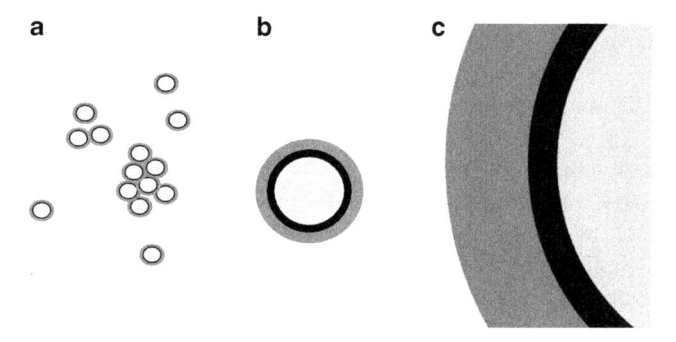

Fig. 9.1 Schematic representation of the range of structuring that may influence the behaviour of a nanodielectric: (**a**) aggregation; (**b**) nanoparticle parameters; (**c**) interfacial factors

structural characteristics include particle size, aspect ratio, crystallinity, bulk chemistry, defect structures, surface chemistry, surface structure and, in the case of its electrical behaviour, the density of electronic states in the interior and at the surface. Second, as introduced above, interactions between the nanoparticles and their environment may lead to a perturbation of the local structure or composition of the surrounding matrix material (Fig. 9.1c). Finally, the range of aggregation states of the primary nanoparticles needs to be considered, together with their distribution throughout the bulk (Fig. 9.1a).

9.2 Nanodielectric Properties

The previous section has introduced in a general way the means by which the introduction of a filler with dimensions in the nanometric range can affect structure and, thereby, lead to novel materials with combinations of properties that are exhibited by neither of the constituents in isolation. In power engineering applications, the technological objective of this is to produce materials with, in addition to improved electrical characteristics, improved mechanical performance, increased thermal conductivity, enhanced thermal endurance, etc. In general, the technological motivation behind nanodielectrics research is to facilitate the design of power plant with reduced footprints, which can operate at higher power densities. The relative importance of these various electrical, mechanical and thermal factors will depend upon the particular application, but improving electrical characteristics alone is rarely sufficient.

As far as dielectrics are concerned, electrical behaviour can be grossly divided into surface properties and bulk properties and these two areas will be briefly discussed in the following sections. While nanodielectrics have exhibited clear promise in both of these domains, our understanding of the fundamental physics and chemistry that underlie the effects that have been reported is, however, often poor.

9.2.1 Surface Electrical Properties of Nanodielectrics

9.2.1.1 Resistance to Corona and Surface Discharges

An increasingly common application area for polymeric dielectrics is as outdoor insulation; that is, on overhead lines or as outdoor bushings. In such circumstances, corona activity and surface discharges can occur, which will progressively degrade the material. Similar processes also occur in large electrical machines. As such, the ability of a dielectric to withstand surface discharge activity is technologically important and considerable efforts have been devoted to formulating materials with improved characteristics. Although corona and surface discharge effects have been evaluated in nanocomposites based upon a range of different base polymers, the technological importance of epoxy resins and silicones has meant that these systems have attracted particular attention.

The effect of nano-alumina on the ability of an epoxy resin to resist surface discharge activity was reported by Anglhuber and Kindersberger (2012). In this study, the nanofiller loading level was varied and the material damage was evaluated by surface profiling. Although damage was found to be substantially reduced by adding 1.5 % of the nanofiller, no further improvements were seen on increasing the amount of nano-alumina present. This work highlights a common theme in nanodielectrics, namely, that while the addition of a small quantity of nanofiller can be beneficial, in general, increasing this beyond a critical level either leads to no further benefits or, as described below, can have detrimental consequences.

From a practical viewpoint, it is not possible to optimise the performance of a material system with respect to just its electrical performance—other factors also need to be considered. In the case of high voltage bushings, for example, the coefficient of thermal expansion (CTE) is of critical importance and, ideally, the insulation and the conductor materials would exhibit the same CTE. To approach this position, significant quantities of an appropriate inorganic filler (e.g. silica) are commonly added to epoxy resins, which has led to the concept of nano/micro-composites. In one study (Iyer et al. 2012), the corona resistance of an epoxy resin containing 65 % microsilica plus 2.5–5 % nano-silica was compared with the same base resin containing either 2.5 % nano-silica or 65 % micrometric silica alone; the nano/micro-composite was found to out-perform both of the other systems. Elsewhere, the beneficial effects of adding ~1 % of an organically modified nanoclay to an epoxy resin heavily filled with silica have been reported, where it was concluded that the nanofiller "substantially improved the surface performance and was capable of resisting the degrading effects associated with partial discharges" (Fréchette et al. 2008). The beneficial consequence of adding nanofillers are not limited to epoxies, similar effects have been reported in silicones. In one study by Raetzke and Kindersberger (2010), a range of silicone rubbers containing a hydrophilic precipitated nano-silica (nominal particle size 20 nm) was subjected to a number of different surface arcing and erosion protocols. In this study, optimal performance was seen in systems containing 2–5 % of nano-silica.

9.2.1.2 Degradation of Nanocomposites: Mechanisms

The empirical evidence demonstrating that surface erosion by electrical activity is reduced when a very small volume fraction of a nanofiller is present seems clear. However, understanding of the underlying mechanisms is less secure and suggestions include changes in the structure of the matrix polymer (Kozako et al. 2004), the existence of interphases that are in some way inherently resistant (Raetzke and Kindersberger 2010), barrier effects resulting from the close proximity of neighbouring nanoparticles in well-dispersed systems (Venkatesulu and Thomas 2010) and increased bulk thermal conductivity (Ramirez et al. 2010).

Electrical activity adjacent at a polymer/gas interface can affect the solid through a multitude of different processes but, essentially, these will either be thermally driven or a consequence of aggressive chemical species generated in electrically active regions. As such, the mechanisms will have much in common with thermal degradation, thermo-oxidative degradation and chemical degradation. A topic where the chemistry of polymer degradation and the effect on this of included nanoparticles has been studied extensively is in connection with flammability and this would seem to have some parallels with the evidence presented above. While numerous factors have been found to influence the thermal stability and flammability of nanocomposites, in general, the beneficial consequences of adding a nanofiller to a polymer are ascribed to two different processes, one physical and one chemical.

The physical effect is related to the formation of a surface barrier layer due to local accumulation of the nanofiller as a consequence of ablation of the polymer. This layer will be rich in the inorganic components of the system, will consequently be thermally stable and will also act to reduce heat transfer into the underlying material. In the case of combustion, its presence will also limit diffusion of degradation products from the polymer to feed the flame. The chemical effect is ascribed to the nanofiller acting to promote the formation of solid rather than gaseous decomposition products (the so-called char), which will then act in the same way as the physical barrier described above. In this latter case, impurity atoms are believed to play an important role (Kashiwagi et al. 2002).

Many experimental studies of the influence of nanofillers on degradation have been conducted, which support these concepts. Saccani et al. (2007) investigated the degradation of polypropylene (PP) containing an organically modified layered silicate by Thermogravimetric Analysis (TGA) and showed that nanocomposites exhibited superior performance. Elsewhere, an increase in thermal stability of more than 50 °C was reported for PP containing an organically modified montmorillonite (MMT) compared with the unfilled polymer (Zanetti et al. 2001); this study is also significant in that it highlights the importance of competing reactions relating to chain scission and cross-linking, where the latter leads to char formation. Indeed, Drozdov (2007) described a combined experimental and theoretical analysis of the degradation of PP/carbon nanotube (CNT) nanocomposites, which indicated that the inclusion of the nanofiller led to the following: an increase in the activation energy for thermal degradation; a reduction in the rate of chain scission due to reduced permeation of oxygen into the system; a reduction in the rate of detachment of side-groups as a consequence of reduced mobility of chain segments in the vicinity of CNTs and reduced mass transport of decomposition products out of the system. Elsewhere (Ramirez-Vargas et al. 2012), a study of the effects of both aluminium hydroxide (ATH) and MMT on the flammability of systems based upon blends of a low density polyethylene (LDPE) and an ethylene-(vinyl acetate) copolymer (EVA) concluded that the enhanced performance of systems containing increased MMT contents should be attributed to the MMT promoting the formation of inorganic residues.

It is reasonable to state that, where surface electrical properties are concerned, the vast bulk of the published literature is consistent and indicates that the addition of even a very small quantity (~1 %) of a nanofiller can significantly improve the performance of polymers. Although different authors have ascribed these effects to different mechanisms, examination of the broader literature provides credible mechanisms, which have a structural origin and are largely associated with a good dispersion of the nanofiller combined with the consequent close proximity of the nanoparticles and the polymer.

9.2.2 Bulk Electrical Properties of Nanodielectrics

The complete topic of the bulk electrical performance of nanodielectrics includes a very wide range of phenomena and properties and, consequently, a comprehensive discussion of this is not possible here. Rather, just two areas are considered below, since these well illustrate the major issues associated with the bulk characteristics of nanocomposites in dielectric applications.

9.2.2.1 Breakdown Strength

A fundamental requirement of electrical insulation is an ability to withstand the applied electric field and a number of different approaches can be used to evaluate this. From a practical perspective, the important criterion is that the material system is able to withstand the applied electric field for the lifetime of the item concerned, but a comprehensive evaluation of this is time consuming, particularly in view of the statistical nature of breakdown. Consequently, a commonly used alternative is to measure the breakdown strength of a material by progressively increasing the applied field until breakdown occurs, repeating this a number of times (10–20 repeats would be typical) and, then, by assuming an appropriate statistical distribution, deriving the required quantities. Conventionally, the statistical analysis of breakdown data assumes a two parameter Weibull distribution (Weibull 1951), whereupon, the parameters of interest are the scale parameter (comparable to the mean in the normal distribution), the shape parameter (conveys information about the spread of the data and, hence, serves a similar function to the normal distribution's standard deviation) plus associated confidence bounds. However, the data that are generated are highly dependent on the chosen experimental protocol and are affected by the rate of increase of the applied field, the sample thickness, the immersion medium, the electrode geometry, etc. Consequently, while the approach is useful in providing relative performance metrics for different material formulations when tested under strictly identical conditions, it is not possible to compare absolute values reported by different authors.

Figure 9.2 presents DC breakdown data (Andritsch 2010) that clearly demonstrate the reason for the current interest in using nanocomposites as high

Fig. 9.2 Effect of particle size on the DC breakdown strength of epoxy-based systems containing 10 % of boron nitride. The *dashed horizontal line* indicates the breakdown strength of the unfilled polymer; data from Andritsch (2010)

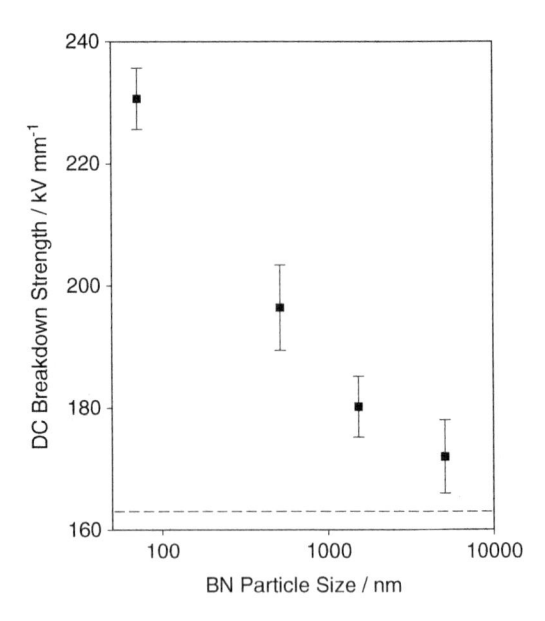

performance dielectric materials. These data were derived from epoxy-based systems containing 10 % of boron nitride (BN), where the size of the BN particles was varied as shown. In this study, all the systems containing BN exhibited increased breakdown strength compared with the unfilled resin and, the smaller the particle size, the greater the increase. Numerous other examples can be found in the literature for the beneficial consequences of adding nanofillers. For example, Okuzumi et al. (2008) considered the effect of nano-magnesium oxide (MgO) on the breakdown strength of PP and reported an increase in DC breakdown strength of about one-third. Elsewhere, the addition of 5 % hydrophobic fumed nano-silica to PP was reported to increase the DC breakdown strength from 511 to 778 kV mm^{-1} (Takala et al. 2010).

The relationships between nanofiller/polymer compatibility and breakdown strength have been studied extensively in connection with many particulate filled systems, where silane chemistry can be used to substitute hydroxyl functionality with other moieties that would be expected to interact more favourably with the matrix material. In an early study of this type, the electrical response of a range of cross-linked polyethylene (XLPE) systems containing 5 % silica (SiO$_2$) was compared (Roy et al. 2005). As is commonly reported, the inclusion of 5 % of micron-sized filler markedly reduced the breakdown strength relative to that of the unfilled XLPE. The performance of the unfilled XLPE and the XLPE containing untreated nano-silica was, however, equivalent, while the inclusion of 5 % of nano-silica in which the surface chemistry had been modified through treatment with a vinylsilane resulted in increased strength compared with the unfilled polymer. In a subsequent study by the same group (Roy et al. 2007), the effect of different silanes on the DC breakdown behaviour of a range of different nanocomposites containing 5 %

nano-silica was reported. While, simplistically, it is not unreasonable to suppose that replacing hydroxyl surface groups with moieties such as alkyl chains could lead to improved compatibility between the nanofiller and PE, in this later study, enhanced performance was seen in a system containing a nano-silica that had been functionalised with N-(2-aminoethyl) 3-aminopropyl-trimethoxysilane (AEAPS), a relatively polar species. This point is discussed further in Sect. 9.3.2 below.

As discussed in the review by Li et al. (2010), if the addition of a small fraction of nanofiller leads to an increase in breakdown strength then, beyond a certain point, this trend is reversed, presumably, as a consequence of increased nanofiller aggregation, which dominates any other factors. Then, the agglomerated nanoparticles behave, effectively, as micron-sized inclusions, the presence of which leads to a reduction in breakdown strength (Roy et al. 2005; Vaughan et al. 2006). This result is consistent with studies in which inclusion of various types were added to polymers (Morshuis et al. 1988), which showed empirically that the presence of sizeable particles degrades breakdown performance.

The studies considered above show that adding a small volume faction of nanoparticles to a polymer can lead to improved breakdown performance. This conclusion is, however, not universal. Hong et al. (2003), for example, reported on the effect of zinc oxide (ZnO) on the DC breakdown strength of LDPE. This work considered a wide range of different loading levels and compared both microscopic and nanoscopic ZnO. A monotonic decrease in strength with increasing filler loading was reported in both cases. Ma et al. (2005) reported a decrease of DC breakdown strength approaching 40 % when titania (TiO_2) nanoparticles were added to LDPE. This reduction in strength could, however, be reduced to about 10 % by drying the TiO_2 nanoparticles. Nevertheless, reacting the nano-TiO_2 with AEAPS—the very same silane that was reported to lead to beneficial effects elsewhere (Roy et al. 2007)—resulted in a decrease in breakdown strength compared with the dried, but unfunctionalised nanofiller.

From the brief account given above, it is clear that published reports on the effect of nanoparticles on the breakdown strength of polymers differ widely. This diversity in behaviour, which is in marked contrast to the consistent picture described in Sect. 9.2.1 in connection with surface electrical performance, suggests that the linkages between material formulation, material structure and breakdown performance are complex and are reliant on factors that are not routinely considered, or at least, not routinely reported.

9.2.2.2 Permittivity

The dielectric response of a material can be represented in terms of the frequency dependence of the permittivity. Permittivity is the macroscopic manifestation of the polarisability of the material constituents, as exemplified by the Claussius–Mossotti equation:

$$\alpha = \frac{3\varepsilon_0 M}{N_A \rho} \cdot \frac{(\varepsilon_r - 1)}{(\varepsilon_r + 2)}$$

In this, M is the molecular mass, ρ is density, N_A is Avogadro's number, ε_0 is the permittivity of free space, ε_r is the relative permittivity/dielectric constant and α is the molecular polarisability. For a full discussion of the dielectric behaviour of polymer-based materials, reference to the excellent works by Kremer (2003) and Jonscher (1983) is highly recommended. Nevertheless, simplistically, permittivity can be thought of in terms of the number and nature of the polarisable species present in the system, plus their dynamics. Since the dielectric response of a given moiety is affected by its environment, dielectric spectroscopy can provide local structural information. In practice, the relative permittivity of a material is a complex quantity:

$$\varepsilon_r = \varepsilon_r' + i\varepsilon_r''$$

where ε_r' is the real part of the relative permittivity, which is related to the energy stored in the system, ε_r'' is the imaginary part of the relative permittivity, which is related to the energy dissipated by the material, and the loss angle, δ, is expressed:

$$\tan \delta = \frac{\varepsilon_r''}{\varepsilon_r'}$$

For the sake of brevity, henceforth, ε_r', ε_r'' and $\tan \delta$ will, in general, be referred to as the real permittivity, the imaginary permittivity and the loss, respectively. In the following discussion, the real permittivity will first be considered, before moving on to consider the imaginary permittivity/dielectric loss.

In the case of a multi-component system such as a nanodielectric, an effective medium approach can be used to define the real permittivity of the whole based upon the composition and the properties of the individual components; Myroshnychenko and Brosseau (2005) provide a good overview of the topic, as applied to binary systems. The Lichtenecker–Rother equation is an example of such a relationship:

$$\log \varepsilon_c' = \phi_p \log \varepsilon_p' + \phi_m \log \varepsilon_m'$$

where ε_c', ε_p' and ε_m' represent the real permittivity of the composite, particles and matrix and ϕ_p and ϕ_m represent the volume fraction of particles and matrix, respectively. While the underlying physical models may vary, the resultant mathematics all share the common feature that the range of possible values is bounded by the real permittivity of each of the two components. Nevertheless, in the case of nanodielectrics, many experimental studies have reported a form of behaviour that differs markedly for this.

Kochetov et al. (2012) described the dielectric response of a range of nanodielectrics based upon particulate nanofillers dispersed within an epoxy matrix. In all cases, with the exception of nano-silica, the inclusion of a low volume fraction (<5 %) of nanofiller resulted in a reduction in the measured real permittivity, below that of the host matrix, despite $\varepsilon'_p > \varepsilon'_m$. In another epoxy-based system, the addition of 5 % of MMT led to a reduction in the real permittivity from ~4.6 to a value around 4.3 (Tagam et al. 2008). In thermoplastics, a study of the dielectric response of LDPE-based nanocomposites containing 5 % SiO_2 revealed evidence of a reduction in the permittivity of the nanocomposite to a value lower than that of the base polymer (Roy et al. 2005). Hui et al. (2010) also considered nanocomposites based upon nano-silica but, in this case, in an EVA matrix; a reduction in permittivity from 3.27 to 3.09 on adding of 5 % of the filler was reported. Elsewhere, a reduction in permittivity from 3.27 to 2.26 was seen in a series of polyimide/silica systems on increasing the percentage of nanofiller from 0.06 to 5.58 % (Srisuwan et al. 2010). Representative data of this form are shown in Fig. 9.3.

Imaginary permittivity data provide a complementary picture of the processes that occur in nanodielectrics and can be particularly informative in probing local interactions; epoxy resins constitute ideal systems for study by dielectric spectroscopy, because of the presence of polar groups that can couple with the applied field. Figure 9.4 shows imaginary permittivity data obtained from a range of epoxy/nano-silica systems. The data shown in Fig. 9.4a were obtained at different temperatures from a single system, in which the nano-silica was thoroughly dried but, otherwise, used as supplied. In this figure, a number of distinct processes can be seen. At low

Fig. 9.3 Variation in the real part of the relative permittivity of nanocomposite systems, ε'_c, as a function of volume fraction of alumina (Al_2O_3) and magnesium oxide (MgO) nanoparticles in an epoxy resin matrix; data from Andritsch (2010)

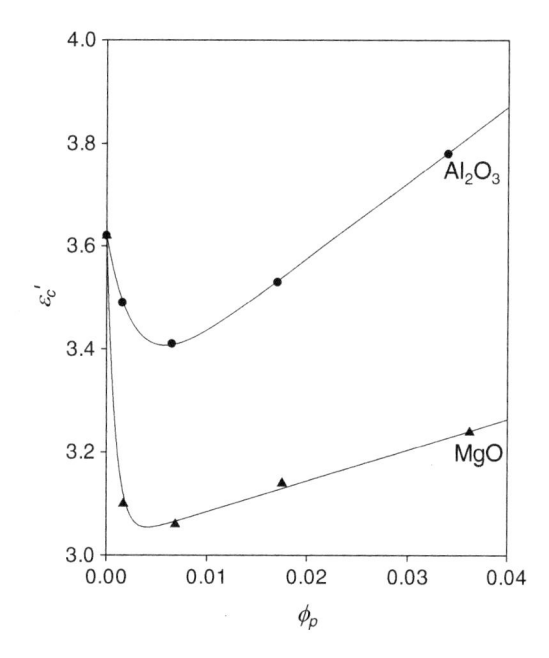

Fig. 9.4 Imaginary permittivity data obtained from nanodielectrics based upon nano-silica in an epoxy resin: (**a**) frequency dependent data obtained at the indicated temperatures from a system containing 2 % of unfunctionalised nano-silica (NC0); (**b**) data obtained from systems where the nano-silica had been functionalised with increasing amounts (NC1 → NC16) of 3-glycidyloxypropyl) trimethoxysilane

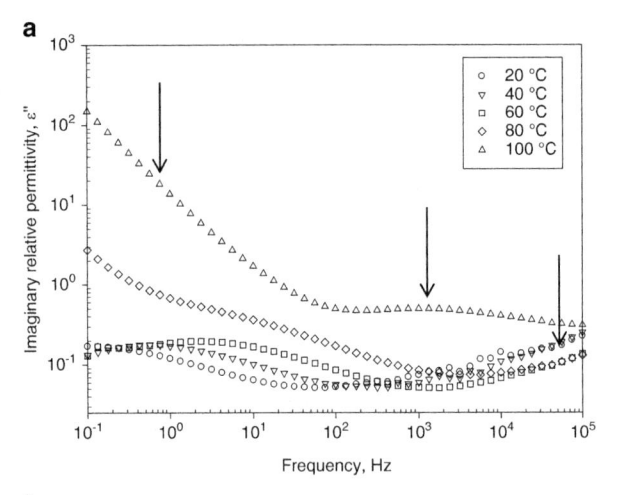

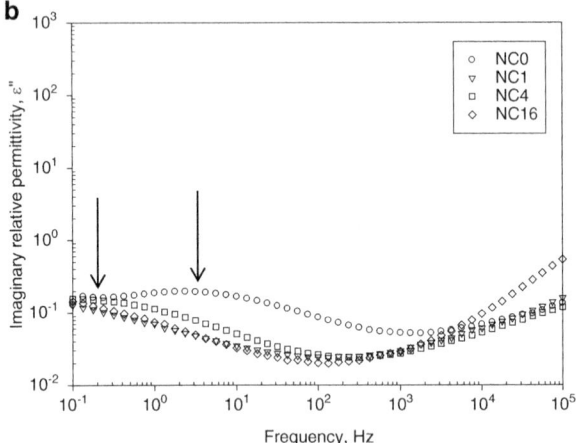

temperatures and high frequencies (arrow at ~5 × 10^4 Hz), the β-relaxation can be seen, which is associated with the hydroxyether groups in the polymer backbone (Kosmidou et al. 2012). Two other processes are evident, which are most easily seen in the data obtained at 100 °C. These are related to the α-relaxation (i.e. the glass transition—arrow at ~2 × 10^3 Hz) and, at low frequencies, a progressive increase in ε'' with decreasing frequency, which is primarily associated with electrical conduction as a result of residual chloride ions in the resin (Hammerton 1997). In Fig. 9.4b, all data were obtained at a fixed temperature (60 °C), but in this case, the surface chemistry of the nano-silica was varied by treating it with different quantities of 3-glycidyloxypropyl)trimethoxysilane (GLYMO). From these data sets, it is evident that surface functionalisation has a marked effect on the α-relaxation. While a detailed interpretation of this is complicated by differences in behaviour between the DSC glass transition and the dielectric α-relaxation, these results nevertheless indicate that changes in nanoparticle surface chemistry do affect local interactions and local structure and that these effects can manifest themselves as changes in bulk properties.

In the case of polymers such as PE and PP, the non-polar nature of the molecule implies that any observed dielectric relaxation effects are either associated with the nano-filler itself, or else, the polymer/nano-filler interface/interphase. In one study of PP containing nano-silica, an invariant dielectric loss was reported at intermediate frequencies, together a marked increase in loss at low frequencies in the nanocomposite containing unfunctionalised nano-silica. This behaviour was interpreted in terms of some quasi-DC conduction process. Zhang and Stevens (2008) contrasted the dielectric response of nanocomposites (0–10 % nano-Al_2O_3) based upon linear low density polyethylene- (LLDPE) and an epoxy resin. Although a complete analysis is not presented, the paper does nevertheless highlight the potential impact of absorbed water at nanoparticle interfaces on the dielectric spectrum. While absorbed water generally has unwanted consequences, it can however be used as a sensitive dielectric probe of nanoparticle interface structures. Indeed, Lau et al. (2013), compared the dielectric response of a range of systems based upon blends of high density polyethylene (HDPE) and LDPE, containing nano-silicas with different surface chemistries, as a function of their exposure time to water. This work showed that the quantity of water accumulated at the nano-silica/polymer interface and its dynamics were both markedly affected by substitution of surface hydroxyl groups with propyl moieties. The implication of this is that changes in the nanoparticle surface chemistry affect the interfacial structure which, in this case, manifests itself in its ability to accommodate water molecules.

9.2.3 Nanoparticles as Structural Modifiers

The primary aim of the preceding sections has been to illustrate the impact that introducing nanoparticles can have on properties. Effective medium theories, of the type described above, are widely used in materials science to estimate a given property, P, of a composite system C composed of materials A and B; in general, limiting values are given as follows:

$$\left(\frac{\phi_A}{P_A} + \frac{\phi_B}{P_B}\right)^{-1} < P_C < \phi_A P_A + \phi_B P_B$$

Applying such concepts to the real permittivity data described above, it is evident that this inequality does not hold. This suggests that the situation in nanodielectrics is more complex, which implies that the presence of the nanoparticles is affecting the overall performance of the system in a much more subtle and complex way.

In the case of breakdown, an immediate issue that needs to be addressed is the potential effect of the nanoparticles on the structure of the matrix polymer, as highlighted by Hoyos et al. (2008), who concluded that, in systems based upon LDPE containing silicate fillers, the final dielectric performance is determined by both the inorganic particles and the semi-crystalline morphology. In PP, in particular, increasing the nucleation density will reduce the spherulite size which, in itself, can lead to markedly increased breakdown strength (Kolesov 1980). This could explain the dramatic beneficial effects reported by some workers on adding nanofillers to PP (Okuzumi et al. 2008; Takala et al. 2010). Although most studies of nanodielectrics have not sought to consider morphological effects in detail, the influence of the added nanofiller on both matrix morphology and breakdown was studied explicitly by Vaughan and co-workers (Vaughan et al. 2006; Green et al. 2008) in systems based upon a polyethylene blend containing organically modified MMT. In the case of systems containing MMT derived from the pre-compounded masterbatch Nanoblend 2101 (PolyOne Corporation), the addition of the MMT was found to have little effect on the morphology of the matrix but to lead to an increase in breakdown strength from 143 to 171 kV mm^{-1} on adding 10 % of the MMT. However, when the same experiments were undertaken using a different masterbatch (Nanocor C30PE), the MMT was found to nucleate the polymer strongly and to impede crystallite evolution, such that reduced overall levels of matrix crystallinity resulted. Although reduced spherulite size would normally be expected to increase the breakdown strength as a consequence of reduced segregation of low molar mass and defective species to spherulite boundaries coupled with reduced specific volume effects, in this case, no improvement in breakdown strength was seen, even when the MMT was well dispersed (see Fig. 9.5). The difference between Nanoblend 2101 and Nanocor C30PE concerns that organic compatibiliser used in the two systems; in the former case it is dimethyl-di(hydrogenated tallow) quaternary amine while the latter contains octadecyl ammonium cations. Elsewhere (Lau et al. 2013), the addition of nano-silica to an HDPE/LDPE blend was also shown to have a strong nucleating effect on

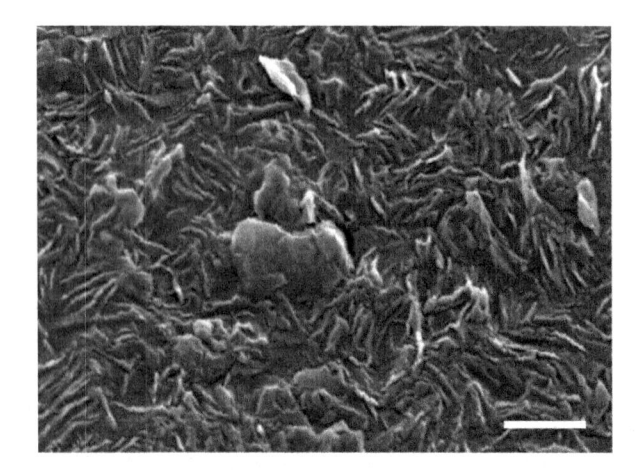

Fig. 9.5 Scanning electron micrograph showing the disordered, highly nucleated structure of a polyethylene blend containing ~5 % of MMT modified using octadecyl ammonium cations. Scale bar: 2 μm

the polymer but, in this case, the presence of the nanofiller did not appear to affect crystal growth and its nucleating ability appeared largely independent of surface chemistry. Evidently, nanofillers can have a major effect on both the matrix morphology and the electrical properties of the resulting nanocomposites, but these factors are not simply linked.

In terms of the real permittivity, the anomalous reduction in this parameter on addition of a low volume fraction (<5 %) of a high permittivity nanofiller can be interpreted in terms of local interactions between the nanofiller and the epoxy serving to immobilise a region of the matrix immediately adjacent to the nanoparticle surface. In such circumstances, the presence of a third component (i.e. an interphase region where the nanoparticles serve to modify the structure and/or dynamics of the matrix) is commonly invoked, such that the Lichtenecker–Rother equation given above should be rewritten.

$$\log \varepsilon'_c = \phi_p \log \varepsilon'_p + \phi_i \log \varepsilon'_i + \phi_m \log \varepsilon'_m$$

Here, the additional term involves the real permittivity (ε'_i) and volume fraction (ϕ_i) of the interphase. Although this provides a qualitative structural interpretation of the data, it is not without its problems, as illustrated by the finite element simulation work of Maity et al. (2010), where it is necessary to assume an unrealistically large interphase thickness of about 200 nm in order quantitatively to reproduce the experimental data. Studies of nitroxide spin-labelled poly(methyl acrylate) containing a synthetic fluoromica suggested a rigid interface region that is just 5–15 nm in thickness (Miwa et al. 2008).

9.3 Theories and Models

A number of different hypotheses have been proposed to account for the structure and consequent properties of nanocomposites, but the dominant concept relates to nanoparticle/matrix interactions. This topic is briefly reviewed below in connection with two issues: nanoparticle dispersion; the formation of structurally distinct interphase regions near nanoparticle surfaces.

9.3.1 Miscibility and Dispersion

Nanocomposites equate to a distribution of nanoparticles within a continuous phase and, consequently, share certain features with colloids, where a range of factors influence the structure of the system. These include: excluded volume effects that are related to the direct interactions between hard particles; where particle surfaces are covered with long chain molecules, additional steric repulsive forces will also exist that are related to local molecular conformations; thermodynamic interactions between the discrete particles and the continuous phase. Finally, in colloids, the particles will generally be charged, which results in the attraction of oppositely charged ions to form an electric double layer, in which the charge density diminishes to zero far from the interface and where the system as a whole is not charged. This structure is termed the Helmholtz–Stern–Gouy–Chapman double layer and is often discussed in terms of the associated zeta potential, which corresponds to the potential at the hydrodynamic slip plane surrounding the charged particle. The zeta potential is an important parameter, since it is often related to the stability of the system. Indeed, a common assumption is that if the zeta potential exceeds about 25 mV, then flocculation is suppressed. However, in view of the range of different interactions that will exist between nanoparticles and their environment, a more generally valid equivalent is to consider the potential energy of particles as a function of their separation. For example, in the case of lyophobic particles, adsorption of lyophilic polymer can act to promote or suppress flocculation, depending on the degree of surface coating. If this is low, then it is possible for a single chain to become adsorbed onto adjacent particles, thereby inducing aggregation. Conversely, if the degree of surface coverage is high, then steric interchain effects will dominate, such that, as neighbouring particles approach, molecular conformations will be perturbed and the particles will be repelled. The concepts of lyophilic and lyophobic are related to enthalpic interactions between the particles and their environment and, as such, this involves comparable physics to that which governs polymer solubility.

In discussing the concept of miscibility of a binary system in the most general terms, for the two components to mix intimately:

$$G_{12} < G_1 + G_2$$

where G_{12} is the Gibbs free energy of the mixed system and G_1 and G_2 represent the Gibbs free energies of the individual components. Alternatively, at a given temperature, T, this can be expressed:

$$\Delta G_m = G_{12} - (G_1 + G_2) = \Delta H_m - T \Delta S_m$$

where ΔH_m is the enthalpy of mixing and ΔS_m is the entropy of mixing; provided $\Delta G_m < 0$, mixing/dissolution will be favoured thermodynamically. In the case of polymer solutions, enthalpic effects are known to dominate and comparable effects are likely to pertain in nanocomposites. Consequently, the topic of surface interaction and nanoparticle dispersion has been considered by a number of workers. Xue et al. (2015), applied coarse grained molecular dynamics (MD) techniques to study molecular conformations in systems based upon polydisperse polymer chains grafted onto nanoparticles. This work adjusted a number of system variables including grafted chain length, grafted chain polydispersity, graft density and nanoparticle size and computed the effect of such parameters on the average radius of gyration of the grafted chains, the average thickness of the film of grafted chains, etc. From this, the authors concluded that enhanced excluded volume interactions occur for larger nanoparticles (reduced curvature) under conditions when the grafted chain polydispersity is low and the grafting density is high. Liu et al. (2011), considered aggregation effects in a non-specific polymer-based nanocomposite and, initially, focused on the effect of varying the nanoparticle/ matrix interactions on the nanoparticle dispersion, as defined by the radial distribution function (RDF). Unsurprisingly, when the nanoparticle/polymer interactions were low ($\varepsilon_{np} < 1$), the simulation led to a peak in the RDF corresponding to direct contact of filler particles while, for $\varepsilon_{np} > 2$, specific peaks in the RDF were reported which suggested a morphology based upon agglomerates in which the nanoparticles were separated from one another by intervening layers of polymer. Interestingly, in a system corresponding to intermediate nanoparticle/matrix interactions ($\varepsilon_{np} = 2$), no strong peaks were observed in the RDF, suggesting a high degree of nanoparticle dispersion. The investigation then examined the effect of grafted chains on the behaviour of the system and showed that, even in the absence of favourable nanoparticle/matrix interactions, this strategy could prevent direct contact aggregation.

9.3.2 Interfaces and Interphases

It is generally accepted that Lewis (1994) first discussed the concept of nanometric dielectrics and introduced the notion of interphases whereby, as filler particle decreases in size, so the specific surface area increases and the total interphase volume progressively increases, as discussed in Sect. 9.1.1.

The most detailed hypothesis of this type is the so-called multi-core model, proposed by Tanaka et al. (2005), which explicitly proposes four local factors. In this, a thin layer of material of the order of 1 nm in thickness immediately adjacent to the nanoparticle surface is thought to be tightly bonded to both the inorganic and organic components, as a consequence of the presence of a suitable coupling agent. Adjacent to the bonded layer is a so-called bound layer (thickness <10 nm), which is composed of a region consisting of polymer chains that are strongly bound and/or strongly interacting with the first layer and/or the surface of the nanoparticle. In this region, chain mobility may be restricted. Also, if the interactions that occur in the bonded and bound layers result in the nanoparticle surface acting to nucleate a semi-crystalline polymer, the overall morphology of the matrix will be altered while, in the case of thermosetting polymers, the presence of reactive moieties will affect the local stoichiometry and, consequently, the network architecture. The final structural element is seen as a region of material that is loosely coupled to the bound, second layer, in which the chain conformations, chain mobility, free volume, crystallinity, stoichiometry, etc., may all differ from that in the unperturbed matrix. Finally, superimposed upon these structurally distinct regions, it is argued that an electric double layer may exist, as in the colloid case described above. A necessary requirement for this is that the nanoparticles are charged, which seems very reasonable, but also that the polymer contains sufficient mobile charge carriers to form an equivalent of the Helmholtz–Stern–Gouy–Chapman double layer.

While the existence of interphases as a consequence of nanoparticle/matrix interactions is eminently reasonable, compelling experimental evidence for these equating to a significant volume fraction of the system remains scarce. As such, it is worth considering other ways in which the presence of interfaces can directly affect the properties of a system. The concept of effective medium theories has been discussed above in Sects. 9.2.2 and 9.2.3, primarily, in connection with the dielectric response of a system. Progelhof et al. (1976), however, have described a multitude of such approaches in connection with the thermal conductivity of mixed systems. Since the dissipation of heat is important in many applications, it is tempting to suggest that the thermal conduction of a polymer can be improved simply by introducing a filler with a high thermal conductivity. However, heat transfer in a disordered dielectric should be considered in terms of phonons that move through the system and undergo frequent scattering, whereupon, the phonon mean free path between scattering events is affected by a combination of phonon/phonon scattering and geometrical scattering from structural irregularities in the material (Berman 1976). As such, internal boundaries within a system can have a major effect on thermal conductivity and can be represented in terms of an interfacial thermal resistance, such that heat transfer through a multicomponent system can be very much less efficient than effective medium theories would suggest.

Electrically, the effect of micro-particles, nanoparticles and nanoparticle surface chemistry on the electrical characteristics of systems based upon XLPE and silica has been reported by Roy et al. (2007). In the systems considered here, it is noteworthy that the nanocomposite containing nano-silica functionalised with AEAPS was characterised by a breakdown strength that was significantly higher than that of the unfilled polymer (see Sect. 9.2.2), despite the fact that is difficult to see how the introduction of polar functionality could lead to enhanced dispersion. However, by conducting an Arrhenius analysis of thermally stimulated current (TSC) data, these workers were able to estimate the depth of the charge carrier traps in their various systems and, from this, they showed the nanocomposite containing the AEAPS-functionalised nano-silica exhibited evidence for the formation of novel trap sites, which the authors ascribed to the polar nature of the surface moieties. That is, the electrical behaviour of nanocomposites may not only be affected by interphases and dispersion effects, but that the local density of interfacial states and the effect of these on charge transport dynamics may also be of great importance.

9.3.3 From Composition to Properties

The preceding sections have considered a wide range of processes by which the addition of nanoparticles to a polymer matrix can affect the structure and, consequently, the properties of the resulting material. In the case of flammability, for example, the mechanisms appear relatively well understood but, in the case of the bulk dielectric response, this is not the case. Evidently, nanoparticles may change the local structure of the polymer or, where they act as nucleation sites, the morphology of the whole system. There is an undoubted need for new dielectric materials to underpin new electrical technologies but, at present, the linkages between the composition, processing, structure and many electrical properties of particulate-filled nanocomposites remain unclear. This lack of certainty and reproducibility currently constitute major impediments to the technological deployment of nanostructured systems based upon inorganic nanoparticles distributed within a polymeric host. Consequently, the following section presents an alternative and less conventional approach to the design of a nanostructured dielectric, where the underlying processes are better understood.

9.4 Advanced Dielectrics Through Morphological Design

It can be argued that the lamellar nature of semi-crystalline polymers means that all such systems contain at least two phases in which at least one characteristic dimension falls into the nanometric range. This section will therefore take a less conventional approach to the topic of nanodielectrics where, here, the concept is broadened to include blend systems that have been intentionally manufactured to exhibit internal structure that are <100 nm in size and which, consequently, exhibit novel properties.

The influence of polymer morphology on the electrical properties of polymers has been studied for many decades. In 1980, Kolesov published a key paper on the effect of spherulite size on the breakdown strength of PE and PP. This work showed that as the spherulite size increased, so the breakdown strength decreased, until the spherulite size became commensurate with the specimen thickness, at which point, it thereafter remained constant. The implications of this are twofold: first, that breakdown is intimately linked with structural defects (spherulite boundaries in this case); second, that optimum performance would be obtained from a material containing no large scale morphological features. Figure 9.6 shows a TEM micrograph of a sample of XLPE, which exhibits a morphology composed of small crystalline units that, even in their lateral extend, appear to be ~100 nm in size.

The weakness of spherulite boundaries in semi-crystalline polymers can be attributed to two effects. First, as spherulites evolve, the lowest molar mass fractions in the system will be rejected from the crystal melt interface and will diffuse away, such that they become concentrated where neighbouring spherulites impinge. Second, the density of crystalline structures is necessarily greater than that of the melt from which they form such that, when crystalline structures form under constant volume conditions, the last regions to solidify will tend to contain increased free volume and inbuilt mechanical stresses. Both of these factors mean that spherulite boundaries are mechanically and electrically weak.

Fig. 9.6 Transmission electron micrograph showing the morphology of a high voltage XLPE cable insulation system (scale bar 200 nm)

9.4.1 Polyethylene Blends

By judiciously blending together different polymers and appropriately controlling subsequent morphological evolution it is, however, possible to engineer out the weakness of spherulitic boundaries; the process is shown conceptually in Fig. 9.7, which is well demonstrated in the case of blends of linear (LPE) and branched (BPE) polyethylenes, where the LPE is the minority component. In PE, the crystalline unit cell means that bulky defects such as branches cannot be incorporated into crystalline structures and, therefore, these are excluded. As a consequence of this, BPE is not able to form lamellae that are as thick as in LPE and, consequently, LPE is able to crystallise at a high temperature than BPE. In the melt, the LPE and the BPE are miscible but, if the temperature is then reduced into the approximate range 110–130 °C, the LPE will begin to crystallise but the BPE will remain in the melt phase. Depending upon the chosen LPE crystallisation temperature, either relatively compact semi-crystalline inclusions of the linear polymer will form within the BPE melt, or else, an array of LPE lamellae will evolve to fill space. This skeleton of LPE crystals are then separated from one another by molten BPE, which acts to solvate and disperse the defective low molar mass fractions which, in a single component system, would become concentrated at spherulite boundaries. In addition, when the BPE subsequently crystallises, associated specific volume effects are also distributed throughout the bulk, with no adverse consequences.

In the above case, control of morphology depends upon three factors: the branch density in the BPE; the temperature at which the LPE component of the blend is crystallised and the fraction of LPE present in the system. Experimentally, it has been found that provided the branch density is not too high, the key features of morphological evolution are insensitive to the nature of the BPE; LDPE, Ziegler Natta LLDPE and metallocene LLDPE behave similarly (Hosier et al. 2000).

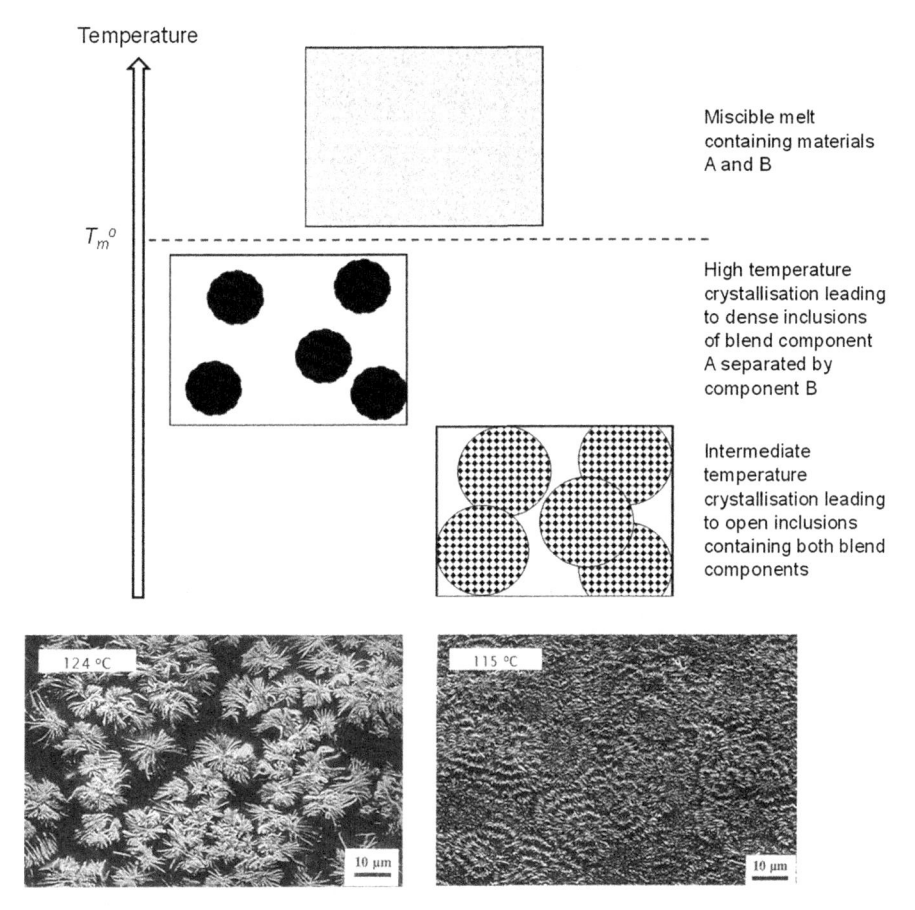

9 Nanodielectrics: The Role of Structure in Determining Electrical Properties   257

Fig. 9.7 Schematic representation showing how isothermal crystallisation at an appropriate, intermediate temperature can be used to instigate a two stage crystallisation process in polyethylene blends, together with two scanning electron micrographs showing the morphology of PE blends crystallised at 124 °C (high temperature) and 115 °C (intermediate temperature)

Figure 9.8 shows the effect of blend composition and crystallisation temperature on breakdown strength (Hosier et al. 1997). From this, it is evident that in the case of quenching, the composition of the blend has little effect on breakdown; within the parameter space considered here, all quenched systems exhibit the same morphology. For both isothermal data sets, the breakdown strength first drops before progressively increasing as the proportion of LPE present in the blend increases. The differences in the behaviour of the three differently crystallised sets of samples clearly demonstrates that the variations in breakdown strength shown are due to morphological effects and that by appropriately controlling morphological evolution, materials with enhanced properties can be generated. Elsewhere, alternative aspects of structure/property relationships in such blends related to their long term breakdown behaviour, mechanical properties and processing have been discussed (Dodd et al. 2003; Green et al. 2013).

Fig. 9.8 Effect of blend composition and crystallisation conditions (quenched, *solid line*; isothermal crystallisation at 115 °C, *dashed line*; isothermal crystallisation at 124 °C, *dotted line*) on the breakdown strength of polyethylene blends

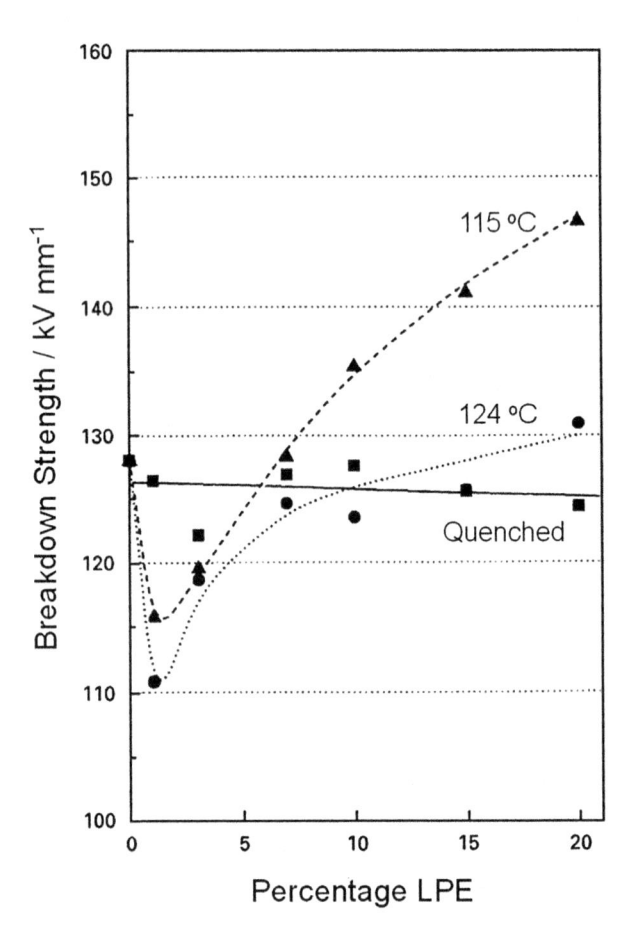

9.4.2 Polypropylene Blends

An attractive extension to the above concept concerns the substitution of LPE and BPE with appropriate propylene-based grades. While this idea has a long history, as judged by the patent literature, control of morphology and hence properties, is very much more difficult in PP-blends. In such systems, the analogue of LPE is an isotactic polypropylene (iPP) homopolymer and the analogue of BPE is a propylene-based copolymer. However, molecular packing in the PP unit cell means that inclusion of defects occurs much more readily than in PE and, consequently, it is much more difficult to control morphological evolution in such systems, such that the desired combination of properties ultimately results. The effect of molecular architecture and composition on the morphology and electrical and mechanical properties of binary blends derived from six propylene-based systems was reported by Hosier et al. (2011). This work set out to develop materials systems that would be characterised by high breakdown strength, low temperature flexibility and good high temperature mechanical integrity. Mechanically, effective medium mixing rules suggest that the required temperature dependence of the modulus can be obtained by combining a relatively stiff iPP homopolymer with a low crystallinity, and therefore softer, copolymer. Then, the copolymer provides the low temperature flexibility while the isotactic homopolymer provides the required high temperature mechanical integrity. However, to achieve desirable electrical performance it is necessary to eliminate distinct spherulite boundaries and to limit phase separation in order to prevent the formation of distinct inclusions rich in the low crystallinity copolymer. Figure 9.9 shows two examples of propylene-based blend morphologies; in Fig. 9.9a, the homopolymer and copolymer undergo neither liquid/liquid nor liquid/solid phase separation such that, even though the majority propylene-ethylene copolymer is unable to crystallise at the chosen crystallisation temperature (130 °C), a fine scale space-filling spherulitic morphology results. In Fig. 9.9b, the appearance is totally dominated by phase separation effects that result in distinct inclusions rich in the propylene-ethylene copolymer ~10 μm in size. It is noteworthy that, in these examples, the system that phase separates contains a copolymer with a very much lower ethylene content than the system that exhibits the

Fig. 9.9 Scanning electron micrographs showing two very different morphologies in homopolymer/copolymer blends: (**a**) pronounced spherulites are evident (scale bar 50 μm); (**b**) the morphology is dominated by phase separation (scale bar 50 μm)

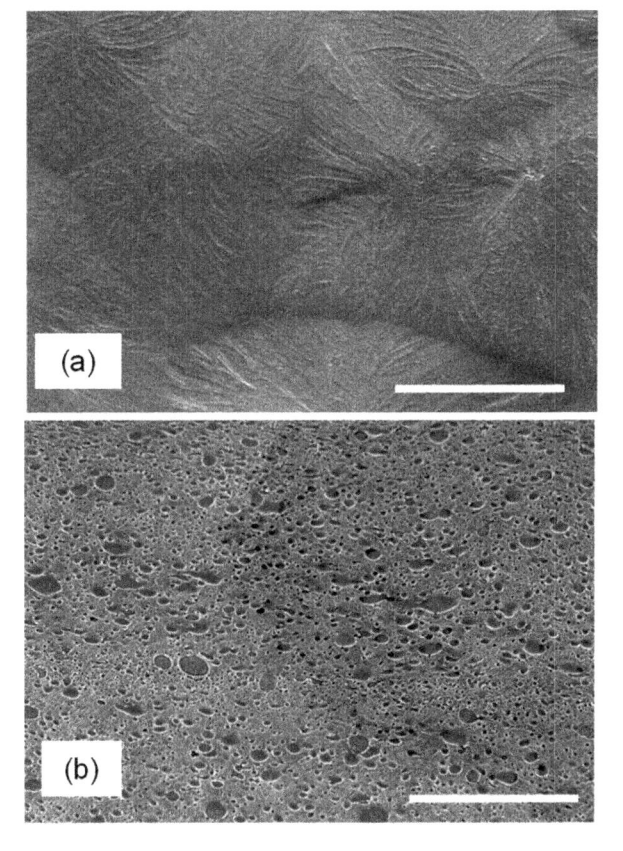

continuous, non-phase separated morphology, which suggests that it is not the gross composition of the copolymer that is important in morphological evolution but, rather, the precise sequences of co-monomers within the backbone.

9.5 Summary

Nanodielectrics undoubtedly remain one of the "hot topics" in dielectrics research. However, despite the 20+ years that have passed since the concept was first introduced, they still remain an enigmatic concept, with many inconsistent and contradictory results being derived from different studies that, apparently, involve very similar systems. Consequently, at present, our understanding of why property enhancements come about and how to achieve them reliably and reproducibly is poor. As such, if these materials are ever to migrate from the research laboratory into our electrical infrastructure, then there is a clear need to develop a better understanding of the key structure/property relationships, suitable techniques for material characterisation and reliable bulk processing routes that will guarantee reliably products. The alternative is to design new materials through polymer blending and, while this strategy is far from simple, the underlying principles appear, at present, to be far better understood.

References

Andritsch T (2010) Epoxy based nanocomposites for high voltage DC applications. Ph.D. Thesis, University of Delft

Anglhuber M, Kindersberger J (2012) Quantification of surface erosion and microscopic analysis of particle distribution in polymer nanocomposites. IEEE Trans Diel Electr Insul 19:408–413

Berman R (1976) Thermal conduction in solids. Clarendon Press, Oxford

Dodd SJ, Champion JV, Zhao Y, Vaughan AS, Sutton SJ, Swingler SG (2003) Influence of morphology on electrical treeing in polyethylene blends. IEE Proc Sci Meas Technol 150:58–64

Drozdov AD (2007) A model for thermal degradation of hybrid nanocomposites. Eur Polym J 43:1681–1690

Fréchette MF, Larocque RY, Trudeau M, Veillette R, Rioux R, Pélissou S, Besner S, Javan M, Cole K, Ton That M-T, Desgagnés D, Castellon J, Agnel S, Toureille A, Platbrood G (2008) Nanostructured polymer microcomposites: a distinct class of insulating materials. IEEE Trans Diel Electr Insul 15:90–105

Green CD, Vaughan AS, Mitchell GR, Liu T (2008) Structure property relationships in polyethylene/montmorillonite nanodielectrics. IEEE Trans Diel Electr Insul 15:134–143

Green CD, Vaughan AS, Stevens GC, Sutton SJ, Geussens T, Fairhurst MJ (2013) Recyclable power cable comprising a blend of slow-crystallized polyethylenes. IEEE Trans Diel Electr Insul 20:1–9

Hammerton I (1997) Recent developments in epoxy resins (Rapra Review Reports). Smithers Rapra Technology, Shrewsbury

Hong JI, Schadler LS, Siegel RW, Martensson E (2003) Rescaled electrical properties of ZnO/low density polyethylene nanocomposites. Appl Phys Lett 82:1956–1958

Hosier IL, Vaughan AS, Swingler SG (1997) Structure property relationships in polyethylene blends: the effect of morphology on electrical breakdown strength. J Mater Sci 32:4523–4531

Hosier IL, Vaughan AS, Swingler SG (2000) On the effects of morphology and molecular composition on the electrical strength of polyethylene blends. J Polym Sci B Polym Phys 38:2309–2322

Hosier IL, Vaughan AS, Swingler SG (2011) An investigation of the potential of polypropylene and its blends for use in recyclable high voltage cable insulation systems. J Mater Sci 46:4058–4070

Hoyos M, Garcia H, Navarro R, Dardano A, Ratto A, Guastavino F, Tiemblo P (2008) Electrical strength in ramp voltage AC tests of LDPE and its nanocomposites with silica and fibrous and laminar silicates. J Polym Sci B Polym Phys 46:1301–1311

Hui S, Chaki TK, Chattopadhyay S (2010) Dielectric properties of EVA/LDPE TPE system: effect of nano-silica and controlled irradiation. Polym Eng Sci 50:730–738

Iyer G, Gorur RS, Krivda A (2012) Corona resistance of epoxy nanocomposites: experimental results and modeling. IEEE Trans Diel Electr Insul 19:118–125

Jonscher AK (1983) Dielectric relaxation in solids. Chelsea Dielectric Press, London

Kashiwagi T, Grulke E, Hilding J, Harris R, Awad W, Douglas J (2002) Thermal degradation and flammability properties of poly(propylene)/carbon nanotube composites. Macromol Rapid Commun 23:761–765

Kochetov R, Andritsch T, Morshuis PHF, Smit JJ (2012) Anomalous behaviour of the dielectric spectroscopy response of nanocomposites. IEEE Trans Diel Electr Insul 19:107–117

Kojima Y, Usuki A, Kawasumi M, Okada A, Fukushima Y, Kurauchi T, Kamigaito O (1993) Mechanical properties of nylon 6-clay hybrid. J Mater Res 8:1185–1189

Kolesov SN (1980) The influence of morphology on the electric strength of polymer insulation. IEEE Trans Electr Insul 15:382–388

Kosmidou TV, Vatalis AS, Delides CG, Logakis E, Pissis P, Papanicolaou GC (2012) Structural, mechanical and electrical characterization of epoxy-amine/carbon black nanocomposites. Express Polym Lett 2:364–372

Kozako M, Fuse N, Ohki Y, Okamoto T, Tanaka T (2004) Surface degradation of polyamide nanocomposites caused by partial discharges using IEC (b) electrodes. IEEE Trans Diel Electr Insul 11:833–839

Kremer F (2003) Broadband dielectric spectroscopy. Springer, Berlin

Lau KY, Vaughan AS, Chen G, Hosier IL, Holt AF (2013) On the dielectric response of silica-based polyethylene nanocomposites. J Phys D Appl Phys 46:095303

Lewis TJ (1994) Nanometric dielectrics. IEEE Trans Diel Electr Insul 1:812–825

Lewis TJ (2004) Interfaces are the dominant feature of dielectrics at the nanometric level. IEEE Trans Dielectr Electr Insul 11:739–753

Lewis TJ (2005) Interfaces: nanometric dielectrics. J Phys D Appl Phys 38:202–212

Li S, Yin G, Chen G, Li J, Bai S, Zhong L, Zhang Y, Lei Q (2010) Short-term breakdown and long-term failure in nanodielectrics: a review. IEEE Trans Diel Electr Insul 17:1523–1535

Liu J, Gao Y, Cao D, Zhang L, Guo Z (2011) Nanoparticle dispersion and aggregation in polymer nanocomposites: insights from molecular dynamics simulation. Langmuir 27:7926–7933

Ma D, Hugener TA, Siegel RW, Christerson A, Martenson E, Önneby C, Schadler LS (2005) Influence of nanoparticle surface modification on the electrical behaviour of polyethylene nanocomposites. Nanotechnology 16:724–731

Maity P, Gupta N, Parameswaran V, Basu S (2010) On the size and dielectric properties of the interphase in epoxy-alumina nanocomposite. IEEE Trans Diel Electr Insul 17:1665–1675

Miwa Y, Drews AR, Schlick S (2008) Unique structure and dynamics of poly(ethylene oxide) in layered silicate nanocomposites: accelerated segmental mobility revealed by simulating ESR spectra of spin-labels, XRD, FTIR, and DSC. Macromolecules 41:4701–4708

Morshuis PHF, Kreuger FH, Leufkens PP (1988) The effect of different types of inclusions on PE cable life. IEEE Trans Electr Insul 23:1051–1055

Myroshnychenko V, Brosseau C (2005) Finite-element modeling method for the prediction of the complex effective permittivity of two-phase random statistically isotropic heterostructures. J Appl Phys 97:044101

Okuzumi S, Murakami Y, Nagao M, Sekiguch Y, Reddy CC, Murata Y (2008) DC breakdown strength and conduction current of MgO/LDPE composite influenced by filler size. Annual Report CEIDP, pp 689–692

Progelhof RC, Throne JL, Ruetsch RR (1976) Methods for predicting the thermal conductivity of composite systems: a review. Polym Eng Sci 76:615–626

Raetzke S, Kindersberger J (2010) Role of interphase on the resistance to high-voltage arcing on tracking and erosion of silicone/SiO$_2$ nanocomposites. IEEE Trans Diel Electr Insul 17:607–614

Ramirez I, Jayaram S, Cherney EA (2010) Analysis of temperature profiles and protective mechanism against dry-band arcing in silicone rubber nanocomposites. IEEE Trans Diel Electr Insul 17:597–606

Ramirez-Vargas E, Sanchez-Valdes S, Parra-Tabla O, Castaneda-Gutierrez S, Mendez-Nonell J, Ramos-deValle LF, Lopez-Leon A, Lujan-Acosta R (2012) Structural characterization of LDPE/EVA blends containing nanoclay-flame retardant combinations. J Appl Polym Sci 123:1125–1136

Roy M, Nelson JK, MacCrone RK, Schadler LS, Reed CW, Keefe R, Zenger W (2005) Polymer nanocomposite dielectrics—the role of the interface. IEEE Trans Diel Electr Insul 12:629–643

Roy M, Nelson JK, MacCrone RK, Schadler LS (2007) Candidate mechanisms controlling the electrical characteristics of silica/XLPE nanodielectrics. J Mater Sci 42:3789–3799

Saccani A, Motori A, Patuelli F, Montanari GC (2007) Thermal endurance evaluation of isotactic poly(propylene) based nanocomposites by short-term analytical methods. IEEE Trans Diel Electr Insul 14:689–695

Srisuwan S, Thongyai S, Praserthdam P (2010) Synthesis and characterization of low-dielectric photosensitive polyimide/silica hybrid materials. J Appl Polym Sci 117:2422–2427

Tagam N, Okada M, Hira N, Ohki Y, Tanaka T, Imai T, Harada M, Ochi M (2008) Dielectric properties of epoxy/clay nanocomposites—effects of curing agent and clay dispersion method. IEEE Trans Diel Electr Insul 15:24–32

Takala M, Ranta H, Nevalainen P, Pakonen P, Pelto J, Karttunen M, Virtanen S, Koivu V, Pettersson M, Sonerud B, Kannus K (2010) Dielectric properties and partial discharge endurance of polypropylene-silica nanocomposite. IEEE Trans Diel Electr Insul 17:1259–1267

Tanaka T, Kozako M, Fuse N, Ohki Y (2005) Proposal of a multi-core model for polymer nanocomposite dielectrics. IEEE Trans Diel Electr Insul 12:669–681

Vaughan AS, Swingler SG, Zhang Y (2006) Polyethylene nanodielectrics: the influence of nanoclays on structure formation and dielectric breakdown. Trans IEE Jpn 126:1057–1063

Venkatesulu B, Thomas MJ (2010) Erosion resistance of alumina-filled silicone rubber nanocomposites. IEEE Trans Diel Electr Insul 17:615–624

Weibull W (1951) A statistical distribution function of wide applicability. J Appl Mech Trans ASME 18:293–297

Xue Y-H, Quan W, Qu F-H, Liu H (2015) Conformation of polydispersed chains grafted on nanoparticles. Mol Simul 41:298–310

Zanetti M, Camino G, Reichert P, Mulhaupt R (2001) Thermal behaviour of poly(propylene) layered silicate nanocomposites. Macromol Rapid Commun 22:176–180

Zhang C, Stevens GC (2008) The dielectric response of polar and non-polar nanodielectrics. IEEE Trans Diel Electr Insul 15:606–617

Chapter 10
Block Copolymers and Photonic Band Gap Materials

Dario C. Castiglione and Fred J. Davis

10.1 Introduction

As discussed in Chap. 1, in general, chemically different polymers do not mix well, particularly as the molecular length of chains is increased. This is largely as a consequence of the unfavourable entropy of mixing. Block copolymers are materials which consist of two or more polymer chains covalently bonded together in a single chain. In a similar way to polymer blends the two blocks may phase separate but the process is substantially different from the phase separation of polymer blends both in terms of the nature of the transitions involved (Liebler 1980) but largely because, by virtue of their chemical connectivity, the macroscopic separation of the blocks is not possible and structures are developed on length scales between 5 and 500 nm; furthermore, the phase separation of the blocks gives rise to periodic structures with arrangements which depend on the nature of the polymer and the size of the blocks (Hamley 2004). Thus such materials offer the potential for a "bottom up" approach to the construction of structures on sub-micrometre scales. Block copolymers have a wide range of uses, and some examples are shown in Table 10.1. However, the ability to produce repeating structure on these length scales has particularly attracted interest in their use as photonic band gap materials (Urbas et al. 1999; Fink et al. 1999) (and more recently phononic band gap materials (Lee et al. 2014)), and it is this area that forms the basis of this contribution. In this chapter we discuss the origin of photonic band gaps, the synthesis of block copolymers, the connections between chemical structure and the large scale

D.C. Castiglione (✉)
AWE, Aldermaston, Reading RG7 4PR, UK
e-mail: dario.castiglione@awe.co.uk

F.J. Davis
Department of Chemistry, The University of Reading, Whiteknights, Reading RG6 6AD, UK

Table 10.1 Exemplar properties and commercial applications of some block copolymers

Properties	Applications	Examples
Dimensional stability, recovery compression set, and utility at high and low temperatures	Mechanical goods like flexible couplings, O-rings, seals, gaskets, and extruded hydraulic and industrial hoses	Brenner et al. (1972)
Low dielectric constants and dissipation factor levels	Electrical and electronic goods like wire and cable insulation and transformer encapsulation	Djiauw and Icenogle (1986)
These materials can be applied either via solution or melt and subsequently develop high strength and recovery characteristics	Sealants, caulks, and adhesives	Crossland and Harlan (1975)
Melt processability, good elastomeric properties, high dynamic coefficient of friction, and excellent abrasion resistance	Footwear industry	Kraton Performance and Inc. (2015)
Block copolymers can contain a high volume fraction of a hard block and a minor concentration of a soft block	Toughened thermoplastic resins	Statz et al. (2004)
Block copolymers having one hydrophobic and one hydrophilic segment can be used as surfactants	Emulsification of aqueous and non-aqueous components, wetting of substrate surfaces and foam stabilisation	Riess (1999)

structures, and the limitations and alternatives to these materials particularly with regard to their interactions with different wavelengths of light.

10.1.1 Phase Behaviour of Block Copolymers

Block copolymers (BCPs) are defined as a class of macromolecules that are produced by joining two or more chemically distinct polymer blocks containing a series of identical monomer units, which may be thermodynamically incompatible. Well-defined block copolymers will have impeccable block integrity and near perfect sequential architecture. These properties are a direct consequence of the limited number of intersegment linkage sites allowed in the block copolymers synthesised. As a result it is possible to predict the properties of the material precisely and use it with greater confidence. Block copolymers with many different molecular architectures have been prepared and some common examples are shown in Fig. 10.1.

The sequential arrangement of polymers can vary from the simplest AB structures, containing two segments only, to ABA block copolymers with three segments, ABC, with three different blocks through to multi-block $(AB)_n$ systems possessing many segments such as those found in some polyurethane block copolymers (Davis and Mitchell 2008). Various chemical coupling strategies permit the synthesis of branched architectures such as those shown in Fig. 10.1; a more detailed description of some of the synthetic routes to such materials is given below in Sect. 10.3.

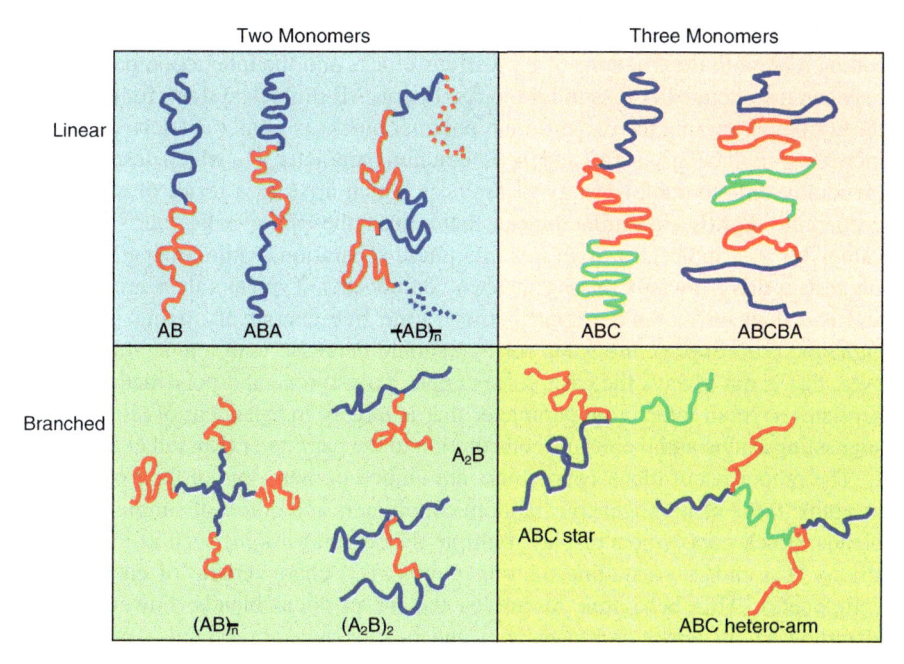

Fig. 10.1 Nomenclature for linear and branched block copolymer architectures of diblock and triblock systems

AB and ABA copolymers typically give rise to four different microphase structures, namely lamellae, double gyroid, cylinders, and spheres and, as described in Chap. 1, these phases are relatively well understood and their presence are predicted by self-consistent mean field theory (Matsen 1994, 1996). The introduction of an additional block as in an ABC type systems opens up the possibility of many more phase structures by self-assembly (segregation) of the blocks (Bates and Fredrickson 1999). The number of structures observed is further increased by the presence of non-equilibrium metastable states which can be formed in conditions of rapid solidification, by, for example, thermal quenching, precipitation, or solvent casting; but therein lies a problem, in that the formation of these states (though potentially of interest) cannot be predicted and may depend on preparation conditions. Furthermore, the presence of such states may compete with the formation of the more predictable equilibrium states. In addition high viscosities are not particularly helpful in terms of the development of equilibrium structures and this is particularly a problem where the formation of photonic band gaps of longer wavelengths is desired, since this requires the formation of rather high molecular weight polymers (*vide infra*) and the development of particular structure is likely to be extremely slow.

In the absence of kinetic effects the variability of phase structure provides a promising route to the control of self-assembly. The main factors that control the phase behaviour have been described by a number of authors (for example, Bates et al. 2012; Stefik et al. 2015; Yoon et al. 2005) and in principle the phase behaviour can be related to the fractions of the various blocks and the interaction parameters between the chemically dissimilar blocks. For an AB diblock system, for example, the entropy of mixing the two different polymer blocks is small (and decreases with increasing molecular weight) and the interaction parameter χ_{AB} which describes the (generally unfavourable) energy differences which arise as a result of mixing of even only slightly dissimilar blocks determines the phase behaviour.[1] Positive values for χ_{AB} indicate a drive towards phase separation, while a negative value suggests a drive towards mixing the two components. Typical values are positive and less than unity. As the contribution to the free energy of mixing is largely enthalpic (*vide supra*), the value for χ_{AB} should decrease with temperature; however, this is not always the case, for example, polystyrene and poly(methyl methacrylate) have an interaction parameter that is largely independent of temperature suggesting a substantial entropic contribution to the parameter (Russell et al. 1990).

The properties of block copolymers are unique in that some of their properties resemble those of homogeneous random copolymers and others of simple physical blends. Block copolymers display multiple thermal transitions, such as glass transitions (T_g) and/or crystalline melting points (T_m) characteristic of each of the components. This behaviour resembles that of physical blends; however, they also show a high degree of transparency and good balance of mechanical properties, which are typical of homogeneous random polymers. This versatility renders block copolymers suitable for many diverse applications. Aside from photonic band gap materials (*vide infra*) block copolymers are particularly well known as thermoplastic elastomers. Such materials have hard/rigid blocks separated by soft flexible blocks, which phase separated and behave as a cross-linked material, with the rigid phase acting as the cross-link points. These materials behave as typical elastomers in that they can be stretched, and will return to their original shape when the stress is released; however, unlike simple elastomers such as vulcanised rubber they can be processed, for example, as a melt at elevated temperature, when the rigid part is heated above the glass transition temperature or melting point. The most common of these include styrene block copolymers and polyurethanes. Styrenic block copolymers include the well-known Kraton® family, initially developed by Shell in the 1950s; these are generally triblock copolymers with styrene blocks surrounding a soft central block formed from butadiene, isoprene, or related materials. Polyurethane thermoplastic elastomers include the ubiquitous Spandex® (also known as Lycra® or Elastane®) originally developed by Du Pont in the 1950s. This consists of a segmented polymer containing a soft block consisting of a polyol [such as poly(tetrahydrofuran)] and a hard block from MDI (4,4′-diisocyanatophenylmethane) [I] and an amine such as ethylenediamine.

Table 10.1 summarises some common applications of block copolymers along with the properties which make them suitable for that particular application

---

[1] Interestingly even protonated and deuterated polystyrene has been found to be incompatible at sufficiently high molecular weights (Bates et al. 1985).

10.2 Photonic Band Gap Materials

The concept of the Photonic Band Gap (PBG) was proposed independently by Yablonovitch (1987) and John (1987). A Photonic Crystal (PC) is any material which exhibits a photonic band gap (*vide infra*) and is therefore able to reflect certain wavelengths of electromagnetic radiation as propagation through it is prohibited.[2]

The structure of any photonic crystal is based upon a regular arrangement of materials of alternating dielectric constant. Any incident waves are reflected by the periodic structure and a band gap will arise if these reflected waves, working in phase with one another, cancel with the incident wave and thus do not allow its propagation through the crystal (Yablonovitch 2001). In many respects, photonic crystal investigations are biomimetic in that they copy well-known phenomena found in nature and scale it to serve the application required (Saito et al. 2006; Starkey 2005). Thus, photonic crystals are often characterised by strong colours and there are numerous examples in nature. A spectacular example of this phenomenon is provided by the wings of certain butterflies (Christensen 2002; Vukisic et al. 1999, 2000) in particular those of the genus *morpho* (Fig. 10.2). Here the distinct colours of the butterfly are probably used as a defence mechanism against predatory attack or as an intraspecies method of communication (Forsman and Merilaita 2003). Unfortunately (for the butterfly) the beautiful iridescence has made this particular genus historically extremely attractive to lepidopterists.

Fig. 10.2 Example of a morpho butterfly; the blue colour arises from the regular arrangement of scales on the wing

---

[2] This differs from the reflections observed from a standard mirror which relates to conductivity of the mirror's tain (highly polished, reflective metal coating) and the abundance of mobile electrons; the electrons respond to an applied electromagnetic field by generating an equal and opposite electromagnetic field of their own, hence reflection is observed.

The interest in photonic crystals has grown significantly since Yablonovitch built the first photonic crystal structure which operated in the microwave regime and, because of the long wavelength, could be built by conventional machining. The action of electromagnetic repulsion/reflection, independent of the frequency at which it occurs, is due to the dielectric materials that make up the crystal and the periodicity (the regular arrangement of materials of alternating dielectric constant). It is these factors that result in the formation of a useful band gap. The width of this band gap [usually expressed either in hertz (Hz) or in electron volts (eV)] depends on the geometry, feature size, spacing, and the materials that make up the crystal.

A photonic band gap covers a range of frequencies where any photon incident on the crystal will be reflected rather than transmitted. The simplest analogy is that of a mirror, as a mirror reflects light that is incident on it. However, while a mirror demonstrates the concept of reflection perfectly, the microscopic behaviour inside a photonic crystal is somewhat more complicated and is described in more detail later in this section.

Just as any periodic structure has dimensionality, so do photonic crystals. Thus, photonic crystals can be 1, 2, or 3-dimensional in terms of their mode of operation. This dimensionality reflects the periodicity which is relevant to the desired optical properties, thus a simple layered structure will show properties in one dimension only (i.e. if light is perpendicular to the surface) whereas a regular 3-dimensional lattice will be dependent on the lattice parameters.

A photonic crystal may consist of a simple square lattice of dielectric cylinders (Fig. 10.3a) but equally, it is also possible to have a square lattice with cylinders consisting of air in a medium of dielectric material (Fig. 10.3b). Both lattices will display a photonic band gap but all other factors being equal, these will not be centred at the same frequency, or as wide as each other. Lattices that make up photonic crystals may be square, cubic, hexagonal, or tetragonal to name a few. Equally the features within the lattices may be cylinders, squares, rectangles, hexagons, or more complex arrays (Liguda et al. 2001; Xia et al. 1999; Yang 2000; Ozin 2001).

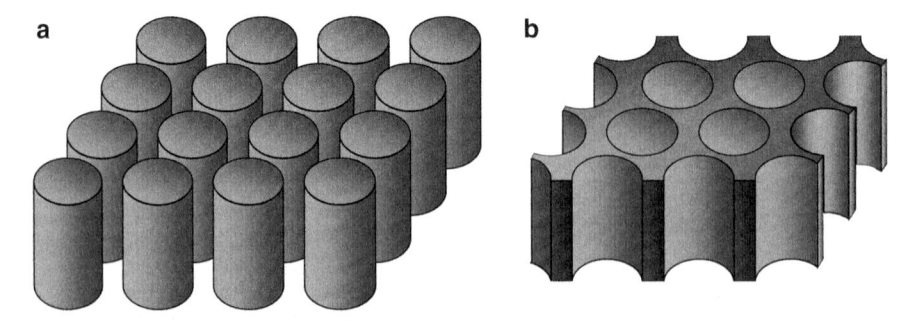

Fig. 10.3 An example of the periodic structure that may be seen in 2-dimensional photonic crystals

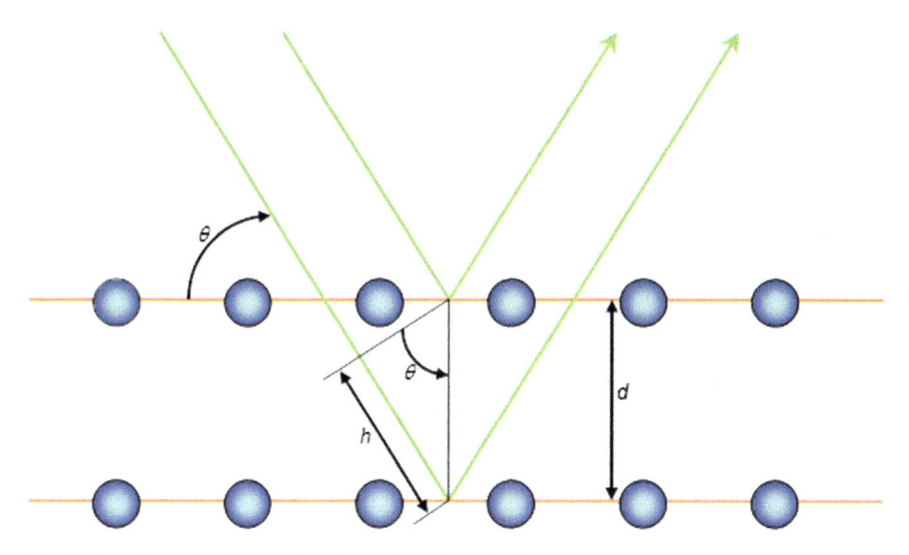

10 Block Copolymers and Photonic Band Gap Materials

Fig. 10.4 Schematic diagram for determining Bragg's Law

The interaction of electromagnetic radiation with a photonic crystal can be explained in part as a problem in diffraction using Bragg's Law, as seen in Fig. 10.4. Figure 10.4 illustrates schematically the interference observed between waves scattering from adjacent rows of atoms in a crystal lattice. The overall effect of scattering from a row is akin to partial reflection from a mirror aligned with the row, and hence the angle of reflection is equal to the angle of incidence for each row. Interference then occurs between the waves reflecting from different rows of atoms in the crystal. For the two adjacent rows shown in Fig. 10.4, simple geometry determines that the path difference between beams is $2d \sin \theta$ which is in turn twice the distance h. For constructive interference h must be an integer number of wavelengths (where the integer is called the order of interference); from this, simple trigonometry gives Bragg's Law of Diffraction (10.1).

$$n\lambda = 2d \sin \theta \quad \text{where} \quad n = 1, 2, 3 \ldots \quad \text{(Bragg's Law)} \quad (10.1)$$

The principles for reflection from two layers can be extended to a plane wave incident on a periodic structure such as the first row of cylinders in a 2-dimensional photonic crystal as shown in Fig. 10.5a. As a normal wave is incident on the structure it may diffract and scatter, but it will also split into forward and reflected waves. As the forward wave continues into the crystal it encounters more obstacles (i.e. periods of alternating dielectric constant) that cause more diffraction, scattering, and forward and reflected waves. Thus as a consequence of the periodic spacing of the cylinders there may be destructive interference between the reflected waves inside the crystal and the incident wave; this occurs if the reflected waves are in phase with each other. The net result is a photonic band gap, ensuring that the

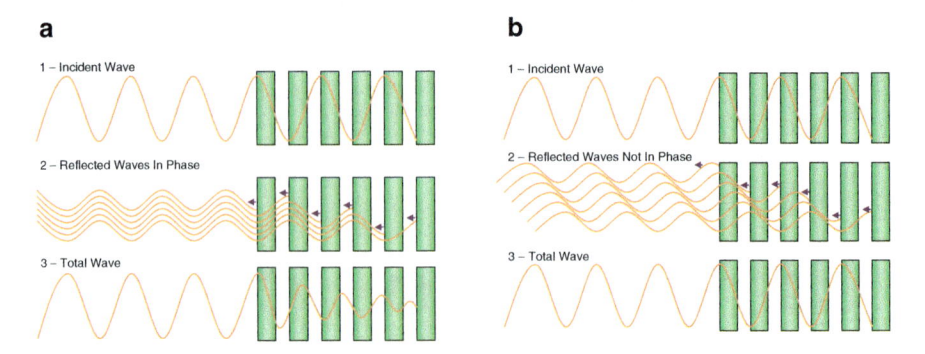

Fig. 10.5 (a) A wave incident on a photonic band gap material partially reflects off each layer of the structure. The reflected waves are in phase with each other and destructively interfere with the incident wave to produce a standing wave that does not propagate through the material. Where there is no destructive interference (b) the wave is propagated (Yablonovitch 2001)

incident wave cannot propagate through the crystal. This does not occur if the wavelength is not in the band gap (Fig. 10.5b).

A crucial factor in the development of photonic band gap materials is the contrast between the two media in terms of their refractive indices. Ideally there should be a large difference in refractive index as the contrast controls reflection through a variety of mechanisms, e.g. total internal reflection and birefringence; without this contrast there will be nothing to distinguish the two layers and the interface would be invisible (refractive index matching). The wavelength of the band gap is then determined by the spacing and refractive indices of the structure (Zhao et al. 2009). In general the wavelength of the band gap is determined by the size of the blocks which should be $\lambda/4n$ where n is the refractive index of the block material (Yoon et al. 2005). The refractive index difference will affect the width of the gap, and will determine the number of layers required to reduce transmission to zero and greater contrast will increase the width of the band gap. This is particularly important when considering applications involving organic polymer systems, as refractive index differences tend to be rather small (Stefik et al. 2015), though this difference may be enhanced by the incorporation of inorganic materials (Bockstaller et al. 2001).

Photonic crystals have considerable potential for applications (Stefik et al. 2015) and have been used widely as the active components in many commercial devices such as optical notch filters (Dowling 1998), dielectric mirrors/optical resonance cavities (Painter et al. 1999), and Bragg gratings (on optical fibres) (Vengsarkar 1996). Photonic crystal fibres provide a route to fibre optic materials and contrast with Bragg filters in terms of the directionality of the periodicity (Russell 2003). Photonic crystals are designed and fabricated to work at a certain frequency, namely the central or mid-band frequency. The range of frequencies around the central frequency over which the crystal is able to operate defines the operational bandwidth of the crystal. Once outside this operation range the photonic band gap nature is no longer seen for that particular photonic crystal.

The means of fabrication of photonic crystals are linked directly to the operational band gap sought. Synthetic photonic crystals may be fabricated by many varied methods. Saado et al. (2001) has demonstrated an array of magnetic particles which are able to self-assemble to form ordered 2D and 3D structures, whereby the lattice parameter may be controlled by an external magnetic field, which exhibit tuneable and non-tuneable microwave and millimetre-wave photonic band gaps. Busch and John (1999) have also demonstrated a tuneable photonic band gap material using liquid crystals. An inverse opal structure is generated by the infiltration of silicon into the air voids of an artificial opal and the template opal structure is etched away to reveal the inverse-opal. An optically birefringent nematic liquid is infiltrated into the voids of the inverse-opal photonic crystal and the tuneable light localisation effects are realised by controlling the orientational disorder in the nematic. Similarly Zhou et al. (2001) prepared a thermally tuneable photonic crystal from a SiO_2 colloid crystal prepared by self-assembly and in filled with $BaTiO_3$. The bare silica colloid crystal shows a dip in transmittance at approximately 570 nm but an in-filled colloid crystal at 400 and 600 °C shows broadening of the dip and also a shift to longer wavelength. Metallic crystals also present a novel route to photonic band gap structures as they can possess a large band gap whilst new electromagnetic phenomena and high temperature applications may be explored. An all metallic three-dimensional photonic crystal featuring a large infrared band gap has been prepared by using chemical vapour deposited (CVD) tungsten to back-fill a polysilicon/SiO_2 structure from which the silicon has been selectively removed (Flemming et al. 2002).

10.3 Synthesis of Block Copolymers

Block copolymers may be synthesised by various methods but two of the common techniques are anionic polymerisation and atom transfer radical polymerisation (ATRP). Living anionic polymerisation, a technique which has been used widely since the 1950s, is now a well-established way of making block copolymers for photonic band gap applications. The first publication and introduction of the term "living polymer" was in 1956 (Szwarc et al. 1956). The work described the making of block copolymers, based on the fact that aromatic hydrocarbons reacted with metallic sodium in suitable solvents yielding coloured and soluble complexes with a composition 1 Na to 1 hydrocarbon which in turn initiated polymerisation of conjugated olefinic hydrocarbons. Paul et al. (1956) showed them to be composed of negative aromatic hydrocarbon ions and Na ions and that they acted as electron transfer agents as shown in Scheme 10.1.

Szwarc et al. postulated that the same type of electron-transfer process was responsible for the initiation of polymerisation by a sodium–naphthalene complex. The monomer anions formed may be represented by (Ia) and (Ib) in Scheme 10.2, and if an excess of monomer is present, polymerisation will proceed with one end growing as a radical and the other as a carbanion. After the addition of the first

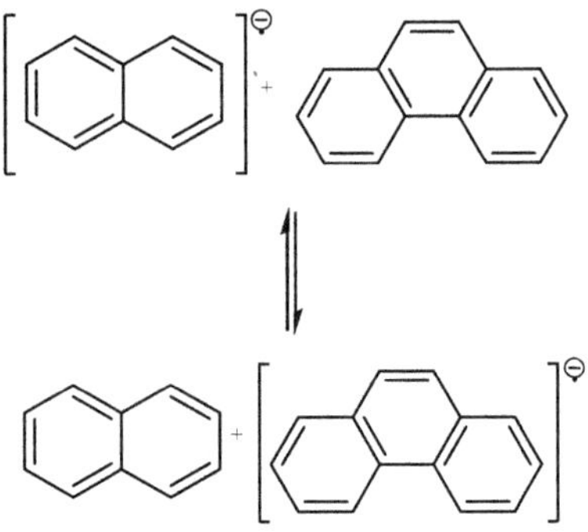

Scheme 10.1 Negatively charged hydrocarbons acting as electron transfer agents (Paul et al. 1956)

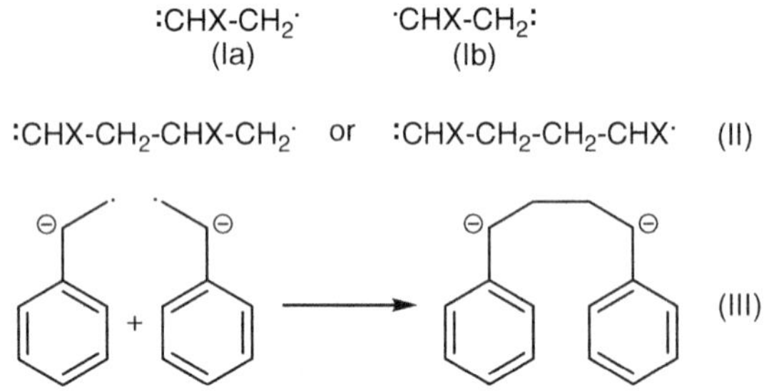

Scheme 10.2 Initial observed reactions of radical anions of styrene generated from a sodium–naphthalene complex

:CHX-CH$_2$·   ·CHX-CH$_2$:
   (Ia)   (Ib)

:CHX-CH$_2$-CHX-CH$_2$·   or   :CHX-CH$_2$-CH$_2$-CHX·   (II)

(III)

monomer unit to either (Ia) or (Ib) a true separation of electrons takes place and species such as (II) in Scheme 10.2 are formed; however, the radical ends have a short lifetime even at low temperature, and they dimerise to form dianions [species (III) in Scheme 10.2]. The experimental details given showed that the characteristic green colour of the sodium–naphthalene complex was immediately replaced by a deep red colour on addition of styrene, due to the styrene ends. At the end of the reaction the red colour persisted indicating that termination had not occurred. Thus, if an additional amount of styrene was added after the initial monomer was converted entirely to polymer, a further increase in molecular weight was observed. If, however, a different monomer was added in place of the second portion of styrene, the living polymerisation still proceed, but in this case, yielding block copolymers of various compositions.

Significant studies have been carried out in the field of living anionic polymerisation over the last 50 years (Uhrig and Mays 2005; Franta et al. 2007; Baskaran and Müller 2007). The "living" anionic polymerisation technique employs anions rather than radicals to propagate the reaction. Alkyllithiums are the initiator used most commonly for this technique. The reaction proceeds in homogeneous solutions in hydrocarbon solvents but it is essential that all active hydrogen impurities (water, alcohols, etc.) be excluded in order to obtain a termination free system. The anionic active chain ends exist primarily in the form of ion pairs when hydrocarbons are used as polymerisation solvents. Polar solvents, such as ethers, result in the appearance of free ion species and reaction rates increase greatly. As there is no

10 Block Copolymers and Photonic Band Gap Materials

Scheme 10.3 General mechanistic scheme for living anionic polymerisation

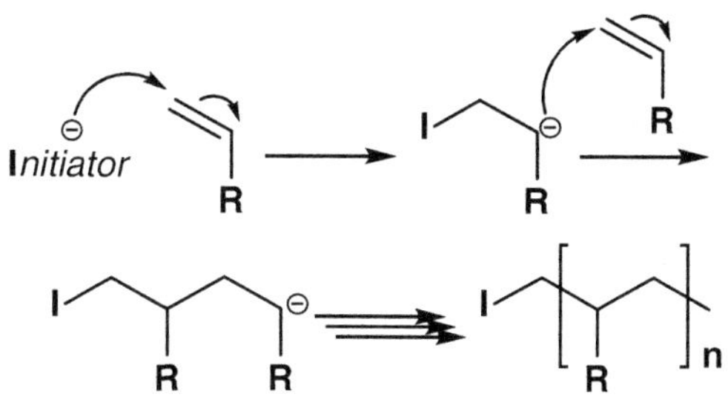

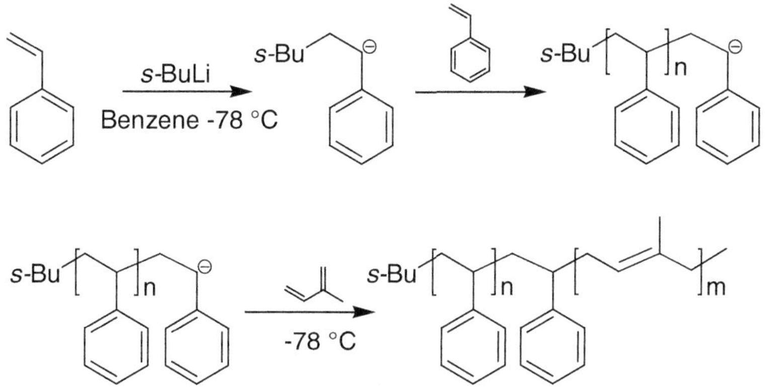

Scheme 10.4 Living anionic polymerisation of styrene and isoprene

unintentional termination, block length, which is dependent only upon the monomer/initiator ratio, is easily controllable and the absence of termination affords an extremely monodisperse polymer sample. A general reaction mechanism is given in Scheme 10.3.

One of the first block copolymer to exhibit a photonic band gap was a styrene isoprene block copolymer. This was prepared by anionic polymerisation using butyllithium as an initiator; the mechanistic details for the copolymerisation of styrene and isoprene are given in Scheme 10.4. Some anionic polymerisations can be undertaken with basic laboratory equipment provided sufficient care is taken (Aragrag et al. 2004). For the formation of photonic band gap materials however, the molecular weights for each block need to be of the order of 200,000; as the number average degree of polymerisation (X_n) is given by the ratio of monomer (M) to initiator (I) concentrations (10.2) the production of the required molecular weight will only be possible with very low initiator concentrations. This places particularly high demands on technique as it will be particularly susceptible to termination of the propagating polymer chain or initiator (Hadjichristidis et al. 2000). For this reason the reactants require extensive purification to remove all traces of water and any other impurities; additionally traces of oxygen must be removed. The preparation requires the highly competent use of well-established vacuum line techniques (Morton 1983) utilising an ultra-high vacuum line, break-seal technology and advanced glassblowing techniques, thus this polymerisation requires a vacuum line equipped with dual pumps; a roughing pump able to evacuate the system to between 1×10^{-2} and 1×10^{-3} Torr and typically a water-cooled silicone oil diffusion pump, which when connected sequentially with the first pump evacuates the system to 1×10^{-6} Torr.

$$X_n = [M]/[I] \qquad (10.2)$$

Using the conditions described above it is possible to produce high molecular block copolymers containing styrene and isoprene. In a typical procedure, degassed styrene was purified by distillation from CaH_2 across a vacuum line into a graduated measuring cylinder containing dibutyl magnesium (DBuMg) and the styrene/DBuMg solution was stirred at room temperature for approximately 1 h until a pale straw colour had developed. The correct volume of styrene required for the reaction was distilled from DBuMg solution into the reactor directly. After ensuring that the reactor was frozen in liquid nitrogen, approximately 500 mL of pre-dried benzene was distilled into the reactor. The mixture was thawed and polymerisation initiated by rupturing the break-seal of an ampoule of *sec*-butyllithium attached directly to the reactor vessel. An instantaneous colour change from colourless to a light yellow colour was observed. After 24 h isoprene was added to the system through a reactor sidearm. The reaction was allowed to proceed for a further 24 h to ensure complete reaction of the monomer although a noticeable change in viscosity was evident after only 3 h. The reaction was terminated using degassed methanol. The polymer was precipitated directly into cold methanol containing 2,6-di-tert-butylated-methylphenol (butylated hydroxytoluene, BHT). The precipitated polymer was dried under vacuum for 48 h. The ^1H NMR spectrum of this polymer is shown in Fig. 10.6; this reveals the composition of the block copolymer to be 40.59 mol% polystyrene to 59.41 mol% polyisoprene (cf. 39.56 mol% styrene to 60.44 mol% isoprene monomer feed ratio). The isoprene polymerised by 1,4-, and 1,2- addition in the ratio; 1,4-, 92.64 %: 1,2-, 7.36 %. In the example shown the molecular weight of the polymer was found to be 750,000 with a polydispersity of 1.21. The polymer was found to be a white solid, but when in solution in toluene showed a vivid blue iridescence indicative of its photonic band gap properties as shown in Fig. 10.7 (Castiglione 2008).

Although anionic polymerisation remains an important technique today for the preparation of block copolymers and other controlled polymer architecture, recent developments in controlled free-radical polymerisation has presented an alternative approach which may complement this methodology.

Atom radical transfer polymerisation (ATRP) has its roots in atom transfer radical addition (ATRA), which involves the formation of 1:1 adducts of alkyl halides and alkenes, and is also catalysed by transition metal complexes. ATRP is a modification of the Kharasch addition reaction (Kharasch et al. 1945) although there may be some differences (Minisci 1975). A general mechanism for ATRP is shown in Scheme 10.5. In ATRP the radicals or the active species are generated through a reversible redox process catalysed by a transition metal complex (Mtn-L/

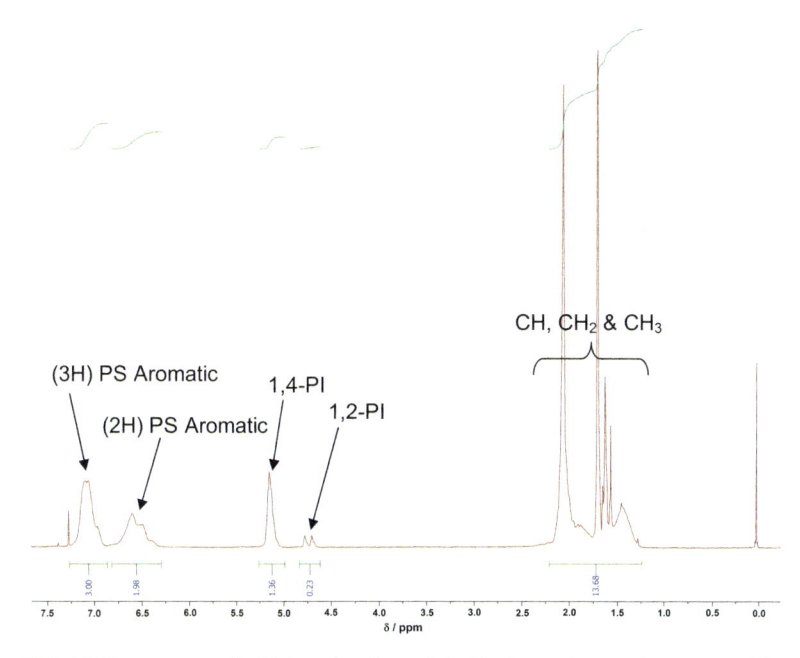

Fig. 10.6 NMR spectrum of a high molecular weight block copolymer of styrene and isoprene produced by anionic polymerisation

Fig. 10.7 High molecular weight styrene-b-isoprene copolymer (**a**) as a solid (**b**) swollen in toluene

Ligand, where L may be another ligand or the counter ion) which undergoes a one-electron oxidation with abstraction of a (pseudo) halogen atom, X, from a dormant species, P-X. This process occurs in a manner similar to a conventional radical polymerisation.

Termination reactions also occur in ATRP, mainly through radical coupling and disproportionation, however, unlike free-radical processes, in a well-controlled ATRP, less than a few percent of the polymer chains undergo termination.

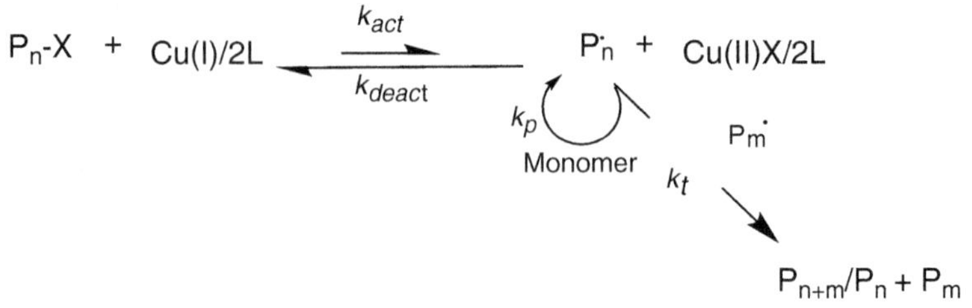

$$P_n\text{-}X + Cu(I)/2L \underset{k_{deact}}{\overset{k_{act}}{\rightleftharpoons}} P_n^{\cdot} + Cu(II)X/2L$$

$$\xrightarrow{k_p \text{ Monomer}} P_m^{\cdot} \xrightarrow{k_t} P_{n+m}/P_n + P_m$$

Scheme 10.5 General mechanism for ATRP

Although other side reactions may additionally limit molecular weights yields and the homogeneity of the products (Fischer 1999), typically no more than 5 % of the total growing polymer chains terminate during the initial, short, non-stationary stage of the polymerisation. A successful ATRP requires both a small contribution of terminated chains and uniform growth of all chains. This is a result of fast initiation and rapid reversible deactivation. Polymers obtained by ATRP have a narrow molecular weight distribution and the degree of polymerisation can be controlled accurately (Hayes and Rannard 2004).

Using ATRP block copolymers can be prepared fairly readily and with some degree of complexity. For example, a commercially available difunctional initiator, dimethyl 2,6-dibromoheptanedioate (DMDBHD [II]), has been used to initiate ATRP of *tert*-butyl acrylate and styrene; in this case the presence of two bromines allows the formation of a tri-block copolymer, with either styrene or *tert*-butyl acrylate units acting as the central block, depending on the order of addition of the monomers. Figure 10.8 shows the NMR spectrum of a styrene-*b*-*tert*-butyl acylate-*b*-styrene triblock copolymer. In this case freshly distilled from calcium hydride and deoxygenated, *tert*-butylacrylate was added to degassed CuBr with DMDBHD being the initiator. After the addition of *tert*-butylacrylate, the ligand PMDETA was added to the mixture and the solution was stirred at 60 °C for a further 6 h. After 3 h the green solution started becoming viscous. The polymer was dissolved in acetone and flushed through a layer of neutral aluminium oxide. Precipitation produced the desired difunctionalised poly(*tert*-butylacrylate) as a white solid. This difunctionalised polymer was then further polymerised with styrene in a similar matter to produce the desired triblock copolymer (Aragrag 2010).

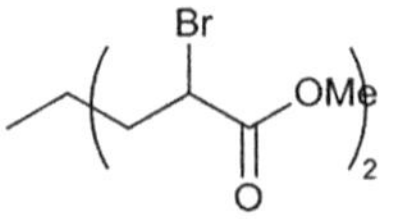

[II]

As inspection of Fig. 10.8 shows the polymer produced is relatively pure, there is a small peak at 3.6 ppm which may reflect the presence of the initiator residue in the central block. The detection of this in the NMR is testament to a lower molecular weight than the anionic polymer described above and GPC measurements showed a

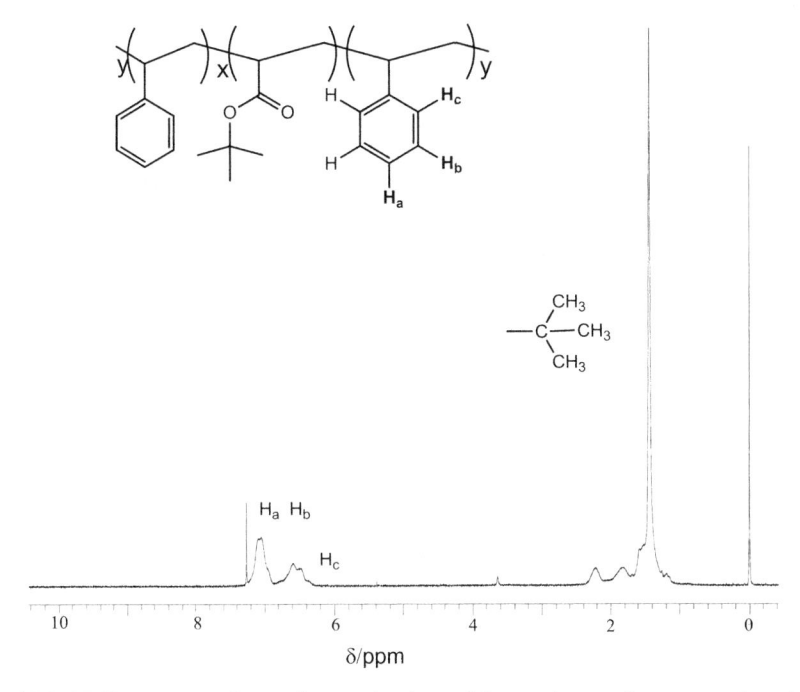

Fig. 10.8 NMR spectrum of a medium molecular weight copolymer of styrene and tert-butyl acrylate produced by ATRP

molecular weight of ca. 11,000 with a polydispersity of 1.55. In general this reflects a problem with atom transfer processes in terms of producing very high molecular weight materials. Thus although Matyjaszewski has been able to produce a very high molecular weight polystyrene under certain conditions (Mueller et al. 2011; Jakubowski et al. 2008), in general such polymerisations are subject to a number of problems, in the case of styrene the initiation rate is too low and there may be thermal self-initiation (Mueller et al. 2011); more generally there is a problem with termination by combination/disproportionation or chain transfer to solvent as with conventional free-radical polymerisations (Simms and Cunningham 2008). To date this limitation has restricted the use of this technique in terms of producing photonic band gap materials with gaps in the visible part of the spectrum, but recent advances by Haddleton et al. has shown the potential of using Cu(0) mediated living radical polymerisation to produce high molecular weight block copolymers (Anastasaki et al. 2013). The use of the related technique RAFT (reversible addition – fragmentation chain transfer) polymerisation is also being developed to produce high molecular weight materials (Gody et al. 2014), however, it is fair to say that in terms of organic polymer systems their potential as photonic band gap materials have been largely explored using anionic polymers.

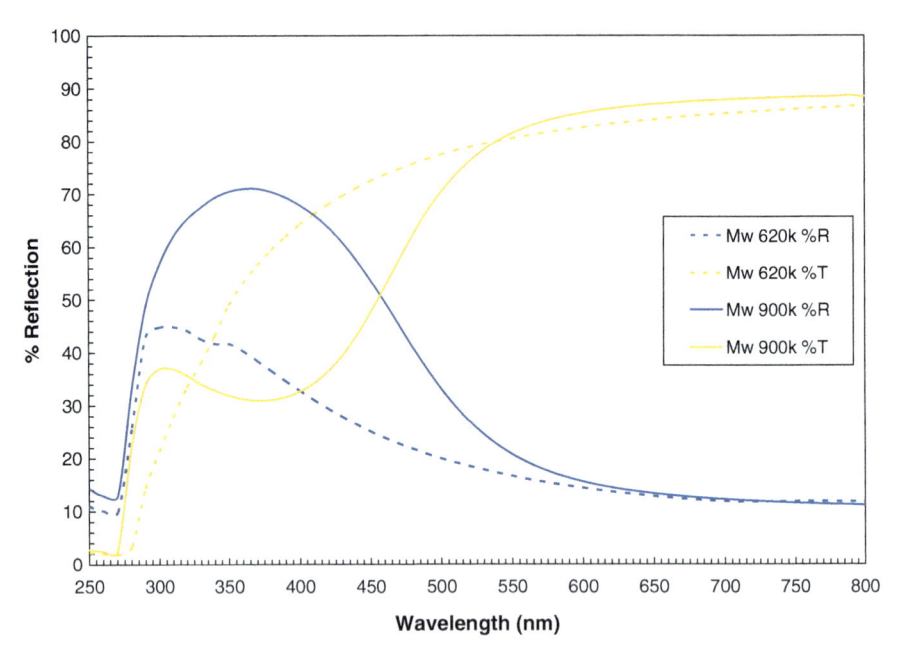

Fig. 10.9 Reflectance (*blue*) and transmission (*yellow*) measurements of poly(styrene-b-isoprene) block copolymer with molecular weights as shown

10.3.1 Characterisation of Photonic Band Gap Materials

The crucial test of a photonic band gap material is the existence of the band gap. For a material with a band gap in the visible range this can be qualitatively gauged by observation of the sample. Quantitative assessment requires the use of UV visible spectroscopy, both in transmittance and reflectance such that any components arising from electronic absorptions can be isolated. For example, Fig. 10.9 shows two poly(styrene-*b*-isoprene) block copolymers of differing molecular weight, the transmission spectrum of the higher molecular weight exhibited a strong trough (300–450 nm) which was also related to the observed reflection of the sample. The difference in appearance between transmission and reflection below 300 nm is due to the absorption of the polymer by virtue of the presence of the styrene chromophores.

To further establish the presence of phase separation, the polymer domains should be analysed by microscopy. In view of the sizes of the domains, this usually means electron microscopy (see Chap. 2, Sect. 2.4), and, Fig. 10.10 shows TEM micrographs of the high molecular weight poly(styrene-*b*-isoprene) block copolymer described earlier. The dark areas relate to the stained polyisoprene and the light areas to the unstained polystyrene (Bockstaller et al. 2001). The polyisoprene domains range from 50 to 80 nm and the polystyrene domains range from 78 to 129 nm. The white spots (24–71 nm) which appear on the dark areas are

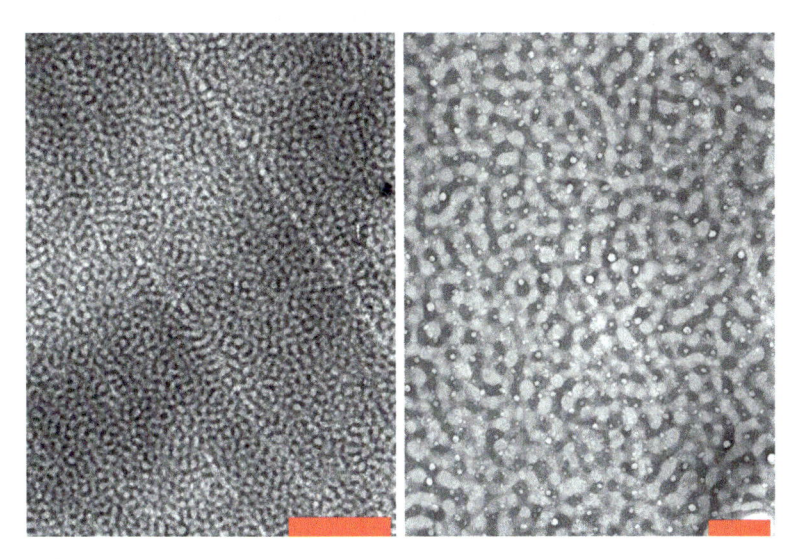

Fig. 10.10 TEM micrograph of poly(styrene-b-isoprene) cast from a 1 % solution in dichloromethane. The *red scale bars* indicate 2 μm (*left*) and 0.5 μm (*right*)

conceivably polystyrene which phase separated quickly. It is possible that annealing the sample would enable the structure to phase separate further and become better defined (Albalak and Thomas 1993).

10.3.2 Extending to Longer Wavelengths

There is a significant desire to increase the range of photonic band gap polymers to the near infrared region, this is prompted particularly by the use of infrared light (of wavelengths around 1.55 μm) in optical communications, which minimises attenuation (crucial over long distances). In addition options for blocking infrared light selectively are limited. A large number of infrared reflecting pigments such as titanium dioxide are available (Bendiganavale and Malshe 2008); such materials presumably block infrared by virtue of their high refractive index rather than through absorption due to an electronic transition. Unfortunately these materials (and metallic films) tend to reflect the visible light also. Highly conjugated organic molecules that show electronic transitions in the infrared are available, some of which may show absorptions above 1000 nm (Fabian et al. 1992); however, in most cases these show additional bands which make them highly coloured and thus partially block visible light.

Infrared reflection is an important consideration in building design. Around 45 % of the sunlight at the earth's surface is in the near infrared (NIR). This can be responsible for considerable build-up of heat within glazed structures; thus removal of infrared is an important consideration. Interestingly one of the best NIR blocking

Fig. 10.11 Schematic representation of the internal structure of a leaf showing reflection of NIR radiation

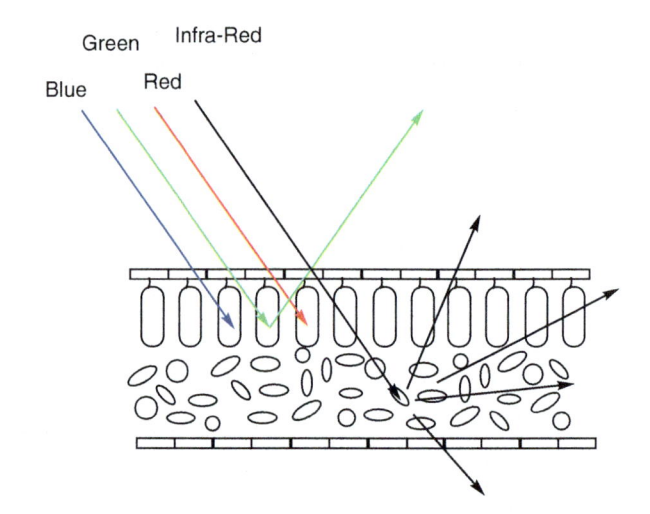

structures in nature is provided by foliage. The cool shade of a tree is due in part to the leaves reducing the amount of NIR reaching the ground. This is not due to the pigments in the leaves, chlorophyll absorption tails off at 700 nm, but rather is due to the leaf structure as illustrated in Fig. 10.11. Thus, a typical leaf structure consists of four layers (Wuyts et al. 2010), the first layers are the epidermis (which is the layer of cells that make up the surface of the leaf) and the palisade parenchyma (which consists of a regular arrangement of elongated cells) and this is where the leaf pigments such as chlorophyll and anthocyanins are to be found. The third layer is known as the spongy parenchyma, and here irregularly shaped cells separated by air spaces are to be found. It is the third layer which is largely responsible for the scattering of infrared light and ray-tracing models have shown how this can result in reflection of a substantial amount of the light passing through the system, depending on the position of the incident light (and hence the time of day) (Jacquemoud et al. 1997). Commercial infrared reflecting films applied to windows to minimise heat build-up are generally rather more efficient (and of course allow the passage of visible light). These tend to be made up from multilayered films, which act by reflecting infrared light but not visible light. An interesting alternative has recently been described by Khandelwal et al. (2014) who use a cholesteric liquid crystal polymer bilayer film to produce an infrared reflective material suitable for building applications.

As the discussion above suggests, at the NIR length, scale top-down fabrication currently may be more effective than self-assembly, in many applications. However the development of larger dimension photonic band gap polymers remains an area of considerable interest. The principle challenges remain the synthetic demands of producing extremely high molecular weight polymers, the poor refractive index contrast generally available between different types of polymer block and thus increased length of the blocks required. There may be some degree of tunability, thus Fig. 10.12 shows significant changes in colour from the high molecular weight

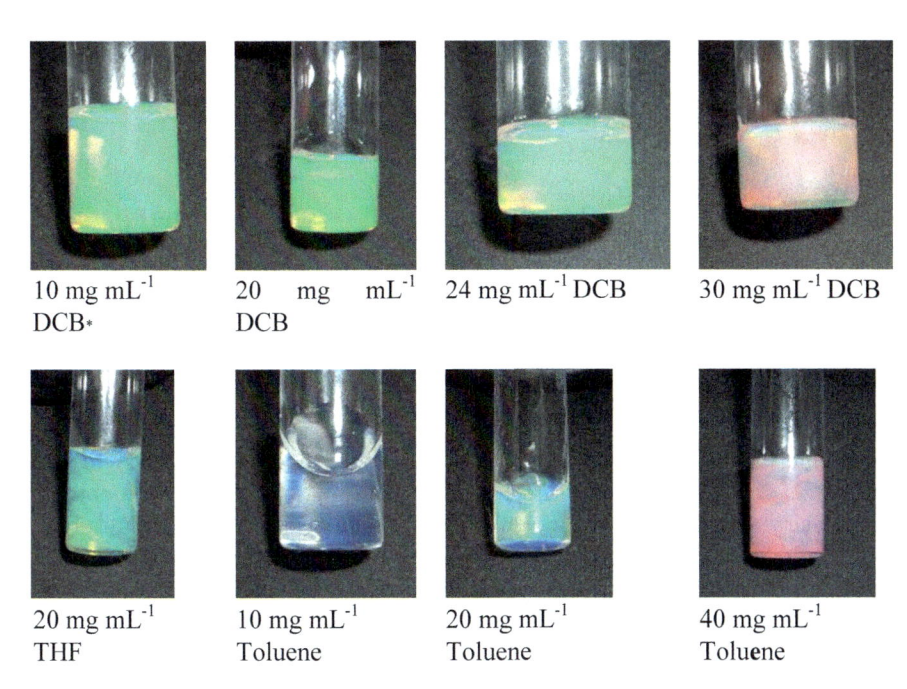

| 10 mg mL^{-1} DCB* | 20 mg mL^{-1} DCB | 24 mg mL^{-1} DCB | 30 mg mL^{-1} DCB |
| 20 mg mL^{-1} THF | 10 mg mL^{-1} Toluene | 20 mg mL^{-1} Toluene | 40 mg mL^{-1} Toluene |

Fig. 10.12 High molecular weight styrene-b-isoprene copolymer (from Fig. 10.7) in different solvents at different concentrations. *DCB* dichlorobenzene

styrene-*b*-isoprene copolymer described earlier arising as a result of swelling in different solvent; in addition the polymer colour is not homogeneous and clearly this is not ideal in terms of processing for applications. However, a number of alternative approaches have been used to utilise phase separation in block copolymers to produce infrared reflecting materials; two examples are given below.

The swollen polymers in Fig. 10.12 show a level of inhomogeneity; this raises another issue with very high molecular weight polymers, namely their high viscosity arising from chain entanglements is likely to slow down the formation of equilibrium structures. Recently a team at Caltech has described block copolymer made of two similar brush copolymer segments, the only difference being a phenyl substituent on one of the brushes (Miyake et al. 2012) as shown in Fig. 10.13. The brush configuration allows the high molecular weight required to form large domains to be achieved without significant entanglement such that NIR reflecting films can be cast directly from solution.

An alternative method to avoid the problems associated with high molecular weight materials has been described by Li et al. (2004). They used an approach where the periodicity was introduced using monodisperse silica microspheres (in this example with a diameter of 700 nm); these spheres were sintered at 950 °C. This process introduces connections between the beads and this provides channels for subsequent removal of the silica by etching. The array thus produced was used as a template and a solution of a thermoplastic elastomer in

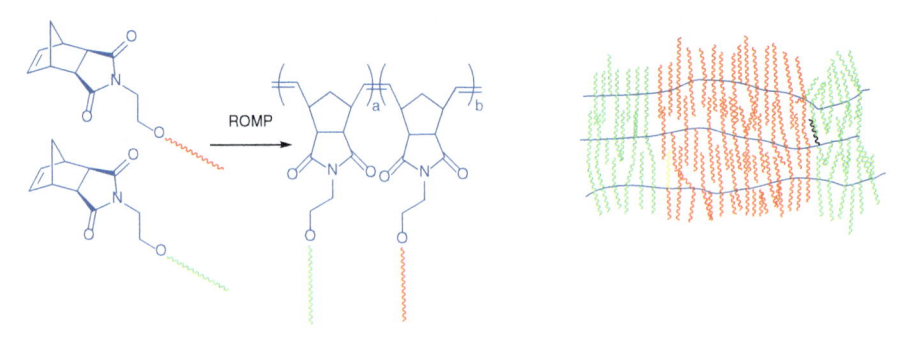

Fig. 10.13 Brush copolymer system to produce NIR reflecting photonic band gap materials

dichloromethane was allowed to fill the voided space. The silica was then removed by etching with HF. This arrangement not only allows for the convenient production of larger photonic band gap wavelengths (in this case in the near infra-red), without the intrinsic time problems resulting from the viscosity of extremely high molecular weight materials, but is also tuneable, as the shape of the pores produced in the structure can be altered by stretching.

10.4 Conclusion

Block copolymers exhibit micro phase separation, which, in the absence of kinetic considerations, results in self-assembly to form a remarkable range of structures. Of particular interest is the potential for such structures to act as photonic band gap materials, where the passage of light is restricted by multiple reflections at the phase boundaries between the different blocks. The development of such materials has faced a number of challenges. The effectiveness of these materials relies on difference in refractive indices and these are often small for organic polymers; the preparation of high molecular weight materials can be synthetically challenging; the high molecular weights required may require significant annealing to form equilibrium phases due to the high viscosity typical of such polymers. That being said recent developments show that there is considerable potential that these materials could be developed into effective products.

References

Albalak RJ, Thomas EL (1993) Microphase separation of block copolymer solutions in a flow field. J Polym Sci B Polym Phys 31:37–46

Anastasaki A, Waldron C, Wilson P, Boyer C, Zetterlund PB, Michael R, Whittaker MR, Haddleton D (2013) High molecular weight block copolymers by sequential monomer addition via Cu(0)-mediated living radical polymerization (SET-LRP): an optimized approach. ACS Macro Lett 2:896–900

Aragrag N (2010) Synthesis and characterisation of thermotropic liquid crystalline elastomers. PhD thesis, University of Reading

Aragrag N, Castiglione DC, Davies PR, Davis FJ, Patel SI (2004) General procedures in chain-growth polymerization. In: Davis FJ (ed) Polymer chemistry: a practical approach. Oxford University Press, Oxford, pp 43–98. ISBN 9780198503095

Baskaran D, Müller AHE (2007) Anionic vinyl polymerization—50 years after Michael Szwarc. Prog Polym Sci 32:173

Bates FS, Fredrickson GH (1999) Block copolymers—designer soft materials. Phys Today 52 (2):32–38

Bates FS, Wignall GD, Koehler WC (1985) Critical-behavior of binary-liquid mixtures of deuterated and protonated polymers. Phys Rev Lett 55(22):2425–2428

Bates FS, Hillmyer MA, Lodge TP, Bates CM, Delaney KP, Fredrickson GH (2012) Multiblock polymers: panacea or Pandora's box. Science 336:434–440

Bendiganavale AK, Malshe VC (2008) Infrared reflective inorganic pigments. Recent Pat Chem Eng 1:67–79

Bockstaller M, Kolb R, Thomas EL (2001) Metallodielectric photonic crystals based on diblock copolymers. Adv Mater 13:1783–1786

Brenner M, Chase FL, Leydon AJ (1972) Gasket-forming solvent-based com-positions containing styrene-butadiene block copolymers. U.S. Patent 3676386

Busch K, John S (1999) Liquid-crystal photonic-band-gap materials: the tunable electromagnetic vacuum. Phys Rev Lett 83:967

Castiglione DC (2008) Block copolymer based photonic band gap materials. PhD thesis, University of Reading

Christensen L (2002) HOME OF: the blue morpho butterfly. New York Times (12 May)

Crossland RK, Harlan Jr, JT (1975) Block copolymer adhesive compositions. U.S. Patent 3917607

Davis FJ, Mitchell GR (2008) Polyurethane based materials with applications in medical devices. In: Bártolo P, Bidanda B (eds) Bio-materials and prototyping applications in medicine. Springer, New York, pp 27–48. ISBN 9780387476827

Djiauw LK, Icenogle RD (1986) Low smoke modified polypropylene insulation compositions. U.S. Patent 4622352

Dowling JP (1998) Mirror on the wall: you're omnidirectional after all? Science 282:1841–1842

Fabian J, Nakazumi H, Matsuoka M (1992) Near-infrared absorbing dyes. Chem Rev 92:1197–1226

Fink Y, Urbas AM, Bawendi MG, Joannopoulos JD, Thomas EL (1999) Block copolymers as photonic bandgap materials. J Lightwave Technol 17(11):1963–1969

Fischer HJ (1999) The persistent radical effect in controlled radical polymerizations. Polym Sci A Polym Chem 37:1885–1901

Flemming JG, Lin SY, El-Kady I, Biswas R, Ho KM (2002) All-metallic three-dimensional photonic crystals with a large infrared bandgap. Nature 417:52–55

Forsman A, Merilaita S (2003) Fearful symmetry? Intra-individual comparisons of asymmetry in signalling versus cryptic colour patterns in butterflies. Evol Ecol 17:491–507

Franta E, Hogen-Esch T, van Beylen M, Smid J (2007) Fifty years of living polymers. J Polym Sci A Polym Chem 45:2576–2579

Gody G, Maschmeyer T, Zetterlund PB, Perrier S (2014) Pushing the limit of the RAFT process: multiblock copolymers by one-pot rapid multiple chain extensions at full monomer conversion. Macromolecules 47:3451–3460

Hadjichristidis N, Iatrou H, Pispas S, Pitsikalis M (2000) Anionic polymerization: high vacuum techniques. J Polym Sci A Polym Chem 38:3211–3234

Hamley IW (2004) Developments in block copolymer science and technology. In: Hamley IW (ed). Wiley, Chichester

Hayes W, Rannard S (2004) Polymer chemistry: a practical approach. 3. Controlled/'living' polymerization methods. Oxford University Press, Oxford, p 116

Jacquemoud S, Frangi J-P, Govaerts Y, Ustin SL (1997) Three-dimensional representation of leaf anatomy—application of photon transport. In: Guyot G, Phulpin T (eds) Physical measurements & signatures in remote sensing. A.A. Balkema, Rotterdam, pp 295–302

Jakubowski W, Kirci-Denizli B, Gil RR, Matyjaszewski K (2008) Polystyrene with improved chain-end functionality and higher molecular weight by ARGET ATRP. Macromol Chem Phys 209:32–39

John S (1987) Strong localization of photons in certain disordered dielectric superlattices. Phys Rev Lett 58:2486

Khandelwal H, Loonen RCGM, Hensen JLM, Schenning APHJ, Debije MG (2014) Application of broadband infrared reflector based on cholesteric liquid crystal polymer bilayer film to windows and its impact on reducing the energy consumption in buildings. J Mater Chem A 2:14622–14627

Kharasch MS, Jensen EV, Urry WH (1945) Addition of carbon tetrachloride and chloroform to olefins. Science 102:128

Kraton Performance Polymers, Inc. http://www.kraton.com/about/history.php. Accessed May 2015

Lee J-H, Koh CY, Singer JP, Jeon S-J, Maldovan M, Stein O, Thomas EL (2014) 25th anniversary article: ordered polymer structures for the engineering of photons and phonons. Adv Mater 26:532–569

Li J, Wu Y, Fu J, Cong Y, Peng J, Han Y (2004) Reversibly strain-tunable elastomeric photonic crystals. Chem Phys Lett 390:285–289

Liebler L (1980) Theory of microphase separation in block copolymers. Macromolecules 13:1602–1617

Liguda C, Böttger G, Kuligk A, Blum R, Eich M, Roth H, Kunert J, Morgenroth W, Elsner H, Meyer HG (2001) Polymer photonic crystal slab waveguides. Appl Phys Lett 78:2434–2436

Matsen MW, Schick M (1994) Stable and unstable phases of a diblock copolymer melt. Phys Rev Lett 72:2660

Matsen MW, Bates FS (1996) Unifying weak- and strong-segregation block copolymer theories. Macromolecules 29:1091

Minisci F (1975) Free-radical additions to olefins in the presence of redox systems. Acc Chem Res 8:165–171

Miyake GM, Raymond A, Weitekamp RA, Victoria A, Piunova VA, Grubbs RH (2012) Synthesis of isocyanate-based brush block copolymers and their rapid self-assembly to infrared-reflecting photonic crystals. J Am Chem Soc 2012(134):14249–14254

Morton M (1983) Anionic polymerization: principles & practice. Academic, New York

Mueller L, Jakubowski W, Matyjaszewski K, Pietrasik J, Kwiatkowski P, Chaladaj W, Jurczak J (2011) Synthesis of high molecular weight polystyrene using AGET ATRP under high pressure. Eur Polym J 47:730–734

Ozin GA, Yang SM (2001) The race for the photonic chip: colloidal crystal assembly in silicon wafers. Adv Funct Mater 11:95

Painter O, Lee RK, Scherer A, Yariv A, O'Brian JD, Dapkus PD, Kim I (1999) Two-dimensional photonic band-gap defect mode laser. Science 284:1819–1821

Paul DE, Lipkin D, Weissman SI (1956) Reaction of sodium metal with aromatic hydrocarbons. J Am Chem Soc 78:116

Riess G (1999) Block copolymers as polymeric surfactants in latex and microlatex technology. Colloids Surf A Physiochem Eng Asp 153:99–110

Russell P (2003) Photonic crystal fibers. Science 299:358–362

Russell TP, Hjelm RP Jr, Seeger PA (1990) Temperature-dependence of the interaction parameter of polystyrene and poly(methyl methacrylate). Macromolecules 23:890

Saado Y, Ji T, Golosovsky M, Davidov D, Avni Y, Frenkel A (2001) Self-assembled heterostructures based on magnetic particles for photonic bandgap applications. Opt Mater 17:1–6

Saito A, Miyamura Y, Nakajima M, Ishikawa Y, Sogo K, Kuwahara Y, Hirai Y (2006) Reproduction of the Morpho blue by nanocasting lithography. J Vac Sci Technol B 24:3248

Simms RW, Cunningham MH (2008) High molecular weight poly(butyl methacrylate) via ATRP miniemulsions. Macromol Symp 2008(261):32–35

Starkey A (2005) The butterfly effect. New Sci 187:46

Statz RJ, Chen JC, Hagman JF (2004) Highly-resilient thermoplastic elastomer compositions. U.S. Patent 6815480

Stefik M, Guldin S, Vignolini S, Wiesner U, Steiner U (2015) Block copolymer self-assembly for nanophotonics. Chem Soc Rev 44:5076–5091

Szwarc M (1956) Living polymers. Nature 178:1168–1169

Szwarc M, Levy M, Milkovich R (1956) Polymerization initiated by electron transfer to monomer. A new method of formation of block polymers. J Am Chem Soc 78:2656

Uhrig D, Mays JW (2005) Experimental techniques in high-vacuum anionic polymerization. J Polym Sci A Polym Chem 43:6179–6222

Urbas A, Fink Y, Thomas EL (1999) One-dimensionally periodic dielectric reflectors from self-assembled block copolymer-homopolymer blends. Macromolecules 32:4748–4750

Vengsarkar AM (1996) Long-period fiber grating shape optical spectra. Laser Focus World (June): 243–247

Vukisic P, Sambles JR, Lawrence CR, Wootton RJ (1999) Quantified interference and diffraction in single Morpho butterfly scales. Proc R Soc Lond B 266:1403

Vukisic P, Sambles JR, Lawrence CR (2000) Structural colour: colour mixing in wing scales of a butterfly. Nature 404:457

Wuyts N, Palauqui J-C, Conejero G, Verdeil J-L, Granier C, Massonnet C (2010) High contrast three-dimensional imaging of the Arabidopsis leaf enables the analysis of cell dimensions in the epidermis and mesophyll. Plant Methods 6:17

Xia Y, Gates B, Park SH (1999) Fabrication of three-dimensional photonic crystals for use in the spectral region from ultraviolet to near-infrared. J Lightwave Technol 17:1956

Yablonovitch E (1987) Inhibited spontaneous emission in solid-state physics and electronics. Phys Rev Lett 58:2059

Yablonovitch E (2001) Photonic crystals: semiconductors of light. Sci Am 285:46

Yang SM, Ozin GA (2000) Opal chips: vectorial growth of colloidal crystal patterns inside silicon wafers. Chem Commun 2507

Yoon J, Lee W, Thomas EL (2005) Self-assembly of block copolymers for photonic bandgap materials. MRS Bull 30:721–726

Zhao J, Li X, Zhong L, Chen G (2009) Calculation of photonic band-gap of one dimensional photonic crystal. J Phys Conf Ser 183:012018

Zhou J, Sun CQ, Pita K, Lam YL, Zhou Y, Ng SL, Kam CH, Li LT, Gui ZL (2001) Thermally tuning of photonic band gap of SiO_2 colloid-crystal infilled with ferroelectric $BaTiO_3$. Appl Phys Lett 78:661–663

Chapter 11
Relationship Between Molecular Configuration and Stress-Induced Phase Transitions

Finizia Auriemma, Claudio De Rosa, Rocco Di Girolamo, Anna Malafronte, Miriam Scoti, Geoffrey R. Mitchell, and Simona Esposito

11.1 Introduction

The physical and mechanical properties of semi-crystalline polymers depend on their molecular architecture, molecular mass and distribution of molecular mass which are fixed in the polymerization step (Heimenz and Lodge 2007; Halary et al. 2011; Billmeyer 1984; Schultz 1984). Furthermore, for any molecular architecture the properties also depend on the processing conditions (Cowie and Arrighi 2008; Richardson and Lokensgard 1996; Bicerano 2002) (Fig. 11.1).

If we confine our attention to linear polymers, the molecular architecture addresses the constitution of the polymer chains, the configuration defined by the sequence of the monomeric units along the chains, presence of branches, and/or defects such as configurational defects, regio-irregularities and constitutional defects in the case of copolymers (IUPAC and Chap. 1). Since the conformation and the intrinsic flexibility of the chains depend on the chain architecture, properties such as glass transition, mechanical strength and rigidity are intrinsically dependent on the chain architecture.

The crystallization properties and polymorphism are dependent on the chain architecture. In fact, chains with a regular constitution and configuration are generally able to adopt low energy conformations with a regular and periodic repetition of the monomeric units along the chain axis, and therefore, they are also more able to crystallize than less regular chains (De Rosa and Auriemma 2013).

F. Auriemma (✉) • C. De Rosa • R. Di Girolamo • A. Malafronte • M. Scoti • S. Esposito
Dipartimento di Scienze Chimiche, Università di Napoli "Federico II", Complesso Monte Sant' Angelo, via Cintia, Naples 80126, Italy
e-mail: finizia.auriemma@unina.it

G.R. Mitchell
Centre for Rapid and Sustainable Product Development, Polytechnic Institute of Leiria, Centro Empresarial da Marinha Grande, Marinha Grande 2430-028, Portugal

Fig. 11.1 Correlations between structure and morphology (dependent on crystallization conditions and processing) and the mechanical properties of semi-crystalline polymers. These properties depend on the chain architecture that in turn depends on the polymerization set-up

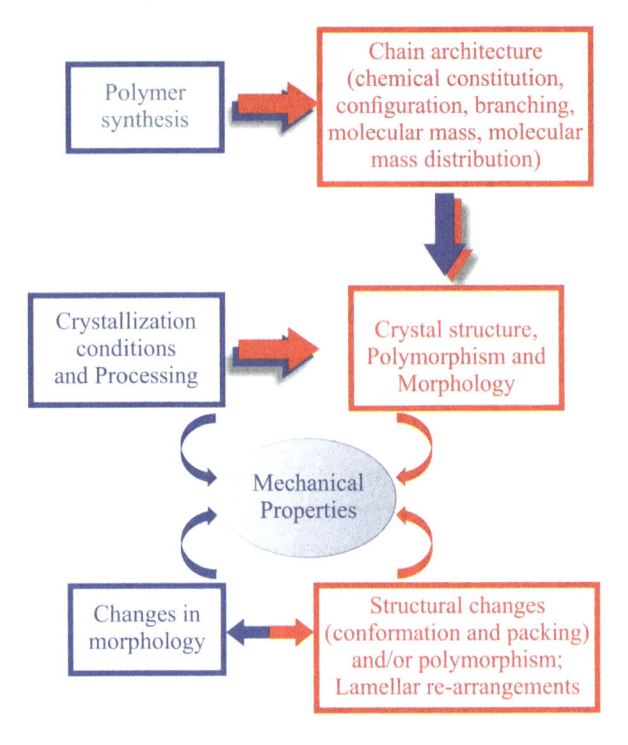

Moreover, the presence of defects along the chains does not necessarily prevent crystallization, even though the degree of crystallinity, the melting temperature, the type and amount of disorder included in the crystals, crystallization kinetics and polymorphism may be strongly influenced by the defects (Wunderlich 1973). Since the properties of a polymer depend on the kind of polymorph present in the material and the degree of disorder in the crystals, small changes in the chain microstructure produce change in the polymorphism and/or crystallization behaviour, resulting in gross changes in material properties and consequent possibility of fine tuning properties through the control of crystallization (De Rosa and Auriemma 2013, Ch. 7).

On the other hand, for any given chain architecture, the final properties of a polymeric material may be also controlled by processing (Richardson and Lokensgard 1996; De Rosa et al. 2014). In fact, small variation in the variables associated with the process, including the maximum temperature achieved in the melt, application of shear forces to the melt (Ward 1997; Somani et al. 2005), cooling rate (Zia et al. 2009; De Rosa et al. 2014; Sorrentino et al. 2007), presence of additives or nucleating agents (Menyhárd et al. 2009), may produce large variations of the segmental dynamics, influence the polymorphism and crystallization ability, or drive a different morphology (Baird 1998). Therefore, a tight control over the manufacturing conditions may allow tailoring the material properties directly in the processing plant, through development of the morphology (Bicerano 2002).

Polymorphic transitions between different crystalline forms can be triggered not only by thermal treatments but also by mechanical deformation (Hay and Keller 1970; Séguéla 2005; De Rosa and Auriemma 2007; Hughes et al. 1997). In general, the deformation mechanism of semi-crystalline polymers involves complex phenomena, and the principal event consists in the transformation of an original isotropic lamellar (generally spherulitic) morphology into a fibrillar one, where polymeric chains become preferentially aligned along the drawing direction (Peterlin 1971; Bowden and Young 1974). The mechanism is complex and involves a hierarchy of transformations, including block slippage within the crystalline lamellae that takes place at low deformations, followed by the stress-induced fragmentation of crystals and recrystallization into fibrils at high values of strain (Lin and Argon 1994; Bowden and Young 1974; Schultz 1974). The exact mechanism of deformation depends on the relative stability of crystals and the density of entanglements in the amorphous network (Men et al. 2003). During the deformation, after yielding and concomitant with the breaking of lamellae, polymorphic transformations may also occur (Hay and Keller 1970; Séguéla 2005; De Rosa and Auriemma 2007; Hughes et al. 1997). Transitions involving polymorphic modifications either characterized by a different conformation of the chains (conformational polymorphism) or a different packing mode, but identical chain conformation (packing polymorphism) may both be involved by effect of strain (De Rosa and Auriemma 2006, 2007, 2013, Ch. 7; De Rosa et al. 2006). These phase transformations dissipate mechanical energy so that the sample breaking is prevented, and the mechanical toughness and ductility are enhanced (De Rosa and Auriemma 2012; De Rosa et al. 2013, 2014). Therefore, not only does the initial morphology of the sample play a key role in the mechanical properties, but also phase transitions are important.

In this chapter, we focus on the effect of the chain microstructure on the mechanical properties of polymers, illustrating how the presence of stereodefects in a stereoregular vinyl polymer may influence polymorphism, morphology, stress-induced phase transitions and mechanical properties in a cascade of events that reflect the hierarchy of structural organization of semi-crystalline polymers. Because of that, the transformations are also hierarchical. In particular, these effects are shown in the case of some selected samples of isotactic polypropylene (iPP) with a well-defined molecular structure that have been synthesized using single site metalorganic catalyst systems. All stretching experiments are performed at room temperature. Since the glass transition temperature of our samples is around $-10\,°C$ to $0\,°C$ (De Rosa et al. 2004, 2005; De Rosa and Auriemma 2007) the effect of the chain microstructure on the deformation behaviour at temperatures above glass transition is illustrated.

In Sect. 11.3, 11.4, 11.5 of this chapter we focus on the polymorphism, morphology and mechanical properties of iPP samples with different concentration of stereo-defects.

In Sect. 11.6, 11.7 and 11.8, the results of X-ray scattering measurements at both wide and small angles recorded in situ during stretching are illustrated.

Therefore, the transformations are related to changes in the packing mode of the chains in the crystals, and those occurring at lamellar length scales. These changes are followed in real time using the high flux of X-rays available at a Synchrotron light source. It is shown how the differences in concentration of stereodefects in iPP samples may influence the critical values of strain at which a given crystalline polymorph starts or ends transforming into another crystalline form throughout the gradual destructions of the original lamellar morphology and transformation into a fibrillar morphology (De Rosa and Auriemma 2006, 2007; De Rosa et al. 2006). The effect of chain microstructure on the critical values of strain at which the original unoriented assembly of lamellae breaks and the chains become oriented in the stretching direction, up to transformation into fibrillae, is also studied. Numerical descriptors of the degree of preferred orientation at atomic and lamellar length scales are also identified. Correlations between mesoscopic phenomena such as buckling instabilities of lamellar stacks and cavitation and microscopic phenomena such as chain orientation and phase transitions are established. The critical value of strain at which changes at the atomic length scale (conformation/packing/polymorphism) occur are then correlated to those at which change at the nanometer length scale (lamellae/lamellar aggregates) occur. The aim is of establishing precise correlations on how stress-induced changes at different length scales are correlated as a function of chain microstructure.

11.2 Isotactic Polypropylene Samples

The development of single-centre metallocene catalysts for the polymerization of olefins has allowed to the synthesis of polyolefins in general and polypropylenes in particular with controlled chain microstructures (Resconi et al. 2000). Polypropylenes characterized by different kinds and amounts of regio- and stereo-irregularities, different distributions of defects and different molecular masses can be prepared through the rational choice of the catalytic system (Resconi et al. 2000). Exploiting the available large library of metallocene catalyst precursors, samples of iPP with different stereoregularity, characterized by only one kind of stereodefect (isolated *rr* triads), and containing no measurable amount of regio-errors, have been synthesized. The samples have been prepared with the complexes of Scheme 11.1, activated with methylalumoxane (MAO) (Ewen 1997, 1998; Brintzinger et al. 1995; Kaminsky 1996, 1998, 1999). The amount of *rr* defects depends on the structure of the catalyst, in particular the ligand substituents and the conditions of polymerization, and the amount of rr defects has been varied in the range between 0.5 and 11 mol% (Nifant'ev et al. 2004). Correspondingly the samples show melting temperatures variable between 162 and 84 °C (Table 11.1) (De Rosa et al. 2004, 2005).

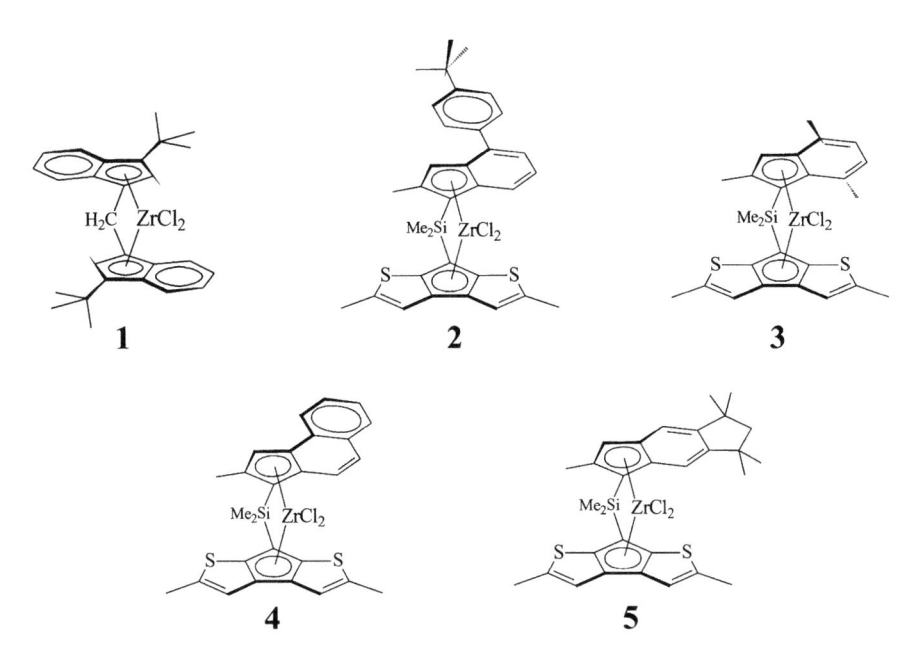

Scheme 11.1 Structure of C_2-symmetric (**1**) and C_1-symmetric (**2–5**) pre-catalysts

Table 11.1 Molecular mass (M_v), melting temperature (T_m) and content of rr triad and $mmmm$ pentad stereosequences of iPP samples prepared with the catalysts of Scheme 11.1[a]

Sample	Catalyst/ cocatalyst	M_v 10⁵ (g/mol)[b]	T_m (°C)[c]	[mm] mol%	[mr] mol%	[rr] mol%	[mmmm] mol%	x_c^d (%)	f_γ^d (%)
iPP1	1/MAO	1.96	162	98.5	1.0	0.49	97.5	70	17
iPP2	2/MAO	1.06	140	92.4	5.1	2.54	87.6	66	60
iPP3	3/MAO	2.02	133	88.9	7.4	3.70	82.2	64	85
iPP4	4/MAO	2.11	114	82.2	11.8	5.92	72.2	55	96
iPP5	5/MAO	1.23	84	66.9	22.0	11.01	51.0	42	100

[a]No or negligible regioerrors (2,1 insertions) could be observed in the ¹³C NMR spectra of the samples (Fritze et al. 2003; Resconi et al. 2005)
[b]From the intrinsic viscosities
[c]The melting temperatures were obtained with a differential scanning calorimeter Perkin Elmer DSC-7 performing scans in a flowing N_2 atmosphere and heating rate of 10 °C/min (De Rosa et al. 2005)
[d]Crystallinity index x_c and relative amount of γ form with respect to the α form f_γ evaluated from X-ray diffraction profiles

11.3 Structural Analysis and Mechanical Properties

The X-ray powder diffraction profiles of melt-crystallized compression moulded specimens of the iPP samples of Table 11.1 are reported in Fig. 11.2a. The samples crystallize from the melt as mixtures of α and γ forms (Fig. 11.2b, c, respectively),

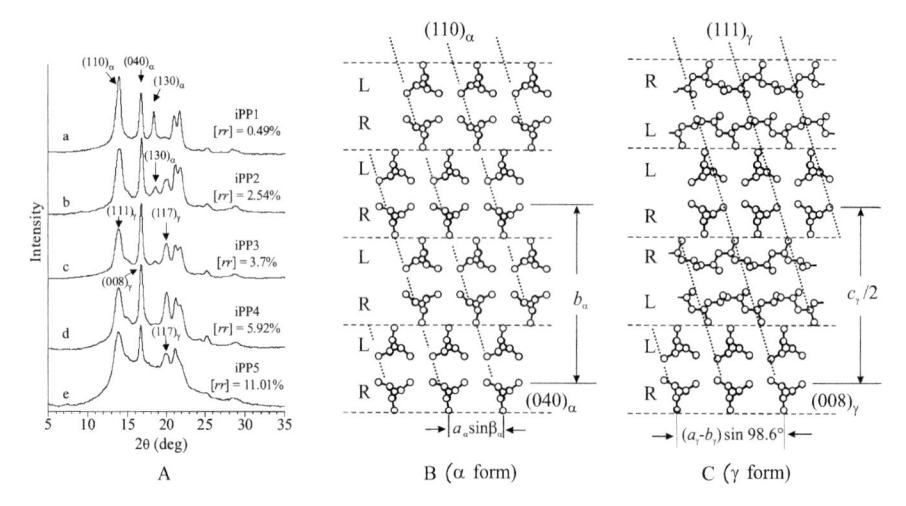

Fig. 11.2 (a) X-ray powder diffraction profiles of iPP samples of Table 11.1 crystallized from the melt by compression moulding and cooling the melt to room temperature at 1 °C/min (De Rosa et al. 2004). The $(110)_\alpha$, $(040)_\alpha$ and $(130)_\alpha$ reflections of the α form (Natta and Corradini 1960) at 2θ (CuKα) = 14, 17 and 18.6° ($d \approx 0.65$, 0.52 and 0.48 nm, respectively) and the $(111)_\gamma$, $(080)_\gamma$ and $(117)_\gamma$ reflections of the γ form (Brückner and Meille 1989) at 2θ (CuKα) = 14, 17 and 20.1° ($d \approx 0.65$, 0.52 and 0.44 nm) are indicated. (**b, c**) Limit ordered models of packing proposed for the α (**b**) and γ (**c**) forms of iPP. The *dashed horizontal lines and dotted oblique lines* are the traces of $(040)_\alpha$ and $(110)_\alpha$ planes of α form in (**b**), respectively, and $(080)_\gamma$ and $(111)_\gamma$, planes of γ form in (**c**), respectively. Subscripts α and γ identify unit cell parameters referred to the monoclinic (Natta and Corradini 1960) and orthorhombic (Brückner and Meille 1989) unit cells of α and γ forms, respectively. ((**a**) redrawn from De Rosa and Auriemma 2007 with permissions)

as indicated by the presence of both $(130)_\alpha$ and $(117)_\gamma$ reflections of α (Natta and Corradini 1960) and γ (Brückner and Meille 1989) forms, respectively, in the diffraction profiles of Fig. 11.2a. The intensity of the $(117)_\gamma$ reflection of the γ form at $2\theta = 20.1°$ increases with increasing the concentration of *rr* defects (De Rosa et al. 2004). The most isotactic sample crystallizes basically in the α form (curve a of Fig. 11.2a), and the concentration of γ form of 17 % is low (Table 11.1). The amount of γ form increases with increasing the content of *rr* defects up to 100 % for *rr* concentrations higher than 6 mol% (Table 11.1) (De Rosa et al. 2004, 2005; De Rosa and Auriemma 2007). Finally, the crystallinity index (Table 11.1) decreases with increasing concentration of *rr* defects from ≈ 70 % for the stereoregular iPP1 sample with $[rr] = 0.49$ mol% to ≈42% for the stereodefective iPP5 sample with $[rr] = 11.0$ mol%.

The small angle X-ray scattering (SAXS) curves of the metallocene-made iPP samples are shown in Fig. 11.3a. The samples have been isothermally crystallized at temperatures T_c lower than the melting temperature of Table 11.1 by 40–50 °C. SAXS data were recorded at the crystallization temperature T_c, on samples first rapidly cooled to room temperature immediately after completion of crystallization process, and successive heating at T_c. The simultaneous recording of the X-ray

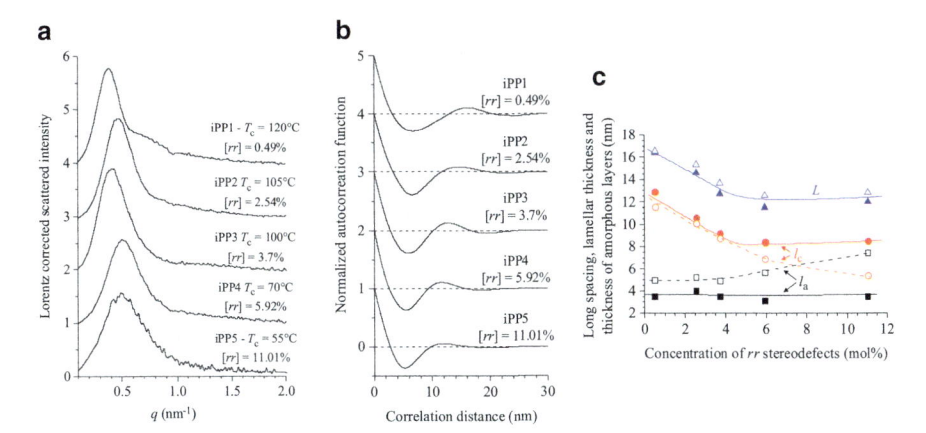

11 Relationship Between Molecular Configuration and Stress-Induced Phase...

Fig. 11.3 (**a**) SAXS profiles of iPP samples of Table 11.1 isothermally crystallized from the melt at the indicated values of temperature T_c (**a**) and corresponding normalized autocorrelation functions. SAXS data are recorded at T_c. Values of long spacing (L), and thickness of crystalline (l_c) and amorphous ($l_a = L - l_c$) layers (**c**). They are evaluated from the autocorrelation function in (**b**) (*full symbols*) and also using the Bragg law for the calculation of L and successive multiplication of L times the crystallinity index (see the text) for l_c (*open symbols*)

scattering intensity in the wide angle (WAXS, data not reported) indicates that in the selected crystallization conditions the samples develop the same crystallinity index as the compression moulded films of Fig. 11.2.

The Lorentz corrected SAXS profiles of all samples (Fig. 11.3a) show a single and broad correlation peak with maximum at $q = q_{peak}$, q being the magnitude of the scattering vector $q = 4\pi \sin\theta/\lambda$ where λ is the wavelength of the incident X-ray beam and 2θ the angle between the incident and scattered beam. The broadness of the scattering peak indicates that all samples present a disordered lamellar morphology, where lamellar crystals (average thickness l_c) alternate to amorphous layers (average thickness l_a) with average periodicity (long spacing) $L = l_c + l_a$. Disorder is due to a broad distribution of both l_c and l_a.

The values of L, l_c and l_a have been evaluated either from the main-correlation triangle in the autocorrelation function (Fig. 11.3b) or with use of the Bragg law, that is $L = 2\pi/q_{peak}$, $l_c \approx L\, x_c$ and $l_a = L - l_c$ where x_c is the crystallinity index obtained from WAXS analysis of Table 11.1. The values obtained are compared in Fig. 11.3c.

It is worth noting that whereas the values of the long spacing calculated with the two methods are in a good agreement within the experimental error (1–2 nm), the agreement for the values of l_c and l_a is less good. In fact the values of l_c and l_a obtained from the autocorrelation function (Fig. 11.3b) are higher and lower, respectively, than those obtained directly, especially for the less stereoregular samples iPP4 and iPP5 with *rr* content of 5.92 and 11.01 mol%, respectively. This is due to the fact that the correlation peak of SAXS profiles (Fig. 11.3a) is more weighted towards the well-organized stacks of lamellar crystals and the amorphous layers containing the less defective crystals. The amorphous phase located in the extralamellar regions and the defective lamellar crystals not participating in the regular stacks, instead, contribute to the SAXS intensity as diffuse scattering in the background and in the tail at low q values of the correlation peak. Therefore, the values of l_c extracted from the autocorrelation function correspond to the thickness of the less defective lamellar crystals in the sample forming regular (periodic) lamellar stacks, whereas the values of l_a correspond to the thickness of intralamellar amorphous layers. The direct method, instead, is based on the use of a value of the crystallinity index (amorphous fraction), that accounts for all kind of crystallites and amorphous phases, located both in the intralamellar and interlamellar regions. Therefore, the values of l_c (and l_a) calculated by the direct method correspond to values that are averaged over both regular and irregular lamellar crystals (both intra- and inter-lamellar amorphous phase). For this reason, the largest difference in the values of l_c and l_a calculated with the two methods occur for the less stereoregular samples, the defective chains of which tend to form a major amount of defective crystals, and/or are rejected in the interlamellar amorphous regions.

The stress–strain curves of compression-moulded films of the iPP samples of Table 11.1 are shown in Fig. 11.4a, b. The values of Young modulus decrease with

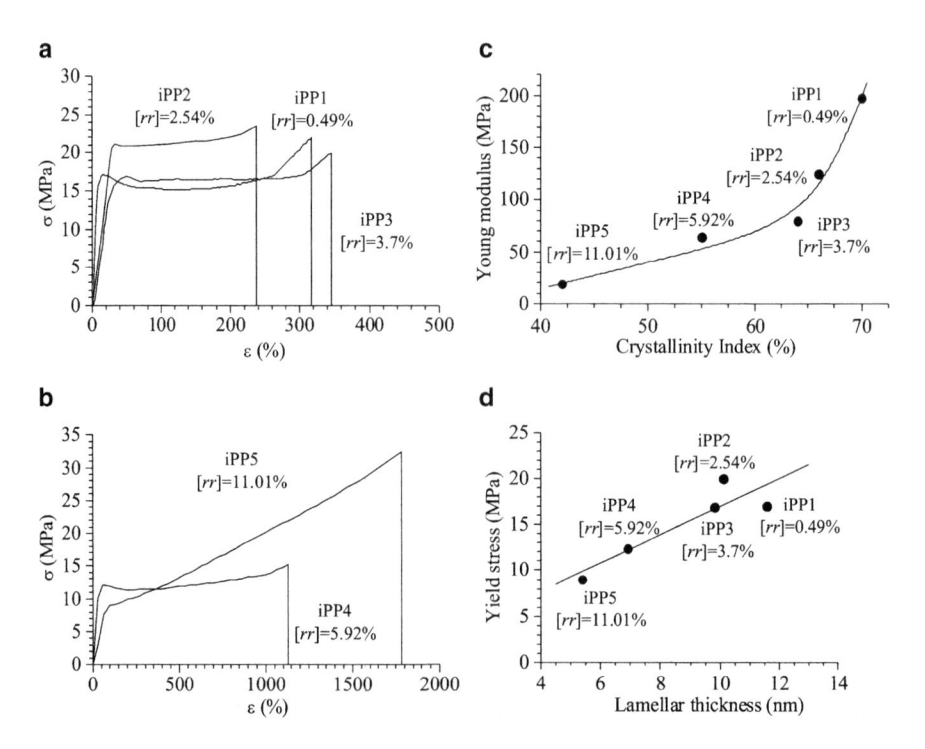

Fig. 11.4 (a, b) Stress–strain curves of unoriented compression moulded films of the samples in Table 11.1 (De Rosa et al. 2004, 2005); (c, d) Average values of Young Modulus as a function of crystallinity index (c) and relationship between the lamellar thickness l_c and the yield stress σ_y (d) of iPP samples with the indicated concentrations of *rr* stereo-defects ((a, b) Reprinted from De Rosa et al. 2005 with permissions)

increasing concentration of rr defects and decreasing the crystallinity (De Rosa et al. 2005; De Rosa and Auriemma 2007) (Fig. 11.4c). The strain at yield, instead, increases with increasing the concentration of defects from values close to 10 % deformation for the sample iPP1 with $[rr] = 0.49$ mol%, to values close to 100 % for the sample iPP5 with $[rr] = 11.0$ mol% (De Rosa et al. 2005; De Rosa and Auriemma 2007). Moreover, the values of yield stress tend to decrease with increasing concentration of defects (De Rosa et al. 2005; De Rosa and Auriemma 2007), whereas they increase almost linearly with the values of the average lamellar thickness (Fig. 11.4d).

It is worth noting that a linear relationship between yield stress and lamellar thickness has been found also in other polymers as for instance in the case of isotactic polypropylene (O'Kane and Young 1995; O'Kane et al. 1995), and random isotactic copolymers of propene with 1-hexene (De Rosa et al. 2009), polyethylene and random copolymer of polyethylene (Kennedy et al. 1994, 1995; Popli and Mandelkern 1987; Crist 1989; Schrauwen et al. 2004), and several different mechanisms have been suggested to explain such behaviour. The proportional relation of yield stress and lamellar thickness has been associated with the intrinsic complex molecular structure and morphology of semi-crystalline polymers and the most popular explanations include occurrence of partial melting and recrystallization (Kennedy et al. 1994, 1995; Popli and Mandelkern 1987; Crist 1989), nucleation and propagation of screw dislocations (Young 1974), the role that thermally activated conformational (chain twist) defects play to overcome energy barriers for nucleation and propagation of dislocations (Séguéla 2002), and the intrinsic stability of the crystalline modifications that develop upon crystallization (De Rosa et al. 2009). It is clear that a definitive and general understanding of this proportionality and of the yield behaviour of semi-crystalline polymers in general has not yet been achieved (Schrauwen et al. 2004). In a recent study it has been pointed out that in order to reach such a more definitive understanding, in addition to the structural and morphological aspects, kinetic factors associated with the segmental relaxation dynamics of the polymer chains should also be included in the theory (Schrauwen et al. 2004).

It is worth noting that in our case, the linear relationship (Fig. 11.4d) holds only for the average values of lamellar thickness evaluated accounting for lamellar crystals included both in the well-developed and regular stacks and those embedded inside the inter-lamellar amorphous regions, and does not hold for the values of l_c deduced from the autocorrelation function (Fig. 11.3b) relative to the best developed lamellar crystals in a sample. Moreover, for the most stereo-regular samples iPP1 and iPP2 with rr content less than 3 mol%, presenting a more regular lamellar organization, the values of l_c present large deviations from the linear relationships holding for the less stereo-regular samples iPP3-iPP5 with rr content higher than 5 mol%. The stereo-irregular samples present a non-negligible population of irregular lamellar crystals not participating to lamellar stacks. This suggest that different mechanisms are involved at yielding in these samples, since different populations of lamellar crystals experience different stress level during the tensile deformation (Hiss et al. 1999; Al-Hussein and Strobl 2002; Men and Strobl 2001;

Men et al. 2003; Hong et al. 2004a, b). In particular the presence of irregular crystals produce values of yield strength, which increase linearly with the average lamellar thickness. Instead, for samples characterized by a negligible population of irregular lamellae, the yield stress seems to level off with the lamellar thickness. In fact, since the most stereoregular samples form also ordered lamellar crystals, we argue that the relaxation mechanisms involving the chains stems in the crystals also play a role at the onset of plastic deformation (Schmidt-Rohr and Spiess 1991). In the stereodefective samples, instead, stem relaxation in the crystals would be damped and/or prevented by a too high concentration of steric defects located at the fold-surface (De Rosa et al. 2009).

At large deformations the samples show cold drawing and values of deformation at break of ≈250–350 %, for concentration of *rr* defects lower than 4 mol% (Fig. 11.4a), and a strong increase of the ductility and toughness, for higher contents of *rr* defects (Fig. 11.4b). Therefore, samples of iPP containing low concentration of *rr* defects (up to [*rr*] = 3.7 %) show high stiffness, coupled with relatively high ductility in spite of the high degree of crystallinity and melting temperature (Fig. 11.4a) (De Rosa et al. 2005; De Rosa and Auriemma 2007). Samples with concentration of *rr* defects around 5–6 % present the behaviour typical of thermoplastic materials with a slightly lower strength but much higher deformation at break ($\varepsilon_b \approx$ 1200 %, Fig. 11.4b). Finally, the less crystalline and stereoregular samples, such as iPP5, with concentration of *rr* defects of 11 mol%, show a strong strain hardening at high deformation, and values of the tensile strength (32 MPa) (Fig. 11.4b) higher than those of the more isotactic and crystalline samples (20–23 MPa, Fig. 11.4a) (De Rosa et al. 2005; De Rosa and Auriemma 2007). The strain hardening of this sample may be somehow related to the fact that it presents uniform stretching behaviour, and involves crystals with low plastic resistance and therefore low cavitation (vide infra) (Pawlak and Galeski 2005). Once crystals have experienced irreversible plastic deformations at low values of deformation, strain hardening is caused by the straightening of the entangled network (Pawlak and Galeski 2005). It is worth to remark, that the sample iPP5 presents also large elastic recovery even at large deformations (vide infra) (De Rosa et al. 2005; De Rosa and Auriemma 2007). These elastic properties are associated with remarkable values of the modulus of nearly 20–30 MPa (Fig. 11.4c), and also high values of tensile strength (Fig. 11.4b) (De Rosa et al. 2005; De Rosa and Auriemma 2007).

11.4 Optical and Atomic Force Microscopy

Polymer films (with a thickness of 90–120 μm) have been prepared for Atomic Force Microscopy (AFM) and Optical Microscopy (OM) experiments by sandwiching a small amount of the samples in between glass cover slips, after melting at ≈200 °C and then crystallizing them at cooling rate of 10 °C/min up to room temperature. The samples crystallize in either the α or γ forms by slow

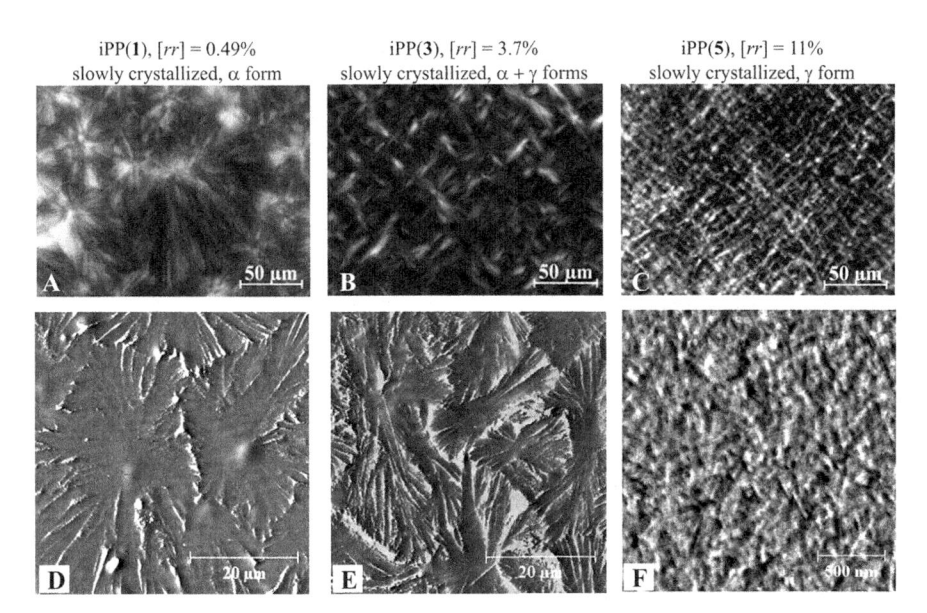

11 Relationship Between Molecular Configuration and Stress-Induced Phase... 297

Fig. 11.5 Polarizing optical microscopy (**a–c**) and high resolution AFM phase (**d–f**) images of samples of iPP of different stereoregularity crystallized from the melt by slow cooling (10 °C/min) to room temperature in α or γ forms. Samples iPP1 with [rr] = 0.49 mol% (**a, d**), iPP3 with [rr] = 3.7 mol% (**b, e**) and iPP5 with [rr] = 11 mol% (**c, f**). The images (**d, e**) are on area of 50 μm × 50 μm, and that in (**f**) is 2 μm × 2 μm (Reprinted from De Rosa et al. 2014 with permissions)

crystallization from the melt and develop different morphologies as shown by the POM and AFM phase images of Fig. 11.5. The different morphologies are variant of the basic lamellar morphology typical of melt crystallized iPP samples, and reflect the crystallization of different amounts of the α and the γ forms (De Rosa et al. 2014), as a result of the different concentration of rr defects. In particular, the micrographs of the highly isotactic sample iPP1 shows lamellae of the α form and space filling spherulitic superstructure (Fig. 11.5a, d). The sample iPP3 with 3.7 mol% of rr defects crystallizes slowly from the melt mainly in the γ form with about 20 % of crystals of α form (WAXS curve c of Fig. 11.2a), and the crystalline supermolecular structure presents bundle-like entities, rather than spherulites, organized in a nearly 90° texture (Fig. 11.5b, e). The sample iPP5 with 11 mol % of rr defects crystallizes slowly from the melt completely in the γ form (WAXS curve e of Fig. 11.2a) and the morphology has a superstructure characterized by single, elongated entities organized in a nearly 90° texture resulting in an interwoven morphology (Fig. 11.5c, f) (De Rosa et al. 2014). The appearance of the three different morphologies shown in Fig. 11.5 and in the corresponding X-ray diffraction profiles of Fig. 11.2 indicates that, with increasing the content of the γ form, the morphology changes from spherulites to bundle-like and elongated entities.

These data are in agreement with early observations by Thomann et al. (1996, 2001) and Stocker et al. (1993) of the morphology of the γ form of iPP, and more recent TEM data by Cao et al. (2009, 2011), and homoepitaxial growth of lamellar branches in the α and γ forms of iPP (Stocker et al. 1993; Compostella et al. 1962; Norton and Keller 1985; Sauer et al. 1965; Khoury 1966; Katayama et al. 1968; Andersen and Carr 1975; Padden and Keith 1966, 1973; Lovinger 1983; Bassett and Olley 1984; Lotz and Wittmann 1986; Zhou et al. 2005). In the case of the α form, the homoepitaxial growth of lamellar branches corresponds to the unique cross-hatched morphology of the α form, which takes place upon crystallization from solution (Sauer et al. 1965; Khoury 1966), melt (Norton and Keller 1985) and in fibres (Compostella et al. 1962; Andersen and Carr 1975). In particular, the cross-hatched structures that develop upon crystallization from solution are so regular that they have been termed "quadrites" by Khoury (Khoury 1966).

For the cross-hatched lamellar structure which develops in spherulites of iPP in the α form (Compostella et al. 1962; Norton and Keller 1985; Sauer et al. 1965; Khoury 1966; Katayama et al. 1968; Andersen and Carr 1975; Padden and Keith 1966, 1973; Lovinger 1983; Bassett and Olley 1984; Lotz and Wittmann 1986), mother lamellae develop from the centre of spherulites according to the radial direction and show on the sides a dense overgrowth of daughter lamellae inclined with their long axis at an angle of ≈80° with respect to the long axis of parent lamellae, according to the self-epitaxial relationship c_α (mother)//a_α (daughter) and a_α (mother)//c_α (daughter), with $a_\alpha = 6.65$ Å and $c_\alpha = 6.5$ Å the nearly identical unit cell parameters of α form forming an angle of $\beta_\alpha = 99.67°$ (Scheme 11.2b) (Compostella et al. 1962; Norton and Keller 1985; Sauer et al. 1965; Khoury 1966; Katayama et al. 1968; Andersen and Carr 1975; Padden and Keith 1966, 1973; Lovinger 1983; Bassett and Olley 1984; Lotz and Wittmann 1986).

In the case of the sample iPP3 that crystallizes in mixture of α and γ forms forming bundle-like entities (Fig. 11.5b, e), the lateral structure of these entities is due to the lateral growth of the γ form on the mother α form lamellae (Scheme 11.2c). In fact, the angle between the polymer chains in the γ form and

Scheme 11.2 Lamellar branching and epitaxial relationships in isotactic polypropylene (Reprinted from Auriemma and De Rosa 2006 with permission)

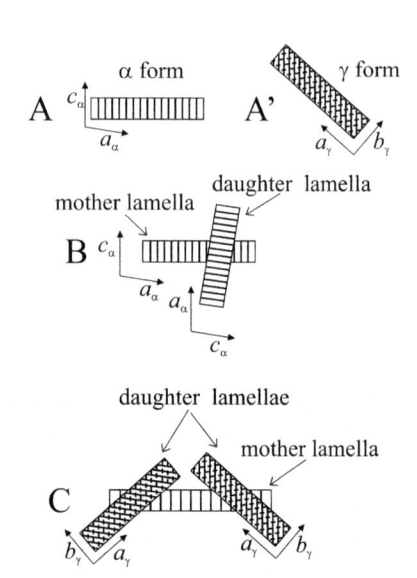

the angle of the daughter lamellae with respect to the mother-lamellae (homoepitaxial growth) in the α form are similar (≈80°). Therefore, the key of the epitaxial growth of γ form on parent lamellae of α form is of structural origin, and lies in the nonparallel arrangement of chain axes in crystals of the γ form at an angle of ≈80°, identical to the complement of angle $β_α$ of 99.67° of the monoclinic unit cell of iPP in the α form. Even in the case of the sample iPP5 that crystallizes nearly completely in the γ form (curve e of Fig. 11.2a), the typical lateral growth is visible, with formation of an interwoven morphology. Straight entities appear frequently arranged in an angle of about 80°–90° to each other, that corresponds to the crosshatching in the α form. This suggests that the interwoven morphology is formed by the γ form growth on crosshatched α form (Thomann et al. 1996, 2001). Small amount of crystals of α form are therefore present in the sample that show X-ray diffraction profile of the pure γ form (profile e of Fig. 11.2a).

It is worth noting that spherulites, bundle-like and elongated needle-like crystals were obtained by Thomann et al. (1996, 2001) for the same sample of metallocene stereoirregular iPP by changing the temperature of isothermal crystallization from the melt and favouring the crystallization of the α form (spherulites) and the γ form (bundle-like and elongated needle-like crystals) at low and high temperatures, respectively. In the case of Fig. 11.5 these morphologies develop under the same crystallization conditions by changing the concentration of *rr* stereodefects, favouring the crystallization of the α and the γ forms at low and high contents of *rr* defects, respectively (De Rosa et al. 2014).

These morphologies are in part responsible of the mechanical properties of iPP samples of different stereoregularity. In particular the morphology accounts for the gradual decrease of rigidity and plastic resistance of these samples with stereoirregularity, ranging from the space filling spherulitic morphology of the rigid sample iPP1 with 0.49 mol% *rr* defect, to the open (bundle like) spherulite morphology of the flexible sample iPP3 with 3.7 mol% *rr* content. Finally the outstanding mechanical properties of high ductility and flexibility of the sample iPP5 crystallized in the γ form (De Rosa et al. 2004, 2005; De Rosa and Auriemma 2006, 2007, 2012) may be accounted for by the interwoven morphology created by small and elongated needle-like crystals of Fig. 11.5c that should be more easily deformable than the better developed lamellar morphology of the stereoregular samples (De Rosa et al. 2014).

11.5 Stress-Induced Phase Transitions in Unoriented Films

The changes in texture and structure occurring during stretching at room temperature in the iPP samples of Table 11.1 are analysed by X-ray fibre diffraction. To this aim, rectangular specimens of initial gauge length $l_0 = 3.0$ mm, initial width $w_0 = 1.6$ mm and initial thickness $t_0 = 0.25$ mm, cut from films prepared by compression moulding in a hot press using a low pressure, are stretched at different deformations. In a series of experiments, diffraction data have been collected while keeping the sample in tension, on samples stretched step by step (ex situ analysis or stretch-and-hold experiments) (De Rosa et al. 2005; De Rosa and Auriemma 2006, 2007). Two benchmarks are drawn in the central part of the specimens at distance l_0 and the local deformation is measured for the region between the benchmarks as $\varepsilon = 100\,(l - l_0)/l_0$ with l the distance between the benchmarks after deformation. In all experiments, the X-ray beam was incident on the central region between the benchmarks. In this way, the structural changes occurring by effect of mere uniaxial deformation with negligible shear components may be probed, provided that the x-ray beam is focused in the center of the specimens and possibly in the central zone of necking regions in case of non homogeneous deformations. Diffraction data are collected 1 h after stretching for at least 40 min. Diffraction data are also collected continuously during stretching at controlled deformation rate of =2.36 mm/min (in situ measurements), using synchrotron radiation light at a time resolution of 1 frame every 5 s. Examples of in situ collected X-ray fibre diffraction patterns are reported in Fig. 11.6 (bi-dimensional diffraction patterns) and Fig. 11.7 (equatorial sections).

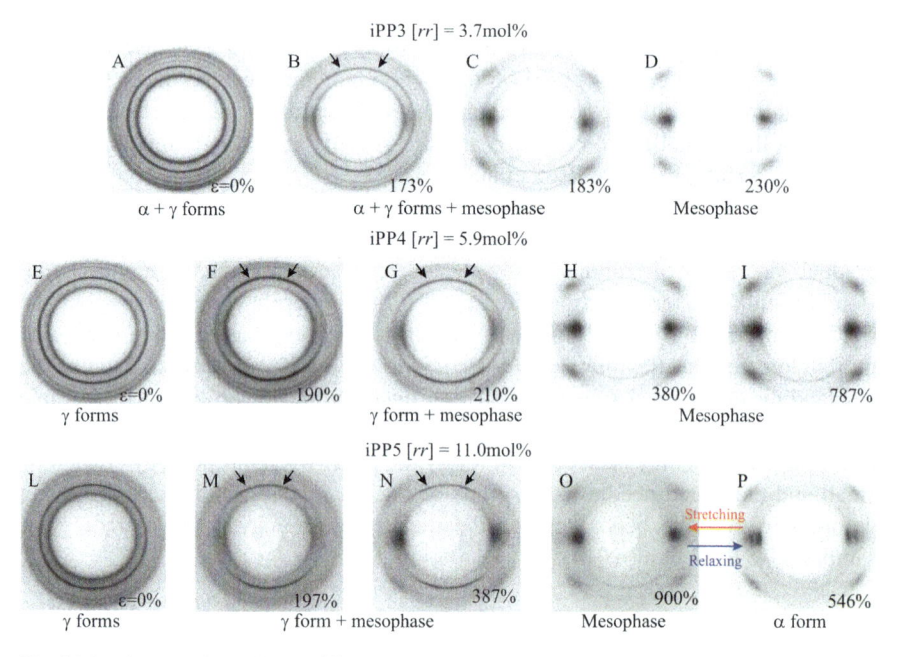

Fig. 11.6 Bi-dimensional X-ray diffraction patterns of fibres of samples iPP3 with $[rr] = 3.7$ mol% (**a–d**) iPP4 with $[rr] = 5.9$ mol% (**e–i**) and iPP5 with $[rr] = 11.0$ mol% (**l–p**), obtained by stretching at room temperature compression moulded films at the indicated values of the strain ε. Data are collected in situ, at the beam-line 16.1 of the Synchrotron Radiation Source in Daresbury (Cheshire, UK), 2.36 mm/min deformation rate, incident wavelength $\lambda = 1.4$ Å, 1 frame/5 s. The stretching direction is vertical. The strong reflections of the unoriented film specimens of iPP are used to determine the correct beam centre coordinates and the sample to detector distance. *Arrows* in **b**, **f**, **g**, **m** and **n** indicate the polarization of the reflection $(040)_\alpha$ of α form, $(008)_\gamma$ of γ form along the meridian, due to the perpendicular chain-axis orientation (cross β, Scheme 11.3a') of crystals in γ form occurring at intermediate deformations, before transformation into the mesophase at the onset of the fibrillar morphology

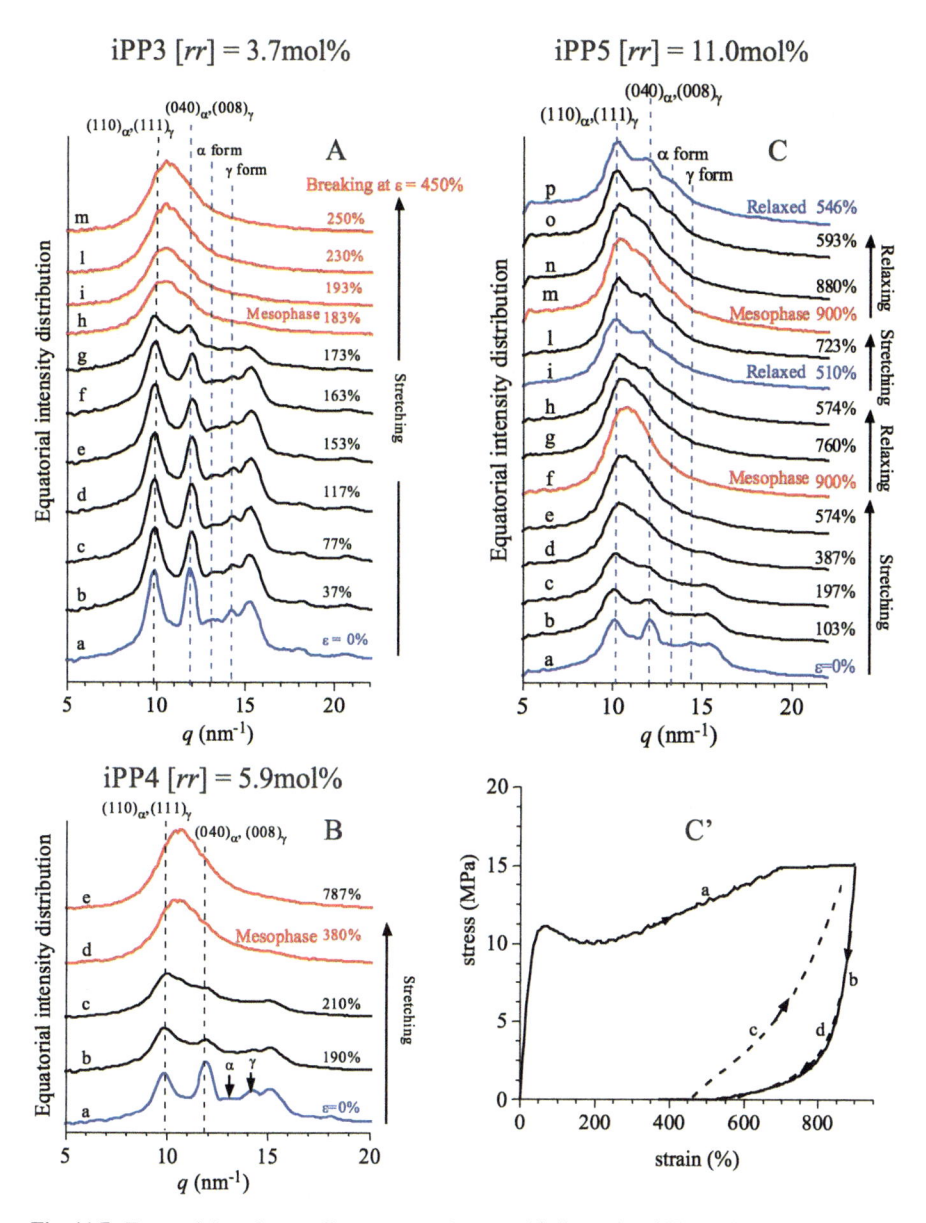

Fig. 11.7 Equatorial section profiles extracted from the bi-dimensional X-ray diffraction patterns of fibres of the samples iPP3 with [*rr*] = 3.7 mol% (**a**), iPP4 with [*rr*] = 5.9 mol% (**b**), and iPP5 with [*rr*] = 11.0 mol% (**c**), obtained by stretching at room temperature compression moulded films at the indicated values of the strain ε. The typical reflections of α and γ forms are indicated. (**c′**) Stress–strain hysteresis cycles of iPP5 ([*rr*] = 11 mol%) recorded during consecutive steps of stretching and relaxation according to the direction of the *arrows* (2.36 mm/min deformation rate) while collecting the diffraction data. In the first cycle (**a**, **b**, *solid lines*) a compression-moulded film of initial length 3 mm is stretched up to 900 % deformation (final length $L_f = 10\,L_0$) and then it is relaxed. In the successive cycles (**c**, **d**, *dashed lines*) the so-obtained relaxed fibre is stretched up to the final length $L_f = 10\,L_0$ and then it is relaxed

The structural changes occurring upon stretching unoriented films of the most stereoregular samples of Table 11.1 with rr content below 5 mol% are shown in Fig. 11.6a–d, in the case of the sample iPP3 with $[rr] = 3.7$ mol%. Compression-moulded films of the sample iPP3 crystallize as mixtures of the α and the γ forms (WAXS curve c of Figs. 11.2a, 11.6a and curve a of Fig. 11.7a) (De Rosa et al. 2005; De Rosa and Auriemma 2007). It is clear that the transformation to the mesomorphic form yields structures which have a very high level of preferred orientation with respect to the stretching direction. This sample behaves as a stiff plastic material even though it can be easily stretched up to 300–400 % deformation at room temperature (Fig. 11.4a). The crystalline forms initially present in the sample starts transforming into the mesomorphic form at ≈ 170 % deformation (Fig. 11.6b) in agreement with the data collected ex situ (De Rosa et al. 2005; De Rosa and Auriemma 2007). The formation of the mesomorphic form is indicated in the X-ray fibre diffraction pattern of the iPP3 sample stretched at 173 % deformation, by the presence of a broad halo in the equatorial regions in the q range 9–14 nm^{-1} (curve g of Fig. 11.7a), typical of the mesomorphic form of iPP, subtending non-oriented reflections of the α or γ forms (Fig. 11.6b). At ≈ 200 % deformation the sample transforms almost completely into the mesomorphic form, achieving a high degree of orientation of the crystals (Fig. 11.6c, d and curves h–m of Fig. 11.7a).

Samples with higher rr content, in the range 4–6 mol%, are highly flexible materials that show high deformation at break (curve d of Fig. 11.4a). For these samples, the γ form, originally present in compression-moulded unstretched films (curve d of Fig. 11.2a), gradually transforms into the mesomorphic form by stretching. The gradual transformation of the γ form into the mesomorphic form at high draw ratios is shown in the case of sample iPP4 with $[rr] = 5.9$ mol% in Figs. 11.6e–i and 11.7b, as an example (De Rosa et al. 2005; De Rosa and Auriemma 2007). The critical deformation at which the mesophase starts appearing is ≈ 200 % and at ≈ 400 % the initial γ form is completely transformed. Also in this case a high degree of orientation of the chain axes in the mesomorphic aggregates along the stretching direction is achieved, typical of fibrillary morphology.

Samples with concentration of rr defects in the range 7–11 % are thermoplastic elastomers with high strength (Fig. 11.4). For these samples, diffraction data collected in stretch-and-hold experiments indicate that the γ form, which is present in unstretched films (Fig. 11.2e, f), transforms by stretching into the α form, up to transformation into the mesomorphic form at very high deformations (De Rosa et al. 2005; De Rosa and Auriemma 2007). As an example, the diffraction data collected in situ during stretching for the sample iPP5 ($[rr] = 11$ %) are shown in Figs. 11.6l–p and 11.7c. In particular, since this sample exhibits elastomeric behaviour, the structural transformations have been followed during mechanical test cycles consisting of consecutive stretching steps up to 900 % deformation, followed by relaxing steps at controlled deformation rate. The stress–strain curves recorded while collecting the bi-dimensional diffraction patterns of Fig. 11.6l–p are shown in Fig. 11.7c′. It is apparent that unoriented samples of γ form (curve e Fig. 11.2a) experience irreversible plastic deformations upon stretching (curve a of

Fig. 11.7c′), and the strain is recovered only in part during the successive relaxing step (curve b of Fig. 11.7c) (De Rosa et al. 2005; De Rosa and Auriemma 2007, 2011). However, after the first mechanical cycle presenting large hysteresis, the sample exhibits elastic recovery in the successive cyclic stretching-relaxation steps (curves c, d of Fig. 11.7c′) (De Rosa et al. 2005; De Rosa and Auriemma 2007).

Diffraction data recorded in situ indicate that the initial γ form at 0 % deformation (Fig. 11.6l, and curve a of Fig. 11.7c) starts transforming into the mesophase at deformations slightly higher than 200 % (Fig. 11.6n and curve e of Fig. 11.7c). In the deformation range 200–500 % (curves b–d of Fig. 11.7c) the diffraction peaks of the γ form become broader and broader and a large amount of diffuse scattering emerges behind the Bragg peaks. This indicates that at 200 % deformation the size of coherently scattering domains decreases, because the lamellar crystals of the initial γ form undergo break through different mechanisms such as lamellar slip, shear banding and/or twinning (Peterlin 1971; Lin and Argon 1994). The transformation into the mesophase is more gradual than in that occurring in iPP3 and iPP4. At $\approx$400 % deformation a small amount of crystals of γ form is still present (Fig. 11.6n) and only at $\approx$900 % deformation the sample is almost completely transformed into the mesophase (Fig. 11.6o and curve f of Fig. 11.7c). A highly oriented fibrillar morphology is achieved at this deformation.

Upon release of the tension, the mesophase gradually transforms into the α form, and the high degree of orientation achieved during stretching is not lost (Fig. 11.6o and curves h–i of Fig. 11.6c). This transition is reversible (Fig. 11.6o, p). In fact, during consecutive mechanical test cycles, the α form transforms back into the mesophase during stretching (curves f–m of Fig. 11.7c) and the mesophase transforms again into the α form during release of the tension (curves m–o of Fig. 11.7c), while elastic recovery occurs (curves c, d of Fig. 11.7c′). Therefore, whereas the development of the fibrillar morphology starting from unoriented specimens entails occurrence of large amount of irreversible plastic deformation coupled with the irreversible transformation of the initial γ form into the mesomorphic form, oriented fibres of the sample iPP5 show elastomeric properties and the reversible phase transition between the α form and the mesophase (De Rosa et al. 2005; De Rosa and Auriemma 2007).

It is worth noting that the diffraction data collected ex situ in a series of stretch-and-hold experiments reveal that the onset of transformation γ form occurs at the same values of deformation identified from the diffraction data collected in situ (i.e. $\varepsilon \approx 200\,\%$). However at variance with the data collected continuously during stretching, the stepwise diffraction experiments evidence that the γ form transforms first into the α form and only at $\approx$500 % deformation the mesomorphic form starts appearing (De Rosa et al. 2005; De Rosa and Auriemma 2006, 2007). The appearance of α form at intermediate deformations is not evident in the stream of WAXS data collected in situ probably because no relaxation occurs. In the stretch-and-hold experiments, instead, relaxation of the samples takes place immediately after the stop of the elongation. Because of relaxation, the mesomorphic aggregates created by the mechanical melting of the γ form by effect of deformation "melt" and then recrystallize again into the α form. The melting/recrystallization of the chains is facilitated by the compliance of the amorphous phase surrounding the crystalline aggregates in the low stereoregular samples iPP4 and iPP5. Similar differences in between the data collected in continuum and ex situ during tensile tests have been evidenced and discussed by Stribeck in the case of commercial iPP samples (see for instance Stribeck et al. 2008). In that case the differences have been attributed to some kind of relaxation effect in the segmental dynamics. In our case, the in situ experiments show that the stress-induced phase transitions of γ form are the results of competition between strain-rate and melting-recrystallization phenomena, and are facilitated by the fast segmental dynamics of the defective iPP chains.

Moreover, the data of Figs. 11.6 and 11.7 lead to the conclusion that the γ form of iPP is mechanically unstable, and gradually transforms into the mesophase (or α form) by stretching through melting and recrystallization phenomena occurring at local scale. This phase transition is associated with the increase in the degree orientation of crystals with chain axes parallel to the stretching direction (De Rosa et al. 2005; De Rosa and Auriemma 2007). In particular, the high degree of orientation achieved by mesophase at high deformation starting from unoriented α and/or γ forms may be taken as the hallmark of melting /recrystallization phenomena induced by stretching (Mitchell 1984).

11.6 Structural Analysis at Microscopic Length Scale

The radial intensity distribution along the azimuthal angle χ in the q regions around 10 and 12 nm^{-1} is reported in Fig. 11.8a–c. The selected q regions correspond to the reflections (110)$_\alpha$ of α form, (111)$_\gamma$ of γ form ($d \approx 0.65$ Å nm), and (040)$_\alpha$ of α form, (008)$_\gamma$ of γ form ($d \approx 0.52$ nm) at low deformations (Scheme 11.3a), to the tails of the mesomorphic halo at high deformations (see Fig. 11.7a–c vertical dashed lines). The azimuthal angle χ is defined by the angle that the **q** radial vector makes with the stretching direction (fibre axis) (Scheme 11.3a). Therefore, the meridian axis corresponds to $\chi = 0°$ and 180° whereas the equator corresponds $\chi = 90°$ and 270°.

In Fig. 11.8a–c the intensity distribution at zero deformation is essentially uniform. With increasing deformation, the intensity at $q \approx 10$ nm^{-1} becomes more intense on the equator (curves a–e of Fig. 11.8b and a–g of Fig. 11.8c). In particular, the distribution of intensity at $q \approx 10$ nm^{-1} shows a broad maximum on the equator up to deformations close to 150–200 % (curves b, c of Fig. 11.8b, c). At deformation higher than 200 % the maximum becomes narrow (curves d, e of Fig. 11.8b, d–h of Fig. 11.8c) marking the development of the fibrillar morphology. In fact, since the normal to the (110)$_\alpha$ planes of α form, (111)$_\gamma$ planes of γ form (poles) are perpendicular to the chain axes direction in α form (Fig. 11.2b) to the direction of one-half of chain axes in the γ form (Fig. 11.2c) the polarization of the intensity at $q \approx 10$ nm^{-1} on the equator marks the tendency of chain-axes to orient parallel to the stretching direction (fibre axis).

This suggests that the development of the fibrillar morphology can be followed using the azimuthal spreading of the intensity at $q \approx 10$ nm^{-1} (curves a–e of Fig. 11.8b and a–g of Fig. 11.8c) to obtain an order parameter $\langle P_2(\cos\chi) \rangle$ defined as:

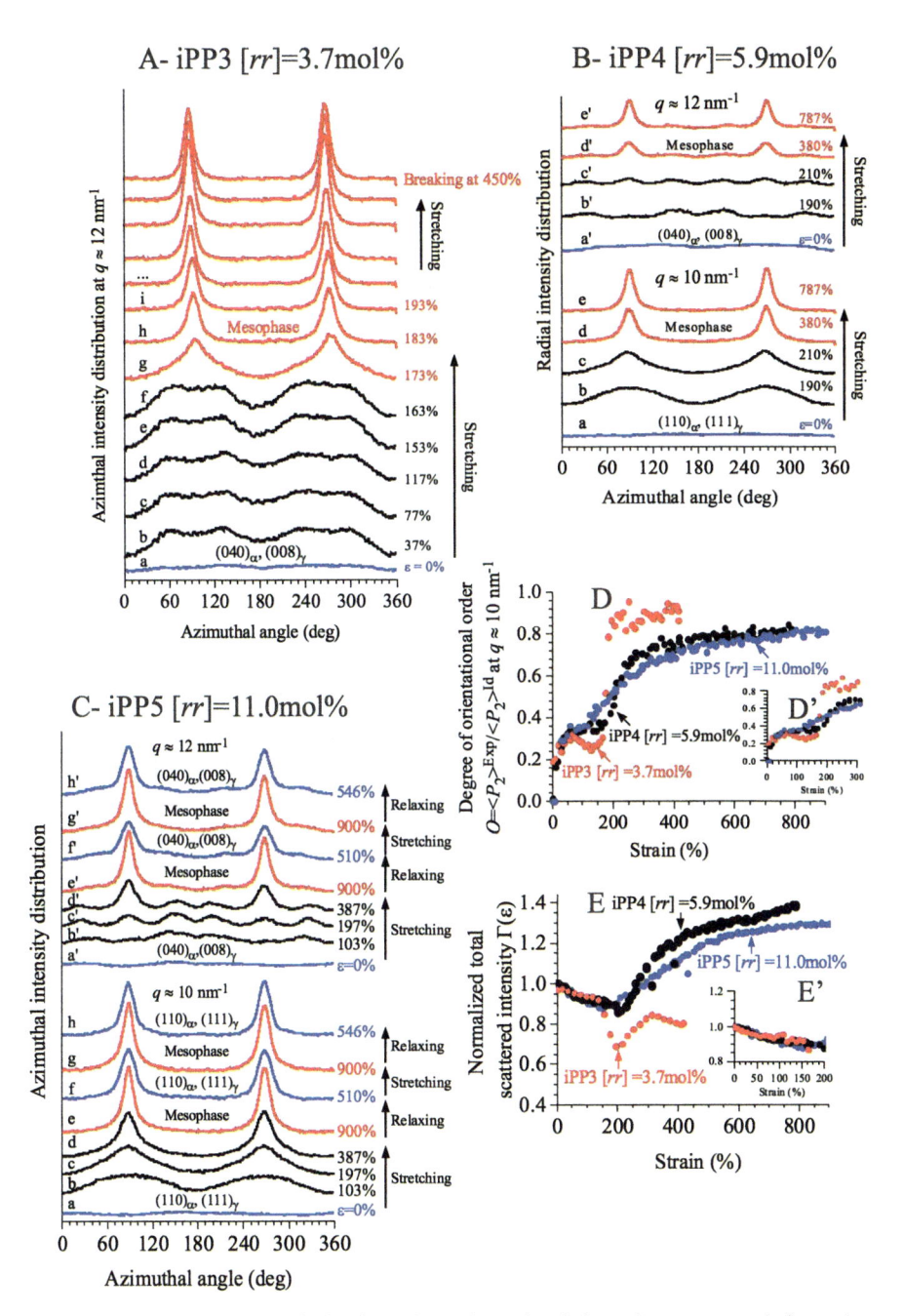

Fig. 11.8 Radial intensity distribution along the azimuthal angle χ extracted from the bi-dimensional X-ray diffraction patterns of Fig. 11.6 relative to the samples iPP3 with $[rr] =$ 3.7 mol% (**a**), iPP4 with $[rr] = 5.9$ mol% (**b**), and iPP5 with $[rr] = 11.0$ mol% (**c**) in the q regions close to 10 (**b**, **c**) and 12 nm^{-1} (**a–c**). The data are obtained by stretching at room temperature compression moulded films at the indicated values of the strain ε. Degree of alignment of chain axes with the stretching direction O (**d**, **d'**) and total scattered intensity normalized for the integrated intensity of the unoriented sample $\Gamma(\varepsilon)$ (**e**, **e'**) as a function of strain. Insets **d'** and **e'** show the parameters O and $\Gamma(\varepsilon)$ at low deformation on an enlarger y-scale. The parameter O is calculated from the radial intensity distribution at $q \approx 10$ nm^{-1} (see the text)

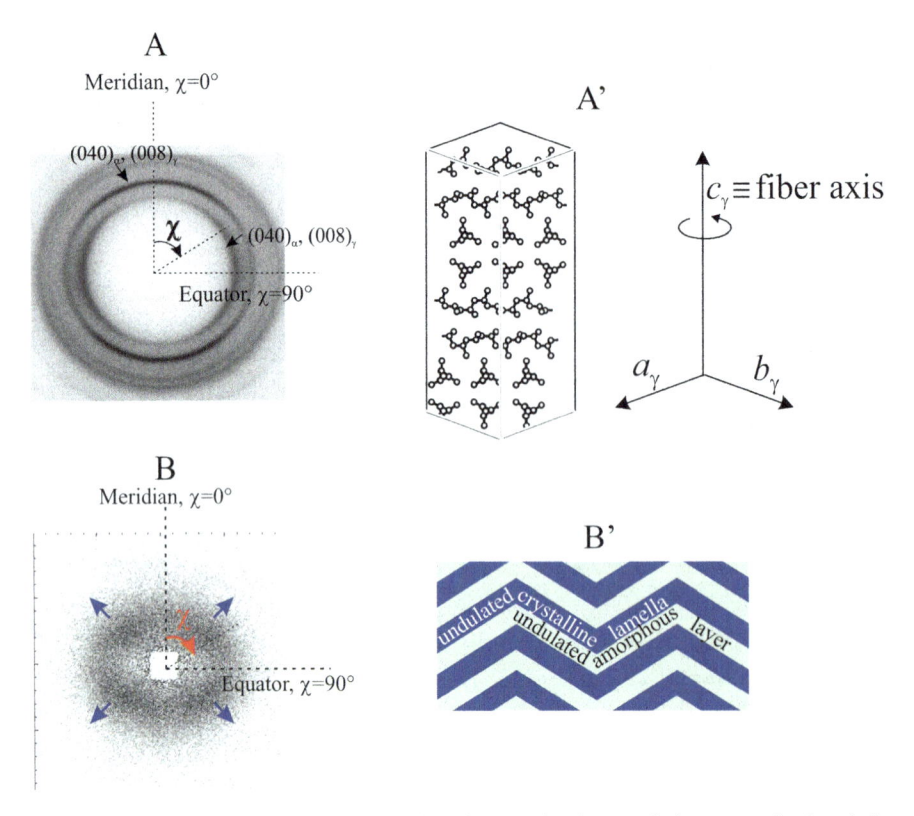

Scheme 11.3 Definition of the azimuthal angle χ and scheme of the perpendicular chain axis orientation (**a**, **a**′) and the chevron-like texture (**b**, **b**′) in the reciprocal (**a**, **b**) and direct space (**a**′, **b**′)

$$\langle P_2(\cos\chi)\rangle = \frac{1}{2}\left(3\langle\cos^2\chi\rangle - 1\right) \qquad (11.1)$$

In Eq. (11.1) $\langle\cos\chi\rangle$ is the average cosine that the poles of a given family of planes make with the preferred (fibre axis) direction and $\langle P_2(\cos\chi)\rangle$ the second order Legendre polynomial of argument $\langle\cos\chi\rangle$. Equation (11.1) is also known as Herman's orientation parameter, where $P_2(\cos\chi) = 1$ corresponds to an ideal case of perfect alignment of the poles toward a preferential direction, $P_2(\cos\chi) = 0$ corresponds to isotropic case and $P_2(\cos\chi) = -0.5$ corresponds to an ideal case of perfect perpendicular orientation. The values of $\langle\cos\chi\rangle$, in turn, are calculated from the azimuthal intensity distribution at $q \approx 10$ nm^{-1} as it follows:

$$\langle\cos^2\chi\rangle = \frac{\int_0^{\pi/2} I(q)\cos^2\chi\sin\chi\,d\chi}{\int_0^{\pi/2} I(q)\sin\chi\,d\chi} \qquad (11.2)$$

Strictly precise determination of the degree orientation of the chain axis in the crystals along the stretching direction should be performed using the azimuthal intensity distribution of a 00l (meridional) reflection or using the azimuthal intensity distribution of three orthogonal reflections (Alexander, 1979). In our case, the chain axis orientation is inferred indirectly, measuring the degree of polarization of the intensity at $q \approx 10$ nm^{-1} on the equator, with respect to the meridional axis. Because of the equatorial polarization of intensity at $q \approx 10$ nm^{-1}, the development of the fibrillar morphology as a function of deformation would be indicated by the decrease of the experimental values of the order parameter $\langle P_2(q)\rangle^{exp}$ from 0 at zero deformation (completely isotropic sample) to values less than zero with increasing the deformation, corresponding to the fibre morphology. Therefore, a measure of the degree of orientation O of the iPP samples as a function of deformation may be obtained by comparing the experimental values of $\langle P_2(q)\rangle^{exp}$ at $q \approx 10$ nm^{-1} with that of an ideal model characterized by an extremely narrow equatorial polarization of intensity at this q (that is $\langle P_2\rangle^{Id} = -0.5$) as:

$$O = \frac{P_2(\cos\chi)^{Exp}}{P_2(\cos\chi)^{Id}} \qquad (11.3)$$

According to Eq. (11.3), the parallel chain axis orientation corresponds to values of $O > 0$. Furthermore, the higher O, the higher the fraction of chain axes in the crystalline aggregates reaching a parallel alignment with the stretching direction during deformation. The values of the parameter O derived for the iPP samples are reported in Fig. 11.8d.

All samples follow a common scheme consisting of three steps. In the first step, for deformations ε lower than 100 %, a steep increment of the degree of orientation occurs, with the parameter O reaching values $O = 0.30$–0.4 (Fig. 11.8d′). Therefore, at low deformations, but higher than yield strain (Fig. 11.4), change in radial intensity distribution along χ of (110)$_\alpha$ of reflection of α form, (111)$_\gamma$ of reflection of γ form at $q \approx 10$ nm^{-1} essentially probes small reorientation of lamellar crystals along the pathway toward the fibrillar morphology. Reorientation of crystals at low deformations is generally due to lamellar or lamellar stacks rotations, and it is associated with interlamellar shear movements (Bowden and Young 1974; Lin and Argon 1994; Humbert et al. 2010; Mourglia-Seignobos et al. 2014).

In the second step, corresponding to the deformation range between 100 and ≈ 150–200 % (Fig. 11.8d, d′), the orientation parameter O reaches a *quasi plateau*. We argue that the leading deformation mechanisms in this region correspond to interlamellar shear that is slip of the crystalline lamellae parallel to each other, and possibly interlamellar separation, both phenomena being assisted above glass transition temperature, by the shear deformation of the compliant interlamellar amorphous phase (Bowden and Young 1974; Schultz 1974; Lin and Argon 1994; Séguéla 2002).

In the third step, starting from deformations close to the critical strain at which stress-induced phase transitions start occurring ($\varepsilon \approx 200$ %, Figs. 11.6 and 11.7), the orientation order O experiences a new increase up to reach a final plateau value at high deformations of $O \approx 0.9$ for the sample iPP3 with $[rr] = 3.7$ mol%, $O \approx 0.8$ for the less stereoregular samples iPP4 and iPP5 (Fig. 11.8d). In this high deformation range, the increase of deformation induces collective shear processes, up to reach the destruction of the pre-existing crystals (mechanical melting) and successive recrystallization with formation of fibrils (Schultz 1974; Lin and Argon 1994). Further increase of plastic strain in the successive steps entails disentanglement of the amorphous network or strain hardening due to the stretching of the amorphous entangled network at high deformations (Schultz 1974; Peterlin 1971; Men et al. 2003).

The analysis of the radial intensity distribution along the azimuthal angle χ in the q region around 12 nm^{-1} of Fig. 11.8a–c reveals that during the development of the fibrillar morphology, for deformations lower than the critical value at which transformation of γ form into mesophase is complete, there is a portion of the crystalline lamellae which tend to orient with the chain axes nearly perpendicular to the stretching direction instead than parallel, contrary to what expected in the standard fibre morphology.

This non-standard crystals orientation is observed in the case of the sample iPP3 up to 170 % deformation (curves b–f of Fig. 11.8a), the sample iPP4 up to 300 % deformation (curves b'–d' of Fig. 11.8b) and the sample iPP5 up to 500 % deformation (curves b'–d' of Fig. 11.8c) as indicated by the polarization of the $(040)_\alpha$ reflection of the α form, $(008)_\gamma$ reflection of the γ form at $d = 5.21$ Å ($q \approx 12$ nm^{-1}), at oblique angles, and as indicated also in Fig. 11.6b, f, g, m, n by arrows. At higher deformations, the diffraction maxima at oblique angles disappear or become very low (Fig. 11.8b, c), and the radial intensity distribution at $q \approx 12$ nm^{-1} becomes peaked on the equator, according to the standard fibre morphology (Fig. 11.8a curves g, i, up to breaking, Fig. 11.8b curves d', e', Fig. 11.8c curves d'–h').

The nearly meridional polarization of the reflection at $q \approx 12$ nm^{-1} (reflections $(040)_\alpha$ of α form, or $(008)_\gamma$ of γ form) in the patterns of Fig. 11.6 and azimuthal profiles of Fig. 11.8a–c indicate that portion of the crystals of γ form, or in disordered modifications intermediate between the γ and α forms more similar to the γ form, assume an orientation with the c_γ-axis of γ form (b_α-axis of α form) nearly perpendicular to the stretching direction (Fig. 11.2b, c) (De Rosa et al. 2005; De Rosa and Auriemma 2007; Auriemma and De Rosa 2006). Since the c_γ-axis of γ form (b_α-axis of α form) are the axes of stacking of the layers of chains (Fig. 11.2b, c) this non-standard mode of orientation of iPP crystals corresponds to lamellae oriented with chain axes nearly perpendicular to the fibre axis (Fig. 11.2b, c, and Scheme 11.3a').

A similar kind of orientation has been well known for many years in some naturally occurring fibrous proteins such as silks (Geddes et al. 1968). The perpendicular orientation of chain axes with respect to the fibre axis, described as cross-β, occurs at low draw ratio and has been explained by the fact that in these soft silks the small crystallites are elongated along the hydrogen bond directions, which run perpendicular to chain axes (Geddes et al. 1968). The cross-β orientation in iPP may be attributed to the simultaneous occurrence of two kinds of slip processes at low deformations, interlamellar and intralamellar (Men and Strobl 2001). Whereas interlamellar shear leads to a location of the $(008)_\gamma$ reflection of γ form $((040)_\alpha$ reflection of α form) on the meridian, the intralamellar shear pushes the chain axes to align parallel with the stretching direction, and thus shifts the position of the reflection toward the equator (De Rosa et al. 2005; De Rosa and Auriemma 2007).

The fact that the cross-β orientation is apparent up to high deformations indicates that not all the crystalline lamellae of γ form originally present in the sample experience simultaneously the uniaxial mechanical stress field. The non-standard mode of orientation of these crystals reflects crystallographic restraints on the slip processes, and topological constrains on the response of crystals to the tensile stress field. In the crystalline domains of γ form, indeed, the chains are oriented along two perpendicular directions, and the crystallites have the shape of elongated entities along the direction normal to the chain axes (De Rosa et al. 2005; Auriemma and De Rosa 2006). Because of the intrinsic structural and morphological characteristics of γ form, at low deformations portion of γ lamellae remains frozen in strained positions of the polymer matrix with the chain axes oriented nearly perpendicular to the stretching direction. By stretching at higher deformations the γ form transforms into the mesophase. Since in the crystals of the mesophase the chains are all parallel, the deformation also induces orientation of crystals with the chain axes oriented along the stretching direction, as in a standard fibre morphology (De Rosa et al. 2005; Auriemma and De Rosa 2006). This mechanism is confirmed by the fact that this non-standard mode of orientation of crystals has been observed only during stretching of iPP samples containing crystals of γ form and has never been observed for iPP samples crystallized in the pure α form.

For samples exhibiting uniaxial symmetry (fibres), the WAXS intensity collected on a bi-dimensional detector (Fig. 11.6) can be integrated over the whole sampled reciprocal space, that is between q_{min} (=3 nm^{-1}) and q_{max} (20 nm^{-1}), obtaining the total scattered intensity in the WAXS region Ω_{waxs}, using (11.4):

$$\Omega_{WAXS}(\varepsilon) = \int_0^{2\pi} \int_{q_{min}}^{q_{max}} I(q,\chi) q^2 \sin\chi \, d\chi \, dq \tag{11.4}$$

In Eq. (11.4), it is implicitly assumed a uniform distribution of intensity along circles of radius $q \sin \chi$ because of the uniaxial symmetry of the fibres. The ratio of the total scattered intensity evaluated at deformation ε and the integrated intensity of the unoriented sample gives the normalized value of the total scattered intensity $\Gamma(\varepsilon) = \Omega_{waxs}(\varepsilon)/\Omega_{waxs}(\varepsilon=0)$. The values of the parameter $\Gamma(\varepsilon)$ of iPP samples are shown in Fig. 11.8e, e' as a function of deformation.

Also in this case the samples follow a common scheme regardless of stereoregularity. In particular, up to ≈200 % deformation the values of $\Gamma(\varepsilon)$ are coincident and decrease with increase of deformation (Fig. 11.8e'), whereas for deformations higher than 200 %, the values of the parameter $\Gamma(\varepsilon)$ increase by 30–40 %, up to reach a quasi-plateau (Fig. 11.8e). Changes occur by effect of deformation because of decrease in thickness, consequent change of absorption factor, change in the relative amount of the phases, stress-induced phase transitions and/or crystallization, appearance of additional phases and consequent change in contrast. Additional effects can also arise from decrease (or increase) of the size of coherent crystalline domains embedded in the amorphous phase, due to lamellar fragmentation (coalescence) of the domains. Since the total scattering intensity is proportional to the squared volume of incoherently scattering objects V_{obj}^2, a decrease (increase) in volume of the scattering objects V_{obj} leads to a decrease (increase) in scattering intensity, even if the total degree of crystallinity, proportional to $N_{obj} V_{obj}$ with N_{obj} the number of scattering objects contributing additively to intensity, remains constant (Alexander 1979; Glatter and Kratky 1982).

In general, the transverse strain of semi-crystalline polymers (perpendicular to the stretching direction under uniaxial elongation) decreases with deformation. For a uniform deformation, the thickness of the sample decreases. For rubbery materials, at low deformations, the thickness t decreases as $t = t_0 (l_0/l)^\nu$, where l_0 and l are the initial gauge length and the gauge length in deformed state, and ν is he Poisson's ratio equal to 0.4–0.5. However, for semi-crystalline polymers the Poisson's ratio changes during deformation, from the maximum value of 0.5 up to drop to values close to zero, by effect of volume expansion due to crazing, cracks and voids (Nitta and Yamana 2012). In the lack of direct measurements of the thickness values during deformation, data in Fig. 11.8e, e′ have not been corrected by this factor. On the other hand, concerning the absorption correction, this factor would not greatly influence the values of $\Gamma(\varepsilon)$, and, at least to a first approximation may be neglected. In fact the linear absorption coefficient of iPP for X-ray wavelength $\lambda = 0.14$ nm is ≈ 3 cm^2/g (Maslen 2006) and the absorption coefficient at 900 % deformation ($l_0/l = 0.1$) would decrease by a factor less than 1.1 assuming a Poisson's ratio of 0.5.

Since in our samples the degree of crystallinity does not greatly change during deformation (De Rosa et al. 2005; De Rosa and Auriemma 2007) the decrease in the normalized total scattering intensity value $\Gamma(\varepsilon)$ of Fig. 11.8e, e′ at deformation less than 200 % can be attributed to a decrease in thickness coupled to a lamellar fragmentation. The increase of the $\Gamma(\varepsilon)$ parameter at deformations higher than 200 %, instead, may be ascribed to structural changes, including appearance of additional phases due to the gradual transformation of γ form, nucleation of voids (due to cavitation, vide infra), small increase in the degree of crystallinity, but also to coalescence of small crystalline mesomorphic aggregates embedded in the mobile amorphous phase into larger aggregates.

11.7 Structural Analysis at Lamellar Length Scale

The structural transformations occurring at lamellar length scale during stretching have been followed by performing time-resolved SAXS measurements at the SRS synchrotron light source at Daresbury, UK. Rectangular specimens of initial gauge length $l_0 = 3.0$ mm, initial width $w_0 = 1.6$ mm and initial thickness $t_0 = 0.25$ mm, prepared by compression moulding, are stretched at controlled deformation rate of 2.36 mm/min (1 frame/20 s, beam diameter 0.1 mm). Scattering data were collected

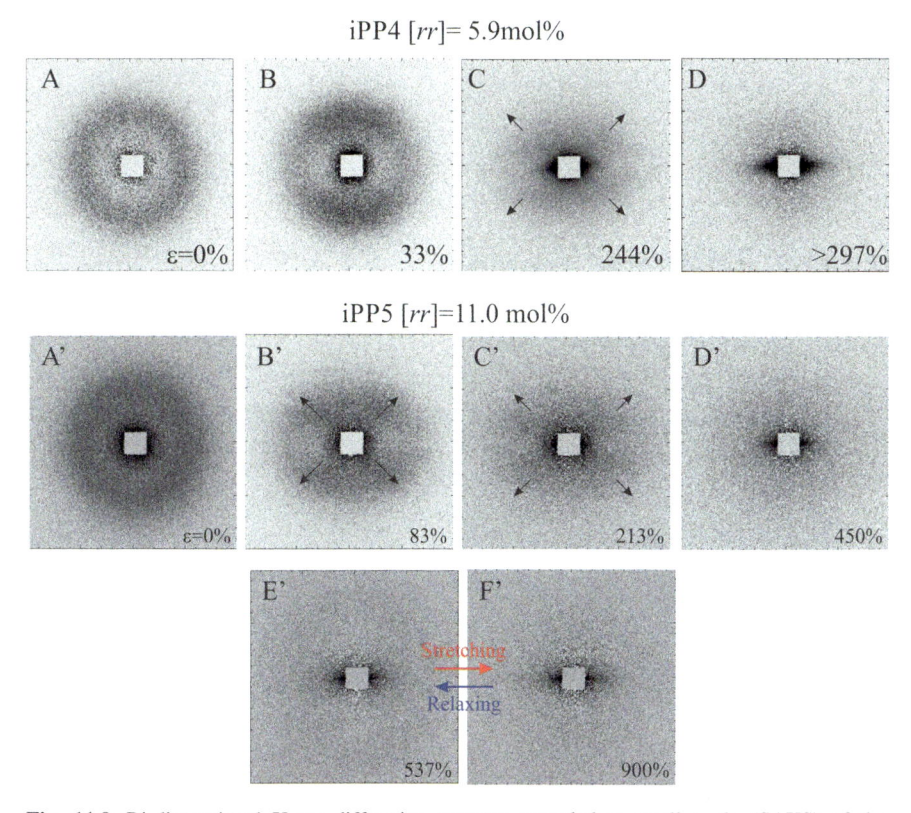

Fig. 11.9 Bi-dimensional X-ray diffraction patterns recorded at small angle (SAXS) of the samples iPP4 with [rr] = 5.9 mol% (**a–d**) and iPP5 with [rr] = 11.0 mol% (**a′–f′**). Data are collected in situ, at the beam-line 16.1 of the Synchrotron Radiation Source in Daresbury (Cheshire, UK), 2.36 mm/min deformation rate, incident wavelength $\lambda = 1.4$ Å, 20 s/frame. The stretching direction is vertical. *Arrows* in (**c**) and (**b′**) and (**c′**) indicate the polarization of intensity off the equator and meridian due to the chevron-like texture (Scheme 11.3b′)

using the Rapid Area Detector system in the q range 0.1–1 nm^{-1}. The scattering geometry was calibrated using a wet collagen fibre.

A selected sequence of the bi-dimensional SAXS patterns of the samples iPP4 and iPP5 ([rr] = 5.9 and 11.0 mol%, respectively) taken during tensile stretching are shown in Fig. 11.9, whereas the corresponding radial distribution along the azimuthal angle χ (Scheme 11.3b) is reported in Fig. 11.10a, b. Both samples show a uniform distribution of intensity along rings centred in the q region 0.3–1 nm^{-1} in the undeformed state (Fig. 11.9a, a′ and curves a of Fig. 11.10a, b).

With increasing deformation the intensity distribution becomes anisotropic, marking the tendency of the lamellae to become oriented with their normal parallel to stretching direction as in fibre morphology.

At deformations lower than ≈300–400 %, the SAXS patterns tend to assume an elliptical geometry characterized by a concentration of intensity above and below the equatorial region (Fig. 11.9b, c, b′, c′). The corresponding azimuthal intensity profiles show broad maxima centred on the meridian ($\chi = 0°$ and 180°) for the

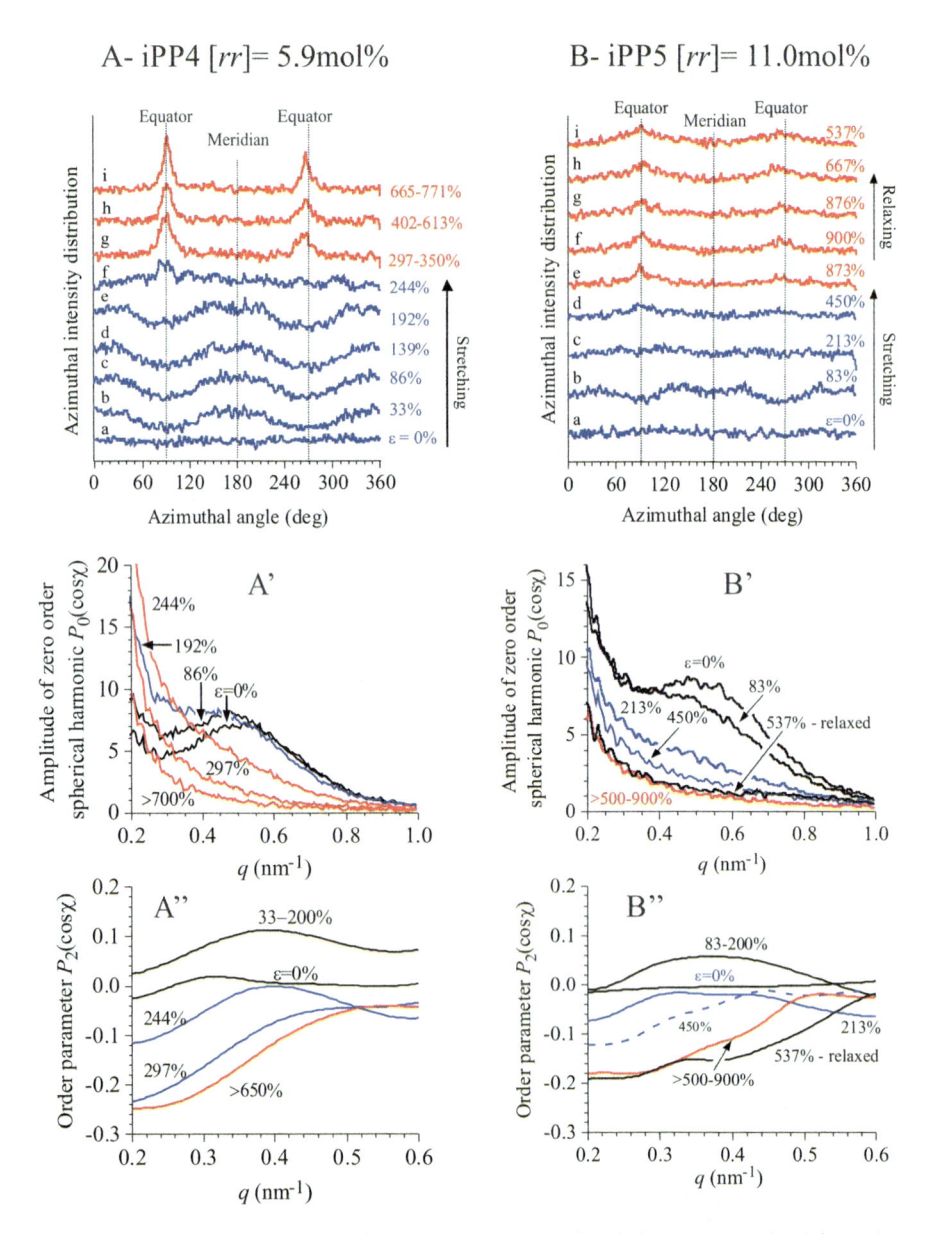

Fig. 11.10 (a, b) Radial intensity distribution along the azimuthal angle χ extracted from the bi-dimensional SAXS patterns of Fig. 11.9 relative to the samples, iPP4 with $[rr] = 5.9$ mol% (**a**), and iPP5 with $[rr] = 11.0$ mol% (**b**). The profiles have been obtained by integrating the intensity in spherical shells of q with q comprised between 0.2 and 0.8 nm^{-1}. (**a′, a″, b′, b″**) Amplitude of zero order spherical harmonics $I_0(q)$ (11.6) (**a′, b′**) and order parameter $P_2(\cos \chi)_q$ (11.7) (**a″, b″**) obtained through interpolation of the bi-dimensional SAXS intensity of Fig. 11.9 with the series expansion using spherical harmonics $P_{2n}(\cos \chi)$ as basis set according to Eq. (11.5)

sample iPP4 at deformations lower than 100 % (Fig. 11.9b and curves b, c of Fig. 11.10a) or four weak maxima displaced by 20°–30° from meridian ($\chi \approx 30°$, 150°, 210° and 330°) at deformations higher than 100 % for the sample iPP4 and in the deformation range 80–220 % for the sample iPP5 (Fig. 11.9c, b′, c′, curves d, e of Fig. 11.10a and curves b, c of Fig. 11.10b). The four maxima off the equator and meridian (evidenced by arrows in Fig. 11.9c, b′, c′) correspond to the buckling of the lamellar stacks according to a chevron-like (undulated) texture as depicted in the Scheme 11.3b′. These undulated textures take place in a large number of semi-crystalline polymers such as polyethylene at low and intermediated deformations (before fibrillation) (Gerasimov et al. 1974; Read et al. 1999; Krumova et al. 2006). According to a general mechanism, lamellar stacks subjected to tensile deformation in the direction normal to their layers experience also compressive stress in the direction transverse to the tensile deformation axis. The crystalline (hard) layers sustain this transverse compression until buckling occurs with consequent formation of undulated structures (Scheme 11.3b′) (Makke et al. 2012).

At deformation higher than the chevron texture (200 % for iPP4, 400 % for iPP5), the four-lobe pattern defined by the intensity distribution is replaced by a diamond-shaped diffuse scattering (Fig. 11.9d, d′, e′). Diffuse scattering appears as streaks located on the equator (Fig. 11.9d, d′, e′), and marks formation of nanocavities in the interlamellar regions during deformation. With increasing deformation, the microcavities undergo elongation in the stretching direction, and the bi-dimensional SAXS patterns become entirely dominated by the equatorial streaks due to cavitation (Pawlak and Galeski 2005). This is evidenced by the azimuthal intensity profiles in Fig. 11.10a, b showing maxima on the equator ($\chi = 90°$ and 270°) superimposed to a flat background (curves f–i of Fig. 11.10a, d, e of Fig. 11.10b). In these curves no meridional maxima are apparent up to very high deformations.

Therefore, the SAXS data evidence that with increasing the deformation the polarization of intensity shifts from positions in between the meridian and the equator, toward the equator, and no meridional maxima become apparent, as instead expected for a well oriented fibre morphology (Figs. 11.9a–d, a′–e′ and 11.10a, b). The absence of meridional maxima in the SAXS intensity distribution at high deformations is in apparent contrast with WAXS results of Figs. 11.6 and 11.7. In fact, WAXS analysis indicate almost complete transformation of the initial γ form into well oriented mesomorphic aggregates already at deformations of $\approx$200 % for the sample iPP4 and 400 % for the sample iPP5. The absence of meridional maxima in the SAXS patterns of the samples iPP4 and iPP5 stretched at high deformations (Fig. 11.9d, d′, e′) is due to the too low contrast in electron density between the mesophase and amorphous components and cavitation. Upon cavitation, indeed, nanocavities act as a third phase with zero electron density, and the scattering intensity becomes entirely dominated, in the small angle, by the large contrast between nanocavities and the surrounding phases.

The damage effect due to cavitation in the elastomeric sample iPP5 stretched at 900 % deformation (Fig. 11.9e and profile f of Fig. 11.10b) is irreversible. In fact, in mechanical cycles of stretching and relaxing the tension no significant changes in the SAXS intensity distribution occur (Fig. 11.9f and profile g–i of Fig. 11.10b). This indicates that even though reversible structural changes occur between the mesophase and α form (see Sect. 3.2), damages due to cavitation are not recovered, and the SAXS intensity distribution keeps dominated by the large contrast between the nanocavities and the surrounding phases (amorphous phase, α or γ form and mesophase).

The SAXS intensity of Fig. 11.9 has been analysed in the framework of series expansion of uniaxially symmetric samples in terms of spherical harmonics $P_{2n}(\cos \chi)$ (Lovell and Mitchell 1981; Mitchell and Windle 1982) as:

$$I(q,\chi) = \sum_{2n=0,2,4,\ldots}^{\infty} I_{2n}(q) P_{2n}(\cos \chi) \qquad (11.5)$$

In Eq. (11.5), only the even terms are required due to the inversion centre intrinsic to X-ray scattering patterns for a non-absorbing sample of uniaxial symmetry. The amplitude of spherical harmonics components in the series expansion of Eq. (11.5) can be obtained by (Lovell and Mitchell 1981; Mitchell and Windle 1982):

$$I_{2n}(q) = (4n+1) \int_0^{\pi/2} I(q,\chi) P_{2n}(\cos \chi) \sin \chi \, d\chi \qquad (11.6)$$

where $P_0(\cos \chi) = 1$, $P_2(\cos \chi) = 1/2(3\cos^2 \chi - 1)$, $P_4(\cos \chi) = 1/8(35 \cos^4 \chi - 30 \cos^2 \chi + 3)$, etc. In practice, the value of the intensity representation with Eq. (11.5) is that the effects of preferred orientation are separated from the dependence of the scattering on the spatial correlations. In particular, the amplitude of the zero order term $I_0(q)$ gives for each q the integrated intensity scattered from an equivalent sample with complete random orientation. The amplitude of the second order spherical harmonics $I_2(q)$ is also relevant. In fact the ratio between $I_2(q)$ and $I_0(q)$ (after normalization by the coefficient $4n+1$) corresponds to the order parameter $P_2(\cos \chi)_q$ as a function of q (11.6):

$$P_2(\cos \chi)_q = \frac{I_2(q)}{5 I_0(q)} \qquad (11.7)$$

At any given q, values of $P_2(\cos \chi)_q$ close to 1 correspond to intensity distribution polarized on the meridian at that q, close to 0 correspond to isotropic distribution of intensity, less than 0 correspond to polarization on the equator.

We have used the amplitude of the zero order spherical harmonics $I_0(q)$ in Eq. (11.6) and the value of $P_2(\cos \chi)_q$ in Eq. (11.7) as descriptors for the structural and textural changes occurring in our samples by effect of deformation. Selected curves are reported in Fig. 11.10a', b', a'', b''.

The isotropic intensity component $I_0(q)$ of undeformed samples ($\varepsilon = 0\%$) presents a correlation peak at $q_{max\,x} \approx 0.5$ nm^{-1} (Fig. 11.10a', b') and isotropic distribution of intensity ($P_2(\cos \chi)_q \approx 0$) (Fig. 11.10a'', b''). The correlation peak is due to the lamellar stacking and corresponds to values of long period $L \approx 12.6$ nm.

Upon stretching, for deformation lower than 203 % the correlation peak of the $I_0(q)$ curves shifts to low q values ($q_{max} \approx 0.45$ nm^{-1}) indicating an increase of the long period to $L \approx 14$ nm (Fig. 11.10a', b') due to interlamellar separation (Bowden and Young 1974; Peterlin 1971; Lin and Argon 1994). The corresponding values of the order parameter are positive and achieve values close to $P_2(\cos \chi)_q \approx 0.1$ for the sample iPP4 and ≈ 0.06 for the sample iPP5 (Fig. 11.10a'', b'') in agreement with the four-lobe patterns observed in the bi-dimensional SAXS patterns (Fig. 11.9b, c, b', c') and the broad distribution of the SAXS intensity profile along the azimuthal angle χ with maxima centred around the meridian (curves b–e of Fig. 11.10a–c of Fig. 11.10b). As discussed before, the four-lobe patterns originate from the formation of the chevron-like undulated structures (Scheme 11.3b') and are formed by lamellar stacks oriented with their normal parallel to the stretching direction, experiencing compressive forces in the transverse direction. Therefore the present analysis evidences that in the deformation range 0–200 % undulation and interlamellar separation take place simultaneously since they possibly involve the same population of lamellae and/or lamellar stacks, oriented with their layers normal parallel to the stretching direction (Bowden and Young 1974).

At deformation higher than 200 % the isotropic intensity component $I_0(q)$ becomes featureless. No correlation peaks are apparent, but only a gradual decay of intensity from low to high q values occurs (Fig. 11.10a', b'). The values of the order parameter $P_2(\cos \chi)_q$ become negative at all q. In particular the values of $P_2(\cos \chi)_q$ are ≈ -0.2 to -0.25 at $q = 0.2$ nm^{-1} and increase only slightly at higher q. This indicates that the polarization is localized on the equator, in agreement with the diamond-like pattern in the bi-dimensional SAXS patterns (Fig. 11.9d, d'–e') due to cavitation. This is also in agreement with the distribution of SAXS intensity along the azimuthal angle χ presenting maxima on the equator and no maxima in the meridional regions (curves f–i of Fig. 11.10a, b). The values of $I_0(q)$ amplitude at these deformations decays according to a power law dependence of q, that is, according to q^{-D} with $D \approx 2$, indicating an ellipsoidal shape of the nanocavities along the stretching direction. Therefore, the analysis in terms of spherical harmonics $P_{2n}(\cos \chi)$ according to (11.5) confirms the formation of nanocavities at deformations higher than a threshold and provides a quantitative insight of the structural and textural evolutions occurring during the stretching of our materials.

In the case of the elastomeric sample iPP5 the values of the scattering $I_0(q)$ and order parameter of $P_2(\cos \chi)$ of the stress-relaxed specimens are also reported in Fig. 11.10a', b', a'', b''. They are similar to those of the fully stretched fibre at 900 % deformation, indicating that the damage effect caused by cavitation is irreversible.

11.8 Determination of Damage Effect

Quantitative information about damage caused by cavitation may be extracted from the SAXS data collected during stretching through calculation of the invariant (Alexander 1979):

$$Q_{SAXS}(\varepsilon) = \int_0^{q_{max}} \int_0^{2\pi} I(q,\chi) q^2 \sin\chi \, d\chi \, dq = \int_0^{q_{max}} I_0(q) q^2 \, dq \quad (11.8)$$

The outer integral in (11.8) is performed after subtraction for the background to $I_0(q)$, and should be classically extended to the whole reciprocal space and not limited to the narrow q range sampled with our experimental set-up, that is $q_{min}(=0.2 \text{ nm}^{-1}) < q < q_{max}(=1 \text{ nm}^{-1})$. However, in our case, since the integrating function is already very low in the sampled region of q close to q_{min} and q_{max}, classic extrapolation methods to extend the inferior and superior integration limits were not applied.

From (11.8), the normalized scattering invariant has been calculated as the ratio between the invariant at a deformation ε to the value of the invariant in the undeformed state:

$$R = \frac{Q_{SAXS}(\varepsilon)}{Q_{SAXS}(\varepsilon = 0)} \quad (11.9)$$

The normalized values of the invariant R relative to the samples iPP4 and iPP5 are reported in Fig. 11.11 as a function of deformation. Curves a and b are not corrected for the decrease in thickness of the specimen during stretching. Curves a' and b', instead, have been corrected for this effect, in the limiting hypothesis that the thickness t decreases according to the power law $t = t_0 \, (l_0/l)^\nu$, with $\nu = 0.5$, in the whole deformation range, as expected for an ideal rubbery materials (Nitta and Yamana 2012).

It is apparent that in the case of the sample iPP4 the normalized scattering invariant (curve a of Fig. 11.11) exhibits a sigmoidal shape. The parameter R decreases slightly from 0 to $\approx$200 % deformation, at $\approx$200 % deformation it decays steeply by $\approx$80 %, and then, in the deformation range from $\approx$400 % up to the breaking, it reaches a new quasi-plateau corresponding to $R \approx 0.20$. In the case of the sample iPP5, instead, (curve b of Fig. 11.11) the decrease of the normalized invariant follows a less pronounced sigmoidal shape. It starts decreasing immediately upon deformation. The rate of decrease becomes more pronounced in the deformation range 200–500 %. Successively, the decrease of the R parameters becomes smooth again and reaches a quasi-plateau in the deformation range 800–900 %. The total decrease of the invariant for the sample iPP5 is similar ($\approx$80 %) to that of the sample iPP4 (final values of $R \approx 0.20$). Considering the effect of thickness contraction according to the rubbery model, instead, (curves a', b' of Fig. 11.11), the invariant would change according to a similar trend,

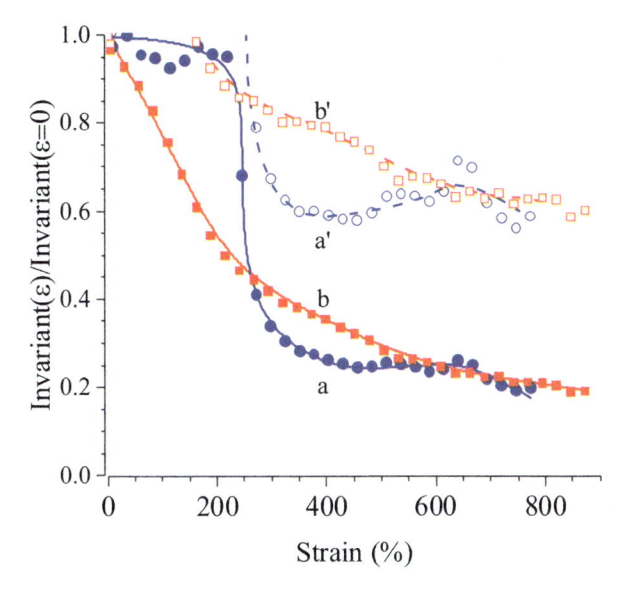

Fig. 11.11 Small angle scattering invariant of the samples iPP4 with $[rr] = 5.9$ mol% (**a**, **a'**) and iPP5 with $[rr] = 11.0$ mol% (**b**, **b'**) normalized for the invariant of the unoriented sample ($\varepsilon = 0$) as a function of strain (R, 11.9). Curves **a'** and **b'** are corrected for the decrease in thickness t by assuming that t decreases according to the power law $t = t_0 \, (l_0/l)^\nu$, with t_0 the thickness of the undeformed specimen and $\nu = 0.5$ as expected for an ideal rubbery material. The correction has been applied for deformations $> 200\%$

corresponding to a total decrease of $\approx 40\%$, and leading to a final value of the R parameter close to 0.6 at high deformations.

The rate of decrease in the invariant changes at deformations of ≈ 200 and $\approx 400\%$ for the sample iPP4 and ≈ 200 and $>500\%$ for the sample iPP5 and these deformations mark the critical values of strain at which the initial γ form starts and ends almost completely to transform into the mesophase (Figs. 11.6, 11.7, and 11.8). In addition (Figs. 11.9 and 11.10) at 200% deformation, also cavitation starts occurring.

This indicates that the change of the invariant with deformation is not due to the mere effect of thickness contraction of the specimens. In fact, cavitation and transformation into mesophase also influence the invariant. In general, the invariant depends on the total sampled volume, the number and relative amount of the phases present in the sample, and on the contrast between the electron density of the phases.

More precisely, the scattering invariant Q_{bin} for an ideal binary phase system without nano-voids is expressed by Eq. (11.10) (Glatter and Kratky 1982):

$$Q_{bin} = k2\pi^2 V \varphi_1 \varphi_2 (\rho_1 - \rho_2)^2 \qquad (11.10)$$

In Eq. (11.10), k is a scaling parameter, V denotes the total sampled volume in the scattering experiment and φ_i and ρ_i denote the volume fraction and the electron density, respectively, of the phase i. The difference in electron density of the two phases define the contrast, the higher the contrast, the higher the invariant. The scaling parameter accounts for the conversion of the intensity to absolute units.

The density of γ form, mesophase and amorphous phase in isotactic polypropylene are 0.939, 0.91 and 0.854 g/cm^3 at room temperature respectively (Brandrup and Immergut 1989), and the corresponding values of electron density are 323, 313 and 294 electrons/nm^3, respectively. Therefore, in a hypothetical two phase model, with no change in the relative amount of the phases and constant thickness of the specimen, the value of invariant by effect of complete transformation of the initial γ form into the mesophase is expected to decrease, because of decrease in contrast. More precisely, using Eq. (11.10), regardless of the relative amount of the two phases, the invariant would decrease by ≈60%, corresponding to a drop in the parameter R from 1 to ≈0.4. This indicates that the transformation of γ form into the mesophase with no change in the degree of crystallinity and no cavitation leads always to a decrease of invariant.

However, in the stretching experiments, we have to consider also cavitation. Cavitation creates an additional phase, characterized by zero electron density. Therefore, in the deformation range between 200% and 400–500% our systems consist of four phases, γ form, mesophase, amorphous component and nanocavities. At strain higher than 400–500% the γ form is almost completely transformed into mesophase, and only three phases (mesophase, amorphous component and nanocavities) should be considered. The scattering invariant for this three-phase system (that is, for deformations ε above the threshold limit at which transformation of γ form is almost complete) Q_ε can be written as (Glatter and Kratky 1982):

$$Q_\varepsilon = k2\pi^2 V_\varepsilon \left[\varphi_{1\varepsilon}\varphi_{2\varepsilon}(\rho_{1\varepsilon} - \rho_{2\varepsilon})^2 + \varphi_{1\varepsilon}\varphi_{\text{voids}}\rho_{1\varepsilon}^2 + \varphi_{2\varepsilon}\varphi_{\text{voids}}\rho_{2\varepsilon}^2 \right] \qquad (11.11)$$

In Eq. (11.11), $\varphi_{i\varepsilon}$ and φ_{voids} denote the volume fractions of the phases and $\rho_{i\varepsilon}$ the corresponding electron densities, whereas V_ε denotes the total sampled volume in the scattering experiment at deformation ε. This volume will change principally by effect of the thickness contraction on stretching and also through phase transitions, cavitation, etc. Therefore, Eq. (11.11) accounts for the change in the scattering invariant because of all effects, volume change, phase transitions and cavitation and it may be utilized to calculate the theoretical invariant at deformation higher than 400–500% that is for a three-phase system. Equation (11.10) instead, accounts for the presence of only two phases and may be utilized to calculate the invariant at zero deformation that is for a two phase system. Equation (11.11) predicts that the main effect of cavitation results in an increase of the invariant already for low values of the volume fraction of voids φ_{voids} because of the increase in contrast.

We have utilized (11.9) and (11.11) to evaluate the volume fraction of voids at high deformations. Since the relative amount of amorphous and crystalline (γ form or mesophase) components does not change by effect of stretching in our samples (De Rosa et al. 2005, 2006; De Rosa and Auriemma 2007), we have assumed that at the maximum deformation the volume fractions of the mesophase and amorphous components ($\varphi_{1\varepsilon}$ and $\varphi_{2\varepsilon}$, respectively) are approximately related to the volume fraction of the initial γ form and corresponding amorphous component (φ_γ and φ_a, respectively) at zero deformation, by (11.12):

$$\varphi_{1\varepsilon} \approx \varphi_\gamma(1 - \varphi_{\text{voids}})$$
$$\varphi_{2\varepsilon} \approx \varphi_a(1 - \varphi_{\text{voids}})$$
(11.12)

Substituting (11.12) into (11.11), and indicating with ρ_m, ρ_γ and ρ_a the electron density of the mesophase, γ form and amorphous component, the theoretical value of the normalized invariant R_ε at deformation higher than a critical value results:

$$R_\varepsilon = \frac{Q_\varepsilon}{Q(\varepsilon = 0)}$$
$$= (1 - \varphi_{\text{voids}}) \frac{V_\varepsilon \left[\varphi_\gamma \varphi_a (1 - \varphi_{\text{voids}})(\rho_m - \rho_a)^2 + \varphi_\gamma \varphi_{\text{voids}} \rho_m^2 + \varphi_a \varphi_{\text{voids}} \rho_a^2 \right]}{V \varphi_\gamma \varphi_a (\rho_\gamma - \rho_a)^2}$$
(11.13)

In practice, considering that the experimental value of the normalized invariant (11.9) at high deformation reaches a quasi-plateau value R of ≈0.2 with no correction (curves a' and b' of Fig. 11.11) and ≈0.6 after correction for the reduction in thickness (curves a' and b' of Fig. 11.11), by setting φ_γ (φ_a) equal to 0.4–0.6 (0.6–0.4) (see Table 11.1) and $V_\varepsilon = V(l/l_0)^{-0.5}$ (as in rubbery materials) we have that in the deformation range 600–900 %,the calculated percentage of voids would reach values close to 1 % of the total volume, for values of R in the range 0.2–0.6.

The low value of the volume fraction for the nanocavities is in agreement with the low intensity of the diamond-like SAXS patterns of these samples (Fig. 11.9c, d, c'–e') compared with the high intensity in the corresponding WAXS patterns of γ- and mesomorphic forms (Fig. 11.6g–i, m–p). In fact, although the presence of voids enhances the contrast with the surrounding environment (11.11), the distribution of SAXS intensity in the bi-dimensional patterns is concentrated in a narrow region of reciprocal space close to the origin and produces only strong diffuse scattering (streaks) on the equator, and no additional significant enhancement. On the other hand, the presence of 1 % of voids does not significantly alter the average electron density and the self-correlation function of fluctuation in electron density at length scale of atomic distances of the environment surrounding the nanocavities. Therefore the WAXS patterns are not affected.

Cavitation corresponds to formation of microvoids in the amorphous layers confined between crystalline lamellae. According to a general view (Pawlak and Galeski 2005), during stretching of semi-crystalline polymers there is a competition between cavitation and activation of crystal plasticity: easiest phenomena occur first, that is cavitation in polymers with crystals of higher plastic resistance, and plastic deformation of crystals in polymers with crystals of lower plastic resistance. The stress level at yield point corresponds to the onset of cavitation and not necessarily to the onset of plastic deformation of crystals. Therefore, cavitation is negligible for polymers having crystals with low plastic resistance.

In our case, cavitation occurs for both samples at deformations higher than the yield point, concomitant with the onset of transformation of the γ form into the mesophase. At this deformation a high level of orientational order is achieved for the mesomorphic aggregates. Cavitation takes place either in the undulated chevron-like structures involving the residual lamellar crystal of γ form or within the amorphous regions of the fibrillar entities including the mesomorphic form. However, the SAXS intensity distribution keeps dominated by the large contrast between the nanocavities and the surrounding phases up to high deformations, and in the case of the sample iPP5 even in the fully relaxed fibre, characterized by presence of crystals of α form, that is at deformations at which the transformation of γ form is almost complete. This suggests that the fibrillar morphology of the mesophase is accomplished by formation of a low fraction of nanovoids, by effect of breaking of the undulated structures including the γ form.

11.9 Concluding Remarks

In this chapter, the complex transformations occurring at different length scales during tensile deformation of semi-crystalline polymers are illustrated in the case of some samples of isotactic polypropylene which present different microstructures of the chains. In particular, the structural, textural and morphological changes occurring at different length scales are studied, by performing time resolved measurements of X-ray intensity scattered at wide and small angle, during stretching. The data are analysed in the framework of our current understanding of the deformation mechanism of polymers at temperatures higher than the glass transition temperature. The main results of the present analysis are depicted in Fig. 11.12.

In particular the WAXS analysis has shown that for all samples, regardless of stereoregularity, the initial crystalline form obtained from the melt, transforms by effect of tensile deformation into the mesomorphic form (Fig. 11.12a–a″). A high degree of orientational order of the mesomorphic aggregates is achieved in the final morphology, with the chain axes parallel to the stretching direction. The values of the critical strain marking the beginning of the polymorphic transitions ε(start) is close to 200 % for all samples, whereas the strain marking the complete transformation ε(end) increases with decreasing the stereoregularity. It has been also shown that at low deformation a non-negligible portion of the crystals in γ form tend to assume a perpendicular chain axis orientation (cross-β) instead then parallel to stretching direction (Fig. 11.12b). The gradual transformation of the initial crystalline form into the mesophase, associated with this non-standard mode of

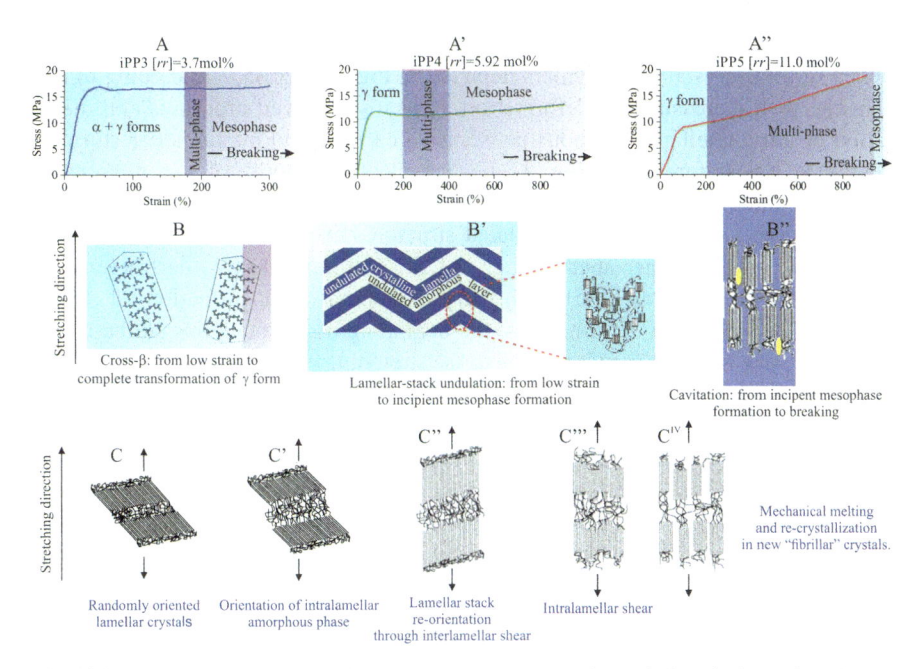

Fig. 11.12 Stress-induced changes in the structure, texture and morphology in isotactic polypropylene samples (iPP) with different content of [rr] stereodefects. All changes occur at critical values of strain according to the general scheme **c–c^{iv}**. Also phase transitions follow a general scheme. The critical values of strain depend on the intrinsic stability of crystals and flexibility of amorphous chains. The latter factors are both dependent on the stereoregularity ((**c–c^{iv}**) redrawn from Schultz 1974)

orientation, indicates that the crystals do not experience simultaneously the phase transition during stretching. If we assume that in semi-crystalline polymers, the strain is homogeneously distributed, whereas the stress is not, the gradual transition entails that with increasing deformation, only the crystals that experience a stress higher than a critical value undergo re-orientation and/or transform into another polymorphic form.

In the present multiscale approach, SAXS data analysis reveals that already at low deformations (lower than ε(start)) the lamellar crystals oriented with their normal parallel to the tensile force, in order to alleviate compressive transversal stress, undergo buckling instability, with consequent formation of undulated chevron-like super-structures (Fig. 11.12b'). However, at deformations close to ε(end) up to the breaking the effect of lamellar stacking and parallel orientation of the lamellar normal to the stretching direction on the SAXS intensity becomes buried by the large diffuse scattering localized on the equator due to cavitation (Fig. 11.12c").

Therefore the values of ε(start) and ε(end) mark not only the deformations at the onset and at the end of phase transitions of γ form but also events involving change in texture and morphology and/or damage. The above cascade of events promoted by tensile strain fits quite well into the general scheme of the deformation mechanism of semi-crystalline polymers depicted in Fig. 11.12c–c^{IV}. According to this scheme, with increasing the deformation, typically involved processes are: orientation of intralamellar amorphous phase, isolated inter and intralamellar slip processes, at low deformation (Fig. 11.12c′); collective activity of slip motions of crystals, including re-orientation of lamellar stacks (Fig. 11.12c″), buckling instability leading to formation of undulated structures at intermediate deformations; collective intralamellar shear (Fig. 11.12c′) followed by destruction of crystals (mechanical melting) and recrystallization with formation of fibrils (Fig. 11.12c^{IV}) at large deformations; the beginning of disentangling of the amorphous network or strain hardening due to the stretching of the amorphous entangled network at deformations close to breaking.

In other terms, the stress-induced phase transitions occurring by effect of plastic deformation are regulated by the same factors that govern the textural and morphological changes during transformation of the initial spherulitic morphology into fibrillar morphology, and reflects crystallographic restraints on the slip processes, and topological constrains on the response of crystals to the tensile stress field transmitted by the interconnected crystallites.

The above considerations support the hypothesis that when polymorphic transformations occur during plastic deformation, in iPP but also in many other polymers, the phase transitions are strain-controlled rather than stress-controlled (De Rosa et al. 2005, 2006; De Rosa and Auriemma 2006, 2007, 2011). As discussed above, the critical values of deformation at which the polymorphic transitions start always correspond to the destruction of a portion of original lamellae of a given crystalline form, and recrystallization with formation of fibrils in a new crystalline form. Moreover these critical values depend on the stereoregularity of the sample (Fig. 11.12). This suggests that the two factors that govern the location of the critical strain corresponding to the formation of fibrils, that is the modulus of the entangled amorphous (Ahmad et al. 2013) and the stability of the crystals, depend on the stereoregularity. Chains of different stereoregularity possess, indeed, different flexibility. In fact, the relative configuration of consecutive stereoisomeric centres along the chain affects the space correlation among skeletal bonds and the rotational energy barriers around the C–C bonds (Flory 1969). Since the dynamics of macromolecular chains is largely controlled by these parameters, which can be defined as "the internal viscosity" (Allegra 1974) different degrees of stereoregularity produce a different entanglement density of the amorphous phase. Moreover, the stereoregularity also affects the stability of crystals, the degree of crystallinity, as well as the relative stability of the different polymorphic forms involved in the structural transformations (De Rosa and Auriemma 2006, 2007, 2011, 2013, Ch. 7; De Rosa et al. 2006).

Acknowledgement The "Ministero dell'Istruzione, dell'Università e della Ricerca" (project PRIN 2010–2011) and Fondazione Cariplo (Cariplo project 2013 "Crystalline Elastomers") are acknowledged for the financial support. The synchrotron based measurements were made at the Science and Technology Facilities Council (UK) Facility at Daresbury and we thank the beamline staff for their help with the experiments. GRM acknowledges the support by the Portuguese Foundation for Science and Technology (FCT) through the Project reference "UID/Multi/04044/2013."

References

Ahmad N, Di Girolamo R, Auriemma F, De Rosa C, Grizzuti N (2013) Relations between stereoregularity and melt viscoelasticity of syndiotactic polypropylene. Macromolecules 46 (19):7940–7946
Alexander LE (1979) X-ray diffraction methods in polymer science. Wiley, New York
Al-Hussein M, Strobl G (2002) Strain-controlled tensile deformation behavior of isotactic poly (1-butene) and its ethylene copolymers. Macromolecules 35(22):8515–8520
Allegra G (1974) Role of internal viscosity in polymer viscoelasticity. J Chem Phys 61 (11):4910–4920
Andersen PG, Carr SH (1975) Formation of bimodal crystal textures in polypropylene. J Mater Sci 10(5):870–886
Auriemma F, De Rosa C (2006) Stretching isotactic polypropylene: from "cross-β" to crosshatches, from α form to γ form. Macromolecules 39(22):7635–7647
Baird DG (1998) Polymer processing: principles and design. Wiley-Interscience, New York
Bassett DC, Olley RH (1984) On the lamellar morphology of isotactic polypropylene spherulites. Polymer 25(7):935–943
Bicerano J (2002) Prediction of polymer properties, 3rd edn. CRC, Boca Raton
Billmeyer FW (1984) Textbook of polymer science, 3rd edn. Wiley Interscience, New York
Bowden PB, Young RJ (1974) Deformation mechanisms in crystalline polymers. J Mater Sci 9 (12):2034–2051
Brandrup J, Immergut EH (eds) (1989) Polymer handbook, 3rd edn. Wiley, New York
Brintzinger HH, Fischer D, Mulhaupt R, Rieger B, Waymouth RM (1995) Stereospecific olefin polymerization with chiral metallocene catalysts. Angew Chem Int Ed 34(11):1143–1170
Cao Y, van Horn RM, Tsai C-C, Graham MJ, Jeong K-U, Wang B, Auriemma F, De Rosa C, Lotz B, Cheng SZD (2009) Epitaxially dominated crystalline morphologies of the γ-phase in isotactic polypropylene. Macromolecules 42(13):4758–4768
Cao Y, van Horn RM, Sun H-J, Zhang G, Wang C-L, Jeong K-U, Auriemma F, De Rosa C, Lotz B, Cheng SZD (2011) Stem tilt in α-form single crystals of isotactic polypropylene: a manifestation of conformational constraints set by stereochemistry and minimized fold encumbrance. Macromolecules 44(11):3916–3923
Compostella M, Coen A, Bertinotti F (1962) Fibers and films of isotactic polypropylene. Preparation and properties with regard to its crystalline structure. Angew Chem 74(16):618–624
Cowie JMG, Arrighi V (2008) Polymers: chemistry and physics of modern materials, 3rd edn. CRC, New York
Crist B, Fisher CJ, Howard PR (1989) Mechanical Properties of Model Polyethylenes: Tensile Elastic Modulus and Yield Stress. Macromolecules 22:1709–1718
De Rosa C, Auriemma F (2006) Structural-mechanical phase diagram of isotactic polypropylene. J Am Chem Soc 128:11024
De Rosa C, Auriemma F (2007) Stress-induced phase transitions in metallocene-made isotactic polypropylene. Lect Notes Phys 714:345–371
De Rosa C, Auriemma F (2011) Single site organometallic polymerization catalysis as a method to probe the properties of polyolefins. Polym Chem 2:2155–2168
De Rosa C, Auriemma F (2012) The deformability of polymers: the role of disordered mesomorphic crystals and stress-induced phase transformations. Angew Chem Int Ed 51:1207–1211

De Rosa C, Auriemma F (2013) Crystals and crystallinity in polymers: diffraction analysis of ordered and disordered crystals. Wiley, New York

De Rosa C, Auriemma F, Di Capua A, Resconi L, Guidotti S, Camurati I, Nifant'ev IE, Laishevtsev IP (2004) Structure-property correlations in polypropylene from metallocene catalysts: stereodefective, regioregular isotactic polypropylene. J Am Chem Soc 126(51):17040–17049

De Rosa C, Auriemma F, De Lucia G, Resconi L (2005) From stiff plastic to elastic polypropylene: polymorphic transformations during plastic deformation of metallocene-made isotactic polypropylene. Polymer 46(22):9461–9475

De Rosa C, Auriemma F, Ruiz De Ballesteros O (2006) A microscopic insight into the deformation behavior of semicrystalline polymers: the role of phase transitions. Phys Rev Lett 96:167801/1

De Rosa C, Auriemma F, Ruiz De Ballesteros O, Dello Iacono S, De Luca D, Resconi L (2009) Stress-induced polymorphic transformations and mechanical properties of isotactic propylene-hexene copolymers. Cryst Growth Des 9:165–176

De Rosa C, Auriemma F, Di Girolamo R, Ruiz De Ballesteros O, Pepe M, Tarallo O, Malafronte A (2013) Morphology and mechanical properties of the mesomorphic form of isotactic polypropylene in stereodefective polypropylene. Macromolecules 46:5202–5214

De Rosa C, Auriemma F, Di Girolamo R, Ruiz De Ballesteros O (2014) Crystallization of the mesomorphic form and control of the molecular structure for tailoring the mechanical properties of isotactic polypropylene. J Polym Sci B Polym Phys 52:611–676

Ewen JA (1997) New chemical tool to create plastics. Sci Am 276(5):86

Ewen JA (1998) Symmetry rules and reaction-mechanisms of Ziegler-Natta catalysts. J Mol Catal A Chem 128:103

Flory PJ (1969) Statistical mechanics of chain molecules. Wiley, New York

Fritze C, Resconi L, Sculte J, Guidotti S (2003) Preparations of dialkylsilylene-bridged and a fused tricyclic compounds and indene derivatives-liganded metallocene catalysts for α-olefin polymerizations PCT Int Appl WO 03/00706. Basell, Italy

Geddes AJ, Parker KD, Atkins EDT, Beighton E (1968) "Cross-β" conformation in proteins. J Mol Biol 32:343

Gerasimov VI, Genin YV, Tsvankin DY (1974) Small-angle x-ray study of deformed bulk polyethylene. J Polym Sci Polym Phys Ed 12:2035–2046

Glatter O, Kratky O (1982) Small angle X-ray scattering. Academic, London

Halary J-L, Laupretre F, Monnerie L (2011) Polymer materials: macroscopic properties and molecular interpretations. Wiley, New York

Hay I, Keller A (1970) Mechanically induced twinning and phase transformations. J Polym Sci C Polym Symp 30:289–296

Heimenz PC, Lodge TP (2007) Polymer chemistry, 2nd edn. CRC, Boca Raton

Hiss R, Hobeika S, Lynn C, Strobl G (1999) Network stretching, slip processes, and fragmentation of crystallites during uniaxial drawing of polyethylene and related copolymers. A comparative study. Macromolecules 32:4390–4403

Hong K, Rastogi A, Strobl G (2004a) A model treating tensile deformation of semicrystalline polymers: quasi-static stress–strain relationship and viscous stress determined for a sample of polyethylene. Macromolecules 37:10165

Hong K, Rastogi A, Strobl G (2004b) Model treatment of tensile deformation of semicrystalline polymers: static elastic moduli and creep parameters derived for a sample of polyethylene. Macromolecules 37:10174

Hughes DJ, Mahendrasingam A, Oatway WB, Heeley EL, Martin C, Fuller W (1997) A simultaneous SAXS/WAXS and stress-strain study of polyethylene deformation at high strain rates. Polymer 38:6427

Humbert S, Lame O, Chenal J, Rochas C, Vigier G (2010) Small strain behavior of polyethylene: in situ SAXS measurements. J Polym Sci B 48:1535–1542

IUPAC. Compendium of chemical terminology, 2nd ed. (the "Gold Book"). Compiled by A.D. McNaught and A. Wilkinson. Blackwell Scientific Publications, Oxford (1997). XML on-line corrected version: http://goldbook.iupac.org (2006) created by M. Nic, J. Jirat, B. Kosata; updates compiled by A. Jenkins. ISBN 0-9678550-9-8. doi: 10.1351/goldbook. Stefano V. Meille,, Giuseppe Allegra,, Phillip H. Geil, Jiasong He, Michael Hess, Jung-Il Jin, Pavel Kratochvíl, Werner Mormann, and Robert Stepto Pure Appl. Chem., vol. 83, No. 10, pp. 1831–1871, 2011. doi:10.1351/PAC-REC-10-11-13

Kaminsky W (1996) New polymers by metallocene catalysis. Macromol Chem Phys 197:3907

Kaminsky W (1998) Mechanistic aspects of the olefin polymerization with metallocene catalysts. Macromol Symp 134:63

Kaminsky W (1999) Metallocene catalysts for olefin polymerization. Stud Surf Sci Catal 121:3

Katayama K, Amano T, Nakamura N (1968) Structural formation during melt spinning process. Kolloid Z Z Polym 226:125

Kennedy MA, Peacock AJ, Mandelkern L (1994) Tensile properties of crystalline polymers: linear polyethylene. Macromolecules 27:5297

Kennedy MA, Peacock AJ, Failla MD, Lucas JC, Mandelkern L (1995) Tensile properties of crystalline polymers: random copolymers of ethylene. Macromolecules 28:1407

Khoury F (1966) The spherulitic crystallization of isotactic polypropylene solution: on the evolution of monoclinic spherulites from dendritic chain-folded crystal precursors. J Res Natl Bur Stand U S A 70A:29

Krumova M, Henning S, Michler G (2006) Chevron morphology in deformed semicrystalline polymers. Philos Mag 86:1689–1712

Lin L, Argon AS (1994) Structure and plastic deformation of polyethylene. J Mater Sci 29:294–323

Lotz B, Wittmann JC (1986) The molecular origin of lamellar branching in the α (monoclinic) form of isotactic polypropylene. J Polym Sci Polym Phys Ed 24:1541

Lovell R, Mitchell GR (1981) Molecular orientation distribution derived from an arbitrary reflection. Acta Cryst A37:135

Lovinger AJ (1983) Microstructure and unit-cell orientation in α-polypropylene. J Polym Sci Polym Phys Ed 21:97

Makke A, Perez M, Lame O, Barrat J-L (2012) Nanoscale buckling deformation in layered copolymer materials. Proc Natl Acad Sci U S A 109:680–685

Maslen EN (2006) X-ray absorption in international tables for crystallography, vol. C, Ch. 6.3

Meille SV, Brückner S (1989) Non-Parallel Chains in Crystalline γ-Isotactic Polypropylene. Nature 340:455–457

Men Y, Strobl G (2001) Critical strains determining the yield behavior of s-PP. J Macromol Sci Phys B40:775–796

Men Y, Rieger J, Strobl G (2003) Role of the entangled amorphous network in tensile deformation of semicrystalline polymers. Phys Rev Lett 91:1

Menyhárd A, Gahleitner M, Varga J, Bernreeitner K, Jääskeläinen P, Øysæd H, Pukánszky B (2009) The influence of nucleus density on optical properties in nucleated isotactic polypropylene. Eur Polym J 45:3138–3148

Mitchell GR, Windle AH (1982) Conformational analysis of oriented non-crystalline polymers using wide angle X-ray scattering. Colloid Polym Sci 260:754

Mitchell GR (1984) A wide-angle X-ray study of the development of molecular orientation in crosslinked natural rubber. Polymer 25:1562–1572

Mourglia-Seignobos E, Long D, Odoni L, Vanel L, Sotta P, Rochas C (2014) Physical mechanisms of fatigue in neat polyamide 6,6. Macromolecules 47:3880–3894

Natta G, Corradini P (1960) Structure and properties of isotactic polypropylene. Nuovo Cim Suppl 15:40–51

Nifant'ev IE, Laishevtsev IP, Ivchenko PV, Kashulin IA, Guidotti S, Piemontesi F, Camurati I, Resconi L, Klusener PAA, Rijsemus JJH, De Kloe KP, Korndorffer FM (2004) C1-symmetric heterocyclic zirconocenes as catalysts for propylene polymerization, 1: Ansa-zirconocenes with linked dithienocyclopentadienyl-substituted cyclopentadienyl ligands. Macromol Chem Phys 205:2275

Nitta K-H, Yamana M (2012) Poisson's ratio and mechanical nonlinearity under tensile deformation in crystalline polymers. In: De Vicente J (ed) Rheology. InTech. ISBN 978-953-51-0187-1. http://www.intechopen.com/books/rheology/poisson-s-ratio-and-mechanical-nonlinearity-undertensile-deformation

Norton DR, Keller A (1985) The spherulitic and lamellar morphology of melt-crystallized isotactic polypropylene. Polymer 26:704

O'Kane WJ, Young RJ (1995) The role of dislocations in the yield of polypropylene. J Mater Sci Lett 14:433

O'Kane WJ, Young RJ, Ryan AJ (1995) The effect of annealing on the structure and properties of isotactic polypropylene films. J Macromol Sci Phys B34:427

Padden FJ, Keith HD (1966) Crystallization in thin films of isotactic polypropylene. J Appl Phys 1966(37):4013

Padden FJ, Keith HD (1973) Mechanism for lamellar branching in isotactic polypropylene. J Appl Phys 44:1217

Pawlak A, Galeski A (2005) Plastic deformation of crystalline polymers: the role of cavitation and crystal plasticity. Macromolecules 38:9688–9697

Peterlin A (1971) Molecular model of drawing polyethylene and polypropylene. J Mater Sci 6:490

Popli R, Mandelkern L (1987) Influence of structural and morphological factors on the mechanical properties of the polyethylenes. J Polym Sci B Polym Phys 25:441

Read D, Duckett R, Sweeny J, Mcleish T (1999) The chevron folding instability in thermoplastic elastomers and other layered material. J Phys D Appl Phys 32:2087–2099

Resconi L, Cavallo L, Fait A, Piemontesi F (2000) Selectivity in propene polymerization with metallocene catalysts. Chem Rev 100:1253

Resconi L, Guidotti S, Camurati I, Frabetti R, Focante F, Nifant'ev IE, Laishevtsev IP (2005) C1-symmetric heterocyclic zirconocenes as catalysts for propylene polymerization. 2. Ansa-zirconocenes with linked diethienocyclopentadienyl-substituted indenyl ligands. Macromol Chem Phys 206:1405

Richardson TL, Lokensgard E (1996) Industrial plastics: theory and applications. Delmar, New York

Sauer JA, Morrow NR, Richarson GC (1965) Morphology of solution-grown polypropylene crystal aggregates. J Appl Phys 36:3017

Schmidt-Rohr K, Spiess HW (1991) Macromolecules 24:5288

Schrauwen BAG, Janssen RPM, Govaert LE, Meijer HEH (2004) Intrinsic deformation behavior of semicrystalline polymers. Macromolecules 37:6069–6078

Schultz JM (1974) Polymer materials science. Prentice Hall, Englewood Cliffs, NJ

Schultz JM (1984) Microstructural aspects of failure in semicrystalline polymers. Polym Eng Sci 24:770–785

Séguéla R (2002) Dislocation approach to the plastic deformation of semicrystalline polymers: kinetic aspects for polyethylene and polypropylene. J Polym Sci B Polym Phys 40:593

Séguéla R (2005) On the strain-induced crystalline phase changes in semi-crystalline polymers: mechanisms and incidence on the mechanical properties. J Macromol Sci C Polym Rev 45:263–287

Somani RH, Yang L, Zhu L, Hsiao BS (2005) Flow-induced shish-kebab precursor structures in entangled polymer melts. Polymer 46:8587–8623

Sorrentino A, De Santis F, Titomanlio G (2007) Polymer crystallization under high cooling rate and pressure: a step towards polymer processing conditions. Lect Notes Phys 714:329–344

Stocker W, Magonov SN, Cantow HJ, Wittmann JC, Lotz B (1993) Contact faces of epitaxially crystallized α- and γ-phase isotactic polypropylene observed by atomic force microscopy. Macromolecules 26:5915–5923

Stribeck N, Nöchel U, Funari SS, Schubert T, Timmann T (2008) Tensile tests of polypropylene monitored by SAXS. Comparing the stretch-hold technique to the dynamic technique. J Polym Sci B Polym Phys 46:721–726

Thomann R, Wang C, Kressler J, Mülhaupt R (1996) On the γ-phase of isotactic polypropylene. Macromolecules 29:8425

Thomann R, Semkea H, Maiera R-D, Thomann Y, Scherblea J, Mülhaupt R, Kressler J (2001) Influence of stereoirregularities on the formation of the γ-phase in isotactic polypropene. Polymer 42:4597

Ward IM (ed) (1997) Structure and properties of oriented polymers. Springer, Netherlands

Wunderlich B (1973) Macromolecular physics, vol 1: crystal structure, morphology, defects. Academic, New York

Young RJ (1974) Dislocation model for yield in polyethylene. Philos Mag 30:85

Zhou J-J, Liu J-G, Yan S-K, Dong J-Y, Li L, Chan C-M, Schultz JM (2005) Atomic force microscopy study of the lamellar growth of isotactic polypropylene. Polymer 46:4077–4087

Zia Q, Radusch HJ, Androsch R (2009) Deformation behavior of isotactic polypropylene crystallized via a mesophase. Polym Bull 63:755

Chapter 12
Summary

Geoffrey R. Mitchell

As long-chain molecular materials, synthetic polymers naturally exhibit structure on a number of scales. We have seen in the earlier chapters, many examples of the consequences of small changes to the chemical configuration on larger scale structure.

This interplay between structural scales places great demands on both the designers of materials and on the development of processing routes to provide shaped objects for end users. This introduces the requirement to define what morphologies are required at different structural scales. This is an area where definitions and standardization have yet to appear.

Such a time lag is typical in technological development, and we can see this with the drive within the European Union for standardization of nanotechnology (EU 2011) and the challenges this presents (Rauscher and Roebben 2014). A number of EU-wide programmes such as Nanodefine (2015) are seeking the development of methods that reliably identify, characterize, and quantify nanomaterials both as substance and in various products and matrices. The development of manufacturing processes to produce defined properties dependent on the control of morphology through the definition of the chemical configuration of the polymer has been a long-held ambition of many polymer scientists and engineers. From a 'design a polymer to deliver the properties' point of view that ambition remains some way in the future. The complications are the interplay between the different scales of structure.

Just as the discussion of the definition of a nanomaterial underlines, any specification must be accompanied by practical steps by which that specification can be verified. There has been a surge in the development of experimental techniques, and details of many of these have appeared in this book. The availability of calibrated

---

G.R. Mitchell
Centre for Rapid and Sustainable Product Development, Institute Polytechnic of Leiria, Marinha Grande, Portugal
e-mail: geoffrey.mitchell@ipleiria.pt

data over a number of structural scales (e.g., see NIMROD in Chap. 2) presents some considerable but interesting challenges in the use of that data to develop a robust understanding of the structure and morphology on the different scales. Clearly, a big computer is no longer sufficient and is unlikely to be sufficient in the foreseeable future. Rather than seeking an ever larger grid computing power, we need to identify how to coarse grain intelligently and effectively so that the interchange of information between the structural scales is effective and efficient. An FCT-funded project to achieve this is underway in the Author's Laboratory (Mitchell 2015).

We have also seen the mix of manufacturing processes, both traditional processes and the emerging area of Direct Digital Manufacturing (Chap. 7) which has the possibility to create complex structures at the nano to macroscopic structural scales. Much of this is driven by applications in the medical field including regenerative medicine (Murphy and Atala 2014). The successes of the twentieth century in delivering mass production so that we could all benefit from the industrial revolution may be replaced by mass customization in this so-called Third Industrial Revolution.

> "A once-shuttered warehouse is now a state-of-the art lab where new workers are mastering the 3-D printing that has the potential to revolutionize the way we make almost everything." Barack Obama during his state of the union address of 2013, Obama (2013).

What is in no doubt is the capacity of polymer-based materials to yield complex targeted structures with specific properties. To achieve this requires an understanding of the interplay between the different scales of structure. We hope that this book has moved that goal nearer to realization.

This work was supported by the Fundação para a Ciência e a Tecnologia (Portugal) through project PTDC/CTM-POL/7133/2014.

References

2011/696/EU (2011) Commission recommendation on the definition of nanomaterial. Official Journal of the European Union L 275/38. 20 Oct 2011
http://www.nanodefine.eu/ (2015)
Mitchell GR (2015) UC4EP Understanding crystallisation for enhanced polymer properties PTDC/CTM-POL/7133/2014
Murphy SV, Atala A (2014) 3D bioprinting of tissues and organs. Nat Biotechnol 32:773–785
Obama B (2013) During his state of the union address of 2013. https://www.whitehouse.gov/state-of-the-union-2013
Rauscher H, Roebben G (eds) (2014) Towards a review of the EC Recommendation for a definition of the term "nanomaterial". Part 1: Compilation of information concerning the experience with the definition. A Science and Policy Report by the Joint Research Centre of the European Commission JRC89369, ISBN 978-92-79-36601-7 (print)

Index

A
Accurate clear epoxy solid (ACES), 185
Acrylonitrile-butadiene-styrene (ABS), 185
Additive manufacturing (AM), 24, 181
Atomic force microscopy (AFM), 53, 167, 176, 296–299

B
Block copolymers
  anionic polymerisation, 271–275
  atom transfer radical polymerisation, 271
Bragg's Law, 269

C
Carbon black, 209, 220, 222–223, 225–228, 230, 232, 238
Carbon nanotubes (CNTs), 7, 74, 76–81, 86–95, 209–210, 214, 221, 242
Composites, 6–7, 16, 20, 46, 70–71, 86–87, 91, 93, 95, 175, 185, 213–214, 218, 238, 246
Computer-aided design (CAD), 181
Computer-aided engineering (CAE)
  STL file format, 182
Confinement effects, 164
Cross-linked polyethylene, 244
Crystal annealing
  melting-recrystallizaiton mechanism, 136
  solid-chain sliding-diffusion mechanism, 136
Crystallinity, 5, 9, 14, 37, 54–56, 89, 130, 136, 195, 197, 230, 253, 292–294, 310

D
Damage effect
  cavitation, 313, 315–320
D-glucitol, 148
Differential scanning calorimetry (DSC)
  high crystallinity polypropylene, 33
  Hoffman–Weeks equation, 30, 31
  isotactic polystyrene, 30
  poly ether ethyl ketone, 32
  recrystallization, 31
Dip coating, 23
Direct digital manufacturing (DDM), 181, 330
3D printing (3DP), 24, 181–204
Dynamic mechanical thermal analysis (DMTA), 36

E
Electrical conductivity
  conduction mechanisms, 219
Electrospinning, 24
Equilibrium melting point
  interaction parameters, 110–111
  molecular weights, 111–113
Etching, 48–50
Ethylene-(vinyl acetate) copolymer (EVA), 242, 247

F
Filler effects
  geometric shape factors, 221
  high aspect ratio, 220–221
Filler localisation, 224–227
Flory–Huggins equation, 108, 116

Free energy barrier, 118–123, 127, 130–133, 135
Fused deposition modelling (FDM), 182, 184–193
Fused filament fabrication (FFF), 185, 187–193

G
Gibbs–Thomson equation, 30, 105, 134
Graphene
    graphene-based materials (GBMs), 79–80

H
High density polyethylene (HDPE), 249, 250

I
Isotactic polypropylene (iPP), 91–92, 258, 289–291, 298–299, 304, 310, 322

L
Lauritzen–Hoffman (LH) theory, 127
Layer-by-layer construction
    deposition speed, 193, 196–198
    extrusion temperature, 193, 201–203
    feed roller velocity, 193, 198–201
    layer patterning, 191
    layer thickness
        slicing, 188
    neck growth, 190
    nozzle, 189, 194, 196, 198–199
LH theory. *See* Lauritzen–Hoffman (LH) theory
Ligand substituents, 290
Light microscopy
    ancillary techniques, 45
    circularly polarized light, 41–42
    interference microscopy
        Nomarski contrast, 43–45
    optical microscopy, 38, 45, 296
    phase contrast microscopy, 42–43
    polarized optical microscopy (POM), 39–40
Low density polyethylene (LDPE), 242
    linear low-density polyethylene (LLDPE), vii, 22, 35, 249

M
Mesogen, 106–108
Mesophase, 106, 107, 133–135, 172, 303–304, 309, 317–320
Multilayer extrusion, 24

N
N-(2-aminoethyl) 3-aminopropyl-trimethoxysilane, 245
Nanocomposites
    dispersion, 86, 92, 232, 251–253
    halloysite, 88–89
    montmorillonite, 82, 84, 85
    nanoclay, 81–85
Nanocrystals, 165, 173–175
Nanodielectrics
    breakdown strength, 243–245
        coefficient of thermal expansion, 241
    miscibility, 251–253
        Helmholtz–Stern–Gouy–Chapman double layer, 252
        steric interchain effects, 252
        zeta potential, 252
    permittivity, 245–249
        Claussius–Mossotti equation, 245
        Lichtenecker–Rother equation, 246
    resistance to corona, 240–241
    surface discharges, 240–241
Nanoparticles
    1-D, 75–76
    2-D, 75–76
    3-D, 75–76
    polymer nanoparticles, 165–167, 173–175
        miniemulsion method, 165, 168, 172–173
        reprecipitation method, 166, 168–169, 172
Nanowires, 164–165, 220
Neutron scattering
    broad Q neutron diffraction, 59–61 (*see also* Wide angle neutron scattering)
    NIMROD, 62
    small angle neutron scattering, 59, 104
    wide angle neutron scattering, 59
Nucleating agents
    dibenzylidene-d-sorbitol (DBS), 146, 148
    dibenzylidene derivative, 148
    d-sorbitol, 148, 149 (*see also* D-glucitol)

O
Order
    amorphous, 8
    crystalline, 7
    liquid crystalline, 7–8

P
PC. *See* Photonic crystals (PC)
PCL. *See* Poly(ε-caprolactone) (PCL)

PDLLA. *See* Poly(D,L-lactic acid) (PDLLA)
PEMA. *See* Polyethylmethacrylate (PEMA)
Percolation
  hard-core systems, 212
  hard-core with soft shell, 212
  soft-core systems, 211
PET. *See* Polyethylene terephthalate (PET)
Phase diagram, 114–117
Phase separation
  metastable, 18
  micelles, 19
  photonic crystals, 19
  reaction injection moulding (RIM), 22
  stable, 18
Photonic band gap, 263–282
Photonic crystals (PC), 19, 267–271
PLLA. *See* Poly(L-lactic acid) (PLLA)
Polarized light microscope (PLM), 126
Poly(D,L-lactic acid) (PDLLA), 168–170, 172–175
Poly(ε-caprolactone) (PCL), 55, 193
Poly(L-lactic acid) (PLLA), 168, 230
Polycarbonate, 50, 185, 210, 211
Polydispersity, 221–223
Polyethylene terephthalate (PET), 9, 14, 111, 163
Polyethylmethacrylate (PEMA), 166, 168
Polymer crystallization
  crystal growth, 125–135
    crystal growth rate, 10–11
    diffusion-controlled mechanism, 126
    interface-controlled mechanism, 126
  fringed micelle model, 9, 33, 103
  Gibbs free energy, 121
  lamellae, 10, 30, 56, 72, 90, 92, 155, 163, 289, 296, 307–309, 319, 322
  lattice models, 108–109, 115, 138, 210
  mean-field lattice theory, 101, 111
  molecular segregation, 132
  nucleation
    bell curve, 13–14
    folding theories, 12–13
    heterogeneous nucleation, 90, 92, 118, 120–122, 125
    Hoffman–Lauritzen theory, 11 (*see also* Lauritzen–Hoffman (LH) theory)
    non-nucleation models, 132–135
    primary nucleation, 11, 118–123
    secondary nucleation, 11, 118, 119, 125–132
    tertiary nucleation, 118, 119
  shish-kebab, 15, 90
  spherulites, 10, 14, 19, 39, 42, 72, 92, 255, 297–299

temperature-dependence curve, 122
thermal fluctuations, 102, 117–118, 120
Polymers
  chain-growth polymers, 2
    free-radical polymerisation, 274, 277
    Ziegler–Natta catalyst, 3
  thermoplastics, 5
  thermosets, 5, 6
Polymer thin films, 164–165, 175
Polymorphism
  polymorphic transitions, 289, 322
Polyphenylenesulfone (PPSU), 185

R
Random-coil model, 102
Refractive index, 39–40, 42, 270, 279–280
Regime-transition phenomenon, 129

S
SALS. *See* Small-angle laser scattering (SALS)
Scanning electron microscopy (SEM)
  contrast, 46–47
  embedding techniques, 47
  sandwiching techniques, 47, 48
SCORIM. *See* Shear-controlled orientation injection moulding (SCORIM)
Selective laser sintering (SLS), 181, 183
Semicrystalline polymers, viii, 9, 101, 103–105, 122, 155, 163–164, 172, 194, 231, 313
Shear, 21, 45, 73, 89–91, 149, 151, 153, 155, 214, 224, 288, 307–309
Shear-controlled orientation injection moulding (SCORIM), 72
SLA. *See* Stereolithography (SLA)
Small-angle laser scattering (SALS), 126
Spectroscopy
  electron spin resonance (ESR), 37
  nuclear magnetic resonance spectroscopy, 37
Spin coating, 23
Stereoirregularity, 299
Stereolithography (SLA), 24, 181, 183
Stress-strain curves, 85, 294, 302
"Switchboard" model, 12, 104

T
Thermal fractionation, 34–36
Thermally stimulated current (TSC), 254
Thermodynamics, 101–117
Thermogravimetric Analysis (TGA), 242

Transmission electron microscopy (TEM)
 inducing contrast, 50–52
Tri-dimensional printing, 184. *See also* 3D
 printing (3DP)
Types of bonds, 1–2

W
Wide angle neutron scattering, 59

X
X-ray scattering

small angle X-ray scattering (SAXS), 17,
 54–55, 125, 150, 153, 154, 193, 292
 azimuthal integration, 194
 correlation function, 60, 195, 198,
  200, 203
 orientation parameter, 56, 74, 159, 194,
  198, 200, 202, 224
 SAXS invariant Ω, 150
wide angle X-ray scattering (WAXS),
 55–57, 61, 171–172, 176
 azimuthal profiles, 151, 152
 orientation distribution function,
  56, 223

CPSIA information can be obtained
at www.ICGtesting.com
Printed in the USA
LVOW01*1431280217
525680LV00003B/18/P